AF353106

Consumer Preferences and Acceptance of Meat Products

Consumer Preferences and Acceptance of Meat Products

Editor

Andrea Garmyn

MDPI • Basel • Beijing • Wuhan • Barcelona • Belgrade • Manchester • Tokyo • Cluj • Tianjin

Editor
Andrea Garmyn
Texas Tech University
USA

Editorial Office
MDPI
St. Alban-Anlage 66
4052 Basel, Switzerland

This is a reprint of articles from the Special Issue published online in the open access journal *Foods* (ISSN 2304-8158) (available at: https://www.mdpi.com/journal/foods/special_issues/consumer_meat).

For citation purposes, cite each article independently as indicated on the article page online and as indicated below:

LastName, A.A.; LastName, B.B.; LastName, C.C. Article Title. *Journal Name* **Year**, *Article Number*, Page Range.

ISBN 978-3-03943-080-2 (Hbk)
ISBN 978-3-03943-081-9 (PDF)

Contents

About the Editor

Andrea Garmyn (Ph.D. in Food Science, Oklahoma State University, USA, 2009) is currently an Academic Specialist in Meat Science Teaching and Outreach at Michigan State University and an Adjunct Professor in the Department of Animal & Food Science at Texas Tech University. Previously, she held roles in the Department of Animal & Food Science at Texas Tech University as Research Assistant Professor (2014–2020), Sr. Research Associate (2012–2014), and Postdoctoral Research Associate (2010–2012), all specializing in Meat Science. Dr. Garmyn's long-term research goal is to better understand the ante- and postmortem factors affecting meat quality, especially palatability. Her research program has focused on beef and lamb eating quality and implementing strategies during animal production or postmortem processing to improve eating quality and ensure positive consumer outcomes. She has authored or coauthored 58 peer-reviewed research articles, 1 book chapter, 81 abstracts, and numerous final research reports, proceedings publications, and technical or popular press articles. She has mentored several graduate and undergraduate students, and has lead research initiatives at the national and international level due to the collaboration between TTU and the red meat industry in New Zealand and Australia.

Preface to "Consumer Preferences and Acceptance of Meat Products"

At the point of purchase, consumers often use extrinsic cues such as color, marbling, leanness, packaging, and price to determine which meat product(s) to buy. The value placed on such cues may vary regionally or even be influenced by the demographic characteristics of the consumer. Tenderness, juiciness, and flavor remain the three pillars of cooked meat palatability, all linked to consumer satisfaction. Historically, tenderness has been the single most important factor affecting beef palatability, yet previous work has shown that flavor becomes the most important aspect of eating satisfaction when tenderness is acceptable. Consumers can distinguish marbling and consequently flavor differences in some muscles, and are willing to pay premiums for the type of flavor they prefer. Several consumer studies over the past two decades have collectively shown that consumer overall acceptability ratings are more highly correlated with flavor ratings than tenderness or juiciness ratings in beef and lamb. However, the role of flavor in the acceptability of muscles from primals other than the rib or loin is not entirely clear. Moreover, consumer acceptance of and preference for flavor potentially shifts when dealing with value-added or processed meats as opposed to fresh meats. This Special Issue includes 12 valuable scientific contributions, including one review article and 11 original research articles, focusing on antemortem and postmortem factors that influence the sensory acceptability of meat products across the major red meat species of beef, lamb, and pork.

Andrea Garmyn
Editor

Editorial

Consumer Preferences and Acceptance of Meat Products

Andrea Garmyn

Department of Animal and Food Sciences, Texas Tech University, Lubbock, TX 79409, USA;
agarmyn@gmail.com; Tel.: +1-785-410-4618

Received: 9 May 2020; Accepted: 26 May 2020; Published: 1 June 2020

At the point of purchase, consumers often use extrinsic cues such as color, marbling, leanness, packaging, and price to determine which meat product(s) to buy. The value placed on such cues may vary regionally or even be influenced by the demographic characteristics of the consumer. Tenderness, juiciness, and flavor remain the three pillars of cooked meat palatability, all linked to consumer satisfaction. Historically, tenderness has been the single most important factor affecting beef palatability, yet previous work has shown that flavor becomes the most important aspect of eating satisfaction when tenderness is acceptable. Consumers can distinguish marbling and consequently flavor differences in some muscles, and are willing to pay premiums for the type of flavor they prefer. Several consumer studies over the past two decades have collectively shown that consumer overall acceptability ratings are more highly correlated with flavor ratings than tenderness or juiciness ratings in beef and lamb. However, the role of flavor in the acceptability of muscles from primals other than the rib or loin is not entirely clear. Moreover, consumer acceptance of and preference for flavor potentially shifts when dealing with value-added or processed meats as opposed to fresh meats. This Special Issue includes 12 valuable scientific contributions, including one review article and 11 original research articles, focusing on antemortem and postmortem factors that influence the sensory acceptability of meat products across the major red meat species of beef, lamb, and pork.

Payne et al. [1] investigated the influence of lamb age class on Australian consumer eating quality scores of eight lamb cuts. Their results showed little difference in the eating quality between "new season" (approx. 5–8 months old) and "old season" (approx. 10–12 months old) lamb, highlighting the market potential for old season lamb products at retail.

Five of the original research articles focused on the influence of postmortem factors, such an enhancement [2,3] or incorporation of non-meat ingredients [4], postmortem aging [5], and packaging [6] on consumer eating quality. Lees et al. [2] assessed the impact of kiwifruit extract (actinidin) on consumer sensory outcomes for beef strip loin and outside flat. In addition, cooking method (grill or roast) and postmortem aging length (10 or 28 days) were also tested. Kiwifruit extract improved all palatability traits (tenderness, juiciness, flavor liking), resulting in greater overall liking for both strip loins and outside flats. Lees et al. [2] suggested actinidin infusion provides an opportunity to improve eating experiences for beef consumers. In another enhancement study, Garmyn et al. [3] explored the eating quality of several beef muscles cooked and served as fajita meat strips after enhancement with either a phosphate-based marinade or a "clean label" alternative ingredient (sodium bicarbonate). Those muscles included the outside skirt, inside skirt, flank, inside round cap, and bottom sirloin flap. Enhanced samples were scored more favorably than non-enhanced samples for all palatability traits; samples enhanced with sodium bicarbonate were more tender and juicier than samples enhanced with phosphate. According to consumers, the inside round cap was the least suitable option for preparation as fajitas. However, creating a "clean label" enhanced fajita product was possible without compromising cooking yield or consumer satisfaction [3].

Consumer response to reformulated burger patties with ingredients that could improve healthfulness [4] is another topic area in this Special Issue. Taylor et al. [4] tested various levels of tempeh inclusion (10%, 20%, and 30%) in beef patties. Their sensory experiments revealed that beef patties could include up to 10% tempeh; however, consumers rated visual appearance lower, along with less flavor and overall acceptability when patties included 20% to 30% tempeh [4].

The final two research articles focusing on postmortem factors that influence consumer eating quality dealt with extended postmortem aging of beef [5] and modified atmosphere packaging of pork [6]. Garmyn et al. [5] investigated the effect of extending the postmortem aging of beef strip loins from 21 to 84 days. Based on their results, samples should not be wet-aged longer than 63 days to prevent negative eating experiences for consumers; however, storage conditions (i.e., temperature) could potentially be adjusted to accommodate longer chilled storage without compromising flavor and overall palatability to the same extent [5]. Peng et al. [6] investigated the effects of high oxygen modified atmosphere packaging (MAP) of pork loins compared to vacuum packaging on eating quality and color following presentation in simulated retail display conditions. Ultimately, retailers should consider vacuum packing the preferred option over high-oxygen MAP, given the inferior consumer acceptability for palatability traits and greater lipid oxidation of MAP samples.

Several research articles in this Special Issue involved the investigation of the consumers' backgrounds and how cultural differences could influence sensory perception and acceptance of meat products. Mena et al. [7] explored the meal and snacking behavior of older adults in Australia and China. In this study, demographics influenced consumer preferences towards food, as older consumers in China and Australia differed in their responses to product traits and segmentation. In another cross-cultural study investigating red meat eating quality, Hastie et al. [8] used a mixed method approach involving both perceptual mapping (qualitative) and sensory (quantitative) methodologies to gain consumer sensory insights into sheep meat and beef. Australian and Asian consumers differed in their perception of 'premiumness' of meat products, which could be related to the traditional meat preparation and presentation styles between those groups of consumers. Moreover, demographic factors, specifically age, influenced eating quality and willingness to pay for sheepmeat and beef. O'Reilly et al. [9] wanted to determine whether demographic factors influenced consumer perceptions of sheep meat eating quality. Their results show consumer age, gender, household size, and income influenced sensory scores, but the impact varied across the three countries where testing took place—Australia, China, and United States. Frequency of lamb consumption is also a relevant factor when assessing eating quality, but again varied between the three countries [9].

Felderhoff et al. [10] aimed to quantify the relative contribution of palatability traits (tenderness, juiciness, and flavor) to beef satisfaction and assess if and to what extent certain demographic variables influence satisfaction. The authors found that flavor was the largest contributor to satisfaction in comparison to tenderness or juiciness, accounting for 59% of the overall rating. The results also indicate that age, income, and gender influenced satisfaction [11].

Arenas de Moreno [11] conducted surveys in three regions of Venezuela to determine buying expectations, motivations, needs, perceptions, and preferences of beef consumers, and their acceptance of domestic and foreign beef. Their results show that two factors explain 74% of the common variance in beef consumption. The first factor focuses on intrinsic factors, such as color, smell, tenderness, flavor, juiciness, and freshness, while the second factor involves more extrinsic factors, primarily product origin. The authors hope to use these results to design and implement strategies to recover and enhance the domestic beef demand in Venezuela [11].

Finally, Miller [12] reviewed the drivers of consumer liking for beef, pork, and lamb, suggesting the drivers are interrelated across species, but differences exist. For example, animal age, animal diet, and marbling influence consumer liking across species. For beef, tenderness has historically been the main driver of consumer liking, but as tenderness has improved and tenderness variation has been reduced, flavor has become a greater determinant to overall liking. Flavor, which is influenced by a number of antemortem and postmortem factors, was explored along with tenderness and juiciness, to determine how changes in palatability traits in response to those factors influence overall liking. Drivers of pork consumer liking can be influenced by pH, color, water holding capacity, animal diet, and presence of boar taint compounds. For lamb, the flavor, which is typically a direct reflection of animal diet and animal age, continues to be the primary driver of consumer liking; however, cultural differences and preferences may exist due to the variable consumption rates in certain countries.

In summary, the Special Issue "Consumer Preferences and Acceptance of Meat Products" demonstrates that the value of different palatability traits has evolved over time. Moreover, consumer acceptance and preference are not solely determined by the inputs of the meat itself, but can also be influenced by various demographic factors. In addition, consumers' views of meat products vary regionally and vary by species.

Conflicts of Interest: The author declares no conflict of interest.

References

1. Payne, C.E.; Pannier, L.; Anderson, F.; Pethick, D.W.; Gardner, G.E. Lamb Age has Little Impact on Eating Quality. *Foods* **2020**, *9*, 187. [CrossRef] [PubMed]
2. Lees, A.; Konarska, M.; Tarr, G.; Polkinghorne, R.; McGilchrist, P. Influence of Kiwifruit Extract Infusion on Consumer Sensory Outcomes of Striploin (*M. longissimus lumborum*) and Outside Flat (*M. biceps femoris*) from Beef Carcasses. *Foods* **2019**, *8*, 332. [CrossRef] [PubMed]
3. Garmyn, A.; Hardcastle, N.; Bendele, C.; Polkinghorne, R.; Miller, M. Exploring Consumer Palatability of Fajita Meat Using Five Australian Beef Muscles Enhanced with Phosphate or Sodium Bicarbonate. *Foods* **2020**, *9*, 177. [CrossRef]
4. Taylor, J.; Ahmed, I.A.M.; Al-Juhaimi, F.Y.; Bekhit, A.-D. Consumers' Perceptions and Sensory Properties of Beef Patty Analogues. *Foods* **2020**, *9*, 63. [CrossRef] [PubMed]
5. Garmyn, A.; Hardcastle, N.; Polkinghorne, R.; Lucherk, L.; Miller, M. Extending Aging of Beef Longissimus Lumborum from 21 to 84 Days Postmortem Influences Consumer Eating Quality. *Foods* **2020**, *9*, 208. [CrossRef] [PubMed]
6. Peng, Y.; Adhiputra, K.; Padayachee, A.; Channon, H.; Ha, M.; Warner, R.D. High Oxygen Modified Atmosphere Packaging Negatively Influences Consumer Acceptability Traits of Pork. *Foods* **2019**, *8*, 567. [CrossRef] [PubMed]
7. Mena, B.; Ashman, H.; Dunshea, F.R.; Hutchings, S.; Ha, M.; Warner, R.D. Exploring Meal and Snacking Behaviour of Older Adults in Australia and China. *Foods* **2020**, *9*, 426. [CrossRef] [PubMed]
8. Hastie, M.; Ashman, H.; Torrico, D.; Ha, M.; Warner, R. A Mixed Method Approach for the Investigation of Consumer Responses to Sheepmeat and Beef. *Foods* **2020**, *9*, 126. [CrossRef] [PubMed]
9. O'Reilly, R.A.; Pannier, L.; Gardner, G.E.; Garmyn, A.J.; Luo, H.; Meng, Q.; Miller, M.F.; Pethick, D.W. Influence of Demographic Factors on Sheepmeat Sensory Scores of American, Australian and Chinese Consumers. *Foods* **2020**, *9*, 529. [CrossRef] [PubMed]
10. Felderhoff, C.; Lyford, C.; Malaga, J.; Polkinghorne, R.; Brooks, C.; Garmyn, A.; Miller, M. Beef Quality Preferences: Factors Driving Consumer Satisfaction. *Foods* **2020**, *9*, 289. [CrossRef] [PubMed]
11. Arenas de Moreno, L.; Jerez-Timaure, N.; Valerio Hernández, J.; Huerta-Leidenz, N.; Rodas-González, A. Attitudinal Determinants of Beef Consumption in Venezuela: A Retrospective Survey. *Foods* **2020**, *9*, 202. [CrossRef] [PubMed]
12. Miller, R. Drivers of Consumer Liking for Beef, Pork, and Lamb: A Review. *Foods* **2020**, *9*, 428. [CrossRef] [PubMed]

Review

Drivers of Consumer Liking for Beef, Pork, and Lamb: A Review

Rhonda Miller

Department of Animal Science, Texas A&M University, College Station, TX 77843-2471, USA; rmiller@tamu.edu; Tel.: +1-979-845-3901

Received: 4 February 2020; Accepted: 17 March 2020; Published: 3 April 2020

Abstract: Tenderness, juiciness, and flavor have been associated with consumer acceptance of beef, lamb, and pork. Drivers of consumer liking are interrelated across these species, but there are differences in consumer preferences. Animal age, animal diet, and subsequent marbling impact consumer liking across species. For beef, consumer research prior to the 1990s showed that tenderness was the main driver of liking. Consumer tenderness and juiciness liking are highly correlated. More recent research has shown that as overall tenderness improved and tenderness variation decreased, flavor has become a more important driver of beef consumer liking. Flavor is affected by consumer preparation methods, familiarity with different flavor presentations, and animal production systems. Animal diet impacts consumer perception of beef tenderness and flavor, especially when comparing forage-fed versus grain-fed beef. Flavor preferences vary across countries more so than preferences for beef based on consumer tenderness preferences and are most likely influenced by the consumption of locally produced beef and the flavor-derived type of beef traditionally consumed. Drivers of pork consumer liking have been shown to be affected by pH, color, water holding capacity, animal diet, and the presence of boar taint compounds. While tenderness and juiciness continue to be drivers of consumer liking for pork, flavor, as impacted by animal diet and the presence of boar taint compounds, continues to be a driver for consumer liking. For lamb, the flavor, as affected by diet, and animal age continue to be the main drivers of consumer liking. Lamb consumers vary across countries based on the level of consumption and preferences for flavor based on cultural effects and production practices.

Keywords: consumer sensory; beef; pork; lamb; tenderness; juiciness; flavor

---

1. Introduction

Meat scientists have understood since the early 1900s that, in assessing and understanding meat eating quality, the end goal is to meet consumer demand and acceptance for meat products. In the early scientific literature, scientists eluded that systems such as the USDA Beef Quality grade system utilized factors that segmented beef carcasses into groups based on expected palatability. Meat scientists used trained descriptive sensory attributes for juiciness, tenderness, and flavor as indicators of consumer acceptability. However, while it was always implied that improvements in juiciness, tenderness, and flavor were associated with consumer acceptance, data were not presented [1–4]. The question was always present, "How are trained descriptive attributes related to consumer liking or acceptability?" While it is intuitive that trained human assessment of juiciness, tenderness, and flavor would most likely be related to consumer liking or disliking, it was not until Francis and others [5] that data were reported to understand consumer acceptance or liking of beef differing in USDA Beef Quality grades [6]. However, consumers were recruited from a farm show, and there was limited interpretation of the data. The main issue was that scientific methods in conducting consumer sensory evaluation that provided repeatable results were evolving. The American Society of Testing Materials published the first guidelines on sensory methods for foods and consumer products [7], and the American Meat Science Association (AMSA) published their first guidelines for cookery and sensory evaluation of meat [8], but these guidelines did not include an assessment of consumer sensory evaluation. It was not until 1995 that AMSA included consumer testing information in their sensory guidelines [9]. As methods evolved within the sensory community, scientifically accepted methodologies and guidelines for conducting and reporting consumer sensory data were developed [9–11]. Disciplines such as psychology, marketing and consumer insights, neural psychology, and sensory science used consumer sensory tools, but it was not until the 1980s that consumer sensory evaluation was used to understand meat product acceptance [12]. Much of the scientific community used trained descriptive panels to evaluate meat eating quality. Today, consumer sensory methodologies are used extensively by researchers to address acceptance of or preference for meat products. While consumer attitudes and familiarity with meat products play an important role in consumer liking and intent to purchase [13–15], these issues will not be directly addressed. Extensive research has also been conducted to examine pre- and post-harvest factors that impact consumer liking and meat eating quality. These papers will not be discussed, as it is not the intent of this paper. The effect of diet on meat flavor and sex on boar-taint in pork will be included in discussions, as these factors impact meat flavor, and pre- and post-harvest factors are important in providing road maps for methods to alter or improve tenderness, juiciness, and flavor of meat. The objective of this paper is to review research that evaluates consumer eating quality acceptance of whole muscle beef, lamb, and pork meat products, and to understand the current drivers of consumer liking. While the effect of meat color and visual assessment play a major role in consumer liking and purchase intent [13,16], this paper will only address the effect of eating quality on consumer liking.

2. Beef Consumers

The first extensive study that included consumers from multiple locations in the US was defined as the National Consumer Retail Beef study (NCRBS) [12,17]. In the NCRBS, the effect of marbling on consumer preferences for consumers in three cities was determined. The study included 540 households that contained two beef eating consumers. While they also conducted trained descriptive sensory evaluation for juiciness, tenderness, and flavor, they concluded that tenderness was the most important trained descriptive attribute driving consumer liking. While they showed that trained descriptive attributes showed similar trends to consumer data, they did not report relationships between trained and consumer sensory attributes. In the second phase of the NCRBS, they concluded that some consumers rated USDA Choice steaks higher for overall liking due to taste, and other consumers rated USDA Select steaks higher for overall liking due to leanness. This study provided evidence that consumer sensory research could provide valuable insight into beef acceptability and that consumer segments existed for whole muscle beef steaks.

The second national study was conducted in 1993 and 1994 and defined as the Beef Customer Satisfaction study. Beef top loin, top sirloin, and top round steaks were evaluated in an in-home placement study involving two beef consumers in each of 300 households in four cities [18–21]). Consumers were asked to cook steaks as they normally would prepare beef and to define the cooking method and degree of doneness. Consumers rated steaks for overall, tenderness, juiciness, and flavor liking. Descriptive sensory attributes and mechanical assessment of tenderness (Warner-Bratzler shear force; WBSF) were evaluated in companion steaks using a standardized cooking method. Relationships were examined between consumer and trained sensory measures and concluded that it was difficult to predict from descriptive sensory data how consumers would rate meat at home [22]. It should be noted that multivariate statistical tools were not used to evaluate relationships. As consumer data are more variable than trained descriptive data, multivariate statistical analysis tools for examining this relationship provide greater insight, and these tools have evolved since the 1990s. Additionally, tenderness and flavor liking were determined to be equal contributors to consumer liking for whole muscle beef steaks [18]. This was pivotal for the beef industry and meat science researchers as much of the focus for improving beef eating quality had been on improvement and assessment of beef tenderness. With the reporting of these results, the contribution of beef flavor to overall consumer liking was recognized.

In the 1990s and beyond, consumer research has become a highly utilized research tool to understand factors that affect meat eating quality and consumer liking. Research by meat scientists around the world has utilized consumer sensory evaluation to understand pre- and post-harvest factors that affect meat eating quality [23–52]. In Australia, consumer research has been the basis for the development of the Meat Standards Australia beef evaluation system [53–57]. This integrated system utilized pre- and post-harvest factors to predict the eating quality of individual beef cuts based on the cooking method. The MSA consumer evaluation methodology has been used to assess consumer liking across countries. Research using European [15,58–69], New Zealand [70], Asian [71–76], South African [77], and United States consumers [47,78–80] using the MSA consumer methods have been conducted. This research has established that consumers respond similarly to differences in tenderness, but flavor is more affected by cultural and environmentally learned behaviors in some countries. Consumer sensory studies that included factors and overall liking scores for beef across countries are presented in Table 1. While other studies have been reported, these selected studies provide evidence of factors affecting consumer overall liking.

A review of the MSA system and the research associated with the development and evaluation of the effectiveness of the system across different countries for beef and lamb has been presented [57]. Most of these studies indicated that tenderness liking is important to consumers, and consumers segmented whole muscle beef based on differences in tenderness. However, beef flavor liking is also a driver of consumer liking, and in some studies, flavor liking was a stronger driver of overall liking than tenderness liking [81–84]. This may be the result of improvements in overall beef tenderness and reductions in variability in beef tenderness. Within the US beef industry, beef in the retail meat case and in the foodservice industry has been monitored since 1989 to assess tenderness differences using WBSF, descriptive sensory attributes, and consumer sensory evaluation [23,85–89]. These surveys, known as the National Beef Tenderness Surveys, have shown that, for most whole muscle beef cuts, beef tenderness assessed using either WBSF, descriptive sensory panels, or consumer sensory panels has improved, and variation in beef tenderness has been reduced. The exception was for beef cuts from the round. With the advent of improved tenderness, it is not surprising that tenderness may not be as strongly related to overall consumer liking as in data from the 1980s, 1990s, and 2000s, where beef was tougher and more variable in tenderness. More recent consumer studies have shown that tenderness and juiciness are closely related, and as long as tenderness is within acceptable thresholds, flavor liking is the most prominent driver of consumer liking. Figure 1 presents a principal component biplot [82]. In these data, beef eaters who consume beef 1–2 times per week, defined as light beef-eaters, evaluated beef from 20 different treatments, where treatments included beef from different cuts cooked to two different degrees of doneness using different cooking methods to create differences in Maillard reaction products and lipid heat denaturation. Additionally, beef cuts were selected from USDA Select and Top Choice beef carcasses. Trained descriptive flavor and texture attributes were evaluated with an expert beef panel and consumers ($n = 239$) from three locations across the US in a central location consumer test. While beef differed in tenderness, least square means varied from 1.8 to 4.2 kg with a root mean squared error of 0.73. The biplot shows that tenderness and juiciness liking were closely related and not as closely related to overall liking as the three consumer flavor attributes were. WBSF was inversely related to consumer and trained descriptive attribute tenderness and juiciness measures, as would be expected, as higher WBSF values indicated increased toughness. Interestingly, trained descriptive flavor attributes of "fat-like" and "overall sweet" were closely related to overall liking or would be classified as positive flavor attributes, whereas trained descriptive flavor attributes of "cardboardy," "liver-like," and "sour aromatic" were negatively associated with overall consumer liking. This trend has been reported in multiple studies indicating that, as long as tenderness is acceptable, beef flavor attributes are currently the major driver for overall consumer liking. However, there is a caveat.

Not all consumers have the same drivers for overall liking, and there are segments or clusters in most consumer data.

Table 1. Selected studies examining consumer overall liking of beef.

Study and Treatments	Country for Consumer Selection	Consumer Liking Rating	
Moreles et al. 2013 [1] [59]	Chili		
Pasture, low marbling		4.9 [ab]	
Pasture, high marbling		4.9 [ab]	
Feedlot, low marbling		4.6 [b]	
Feedlot, high marbling		5.2 [a]	
Garcia-Torres et al. 2016 [2] [77]	Spain		
Organic fed on grass		5.95 [b]	
Organic fed on concentrate		6.74 [a]	
Conventional production fed on concentrate		6.89 [a]	
Realini et al. 2013 [3] [90]			
Spain	Spain		
Grass-fed		5.66	
Grass plus concentrate (0.6%) fed		5.83	
Grass plus concentrate (1.2%) fed		5.59	
Concentrate plus hay fed		5.43	
France	France		
Grass-fed		5.53 [ab]	
Grass plus concentrate (0.6%) fed		5.63 [a]	
Grass plus concentrate (1.2%) fed		5.69 [a]	
Concentrate plus hay fed		5.11 [b]	
United Kingdom	United Kingdom		
Grass-fed		5.48 [a]	
Grass plus concentrate (0.6%) fed		5.67 [a]	
Grass plus concentrate (1.2%) fed		5.62 [a]	
Concentrate plus hay fed		4.98 [b]	
Killinger et al. [4] [35]	United States		
High marbled beef		5.4 [a]	
Low-marbled beef		5.1 [b]	
Sepulveda et al. 2019 [5] [91]	United States		
Prime		67.8 [a]	
Top Choice		65.0 [ab]	
Low Choice		61.2 [bc]	
Select		59.6 [c]	
Bueso et al. 2018 [4] [92]	Hondurus	5.2 [a]	
	United States	4.0 [b]	
		US.	Honduran
Grain fed, Select US beef	Values estimated	5.2 [b]	5.0 [b]
Grain fed Top Choice US beef	From graph	5.8 [a]	5.1 [b]
Honduran grass-fed, Bos indicus		5.0 [b]	3.5 [d]
Honduran grain-fed		4.4 [c]	3.1 [d]
Corbin et al. 2015 [5] [78]	United States		
Australian Wagyu (26.64%)		70.15 [a]	
American Wagyu (18.37%)		73.22 [a]	
Prime (14.67%)		71.58 [a]	
High Choice (8.99%)		61.24 [b]	
Top Choice, Holstein (8.54%)		62.67 [b]	
Low Choice (5.56%)		62.93 [b]	
Grass-finished (3.81%)		43.31 [d]	
Select, Holstein (3.45%)		50.40 [c]	
Select (3.31%)		50.95 [c]	
Standard (1.96%)		45.20 [cd]	

Table 1. *Cont.*

Study and Treatments	Country for Consumer Selection	Consumer Liking Rating
Van Wezemail et al. 2014 [3] [61]	Norway and Belgium	
WBSF 19–29.99 N		6.04
WBSF 30–40.99 N		6.08 [a]
WBSF 41–51.99 N		5.16 [ab]
WBSF 52–62.99 N		5.28 [ab]
WBS 63–73.99 N		4.18 [b]
McCarthy et al., 2017 [5] [68]	Republic of Ireland	
Irish beef		58.7
Australian beef		62.2
Chong et al. 2019 [5] [69]	Northern Ireland	55.6 [a]
	Republic of Ireland	55.7 [a]
	Great Britain	59.6 [b]
Hwang et al. 2008 [5] [72]		
Grilling cooking method	Australian	63.5
Barbeque cooking method		66.2
Grilling cooking method	Korean	55.9
Barbeque cooking method		61.8
Bonny et al. 2017 [5] [66]	France	56.3
	Ireland	54.0
Only for consumers preferring medium degree of doneness	Northern Ireland	51.2
	Poland	55.6
Sitz et al. 2005 [4] [38]	Australian	4.34
	United States	5.37
	Canadian	5.49
	United States	5.79

[abc] Within a study and column, means with the same letter are not significantly different ($p > 0.05$). Note that not all studies provided mean separations. [1] 1 = dislike extremely; 7 = like extremely. [2] 1 = dislike extremely; 10 = like extremely. [3] 1 = dislike extremely; 9 = like extremely. [4] 1 = dislike extremely; 8 = like extremely. [5] 1 = dislike extremely; 100 = like extremely. WBSF = Warner-Bratzler shear force.

Beef is traditionally produced using either forages (defined as forage- or grass-fed), forages with grain supplementation, or high energy gran-based diets fed in the last 60 or more days prior to slaughter. For some consumers, the production system affects beef preferences based on personal beliefs or emotions in relationship to animal welfare, environmental issues, health, sustainability, and other factors [13,15,16]. The intent of this article is not to address psychological or marketing issues affecting consumer intent to purchase, but to concentrate on the meat sensory factors impacting consumer liking. The question is whether beef production systems, forage- or grass-fed versus grain-fed beef systems, impact the consumer perception of beef's overall liking and the assessment of tenderness, juiciness, and flavor. Grass- or forage-based production systems are extensively used in Europe, Mexico, Central and South America, Africa, Australia, and Asia. While grain-based systems are the prevailing production systems for commodity beef in North America, grass-based beef production systems are evolving and becoming more prominent. Grass- or forage-based systems vary in the type of forage and quality of forage available for consumption, and extensive research has been conducted to understand beef production characteristics associated with quality of forage [93]. It has been well established that the quality of forage influences animal growth and the subsequent beef carcass characteristics. The question is whether beef derived from forage- or grain-based beef production systems impacts consumer liking.

It has been well established that beef produced on forage-based diets results in beef that is generally lower in total lipids [94,95]. Additionally, fatty acid composition can also be impacted; however, breed type and dietary forage can influence fatty acid composition [94,95]. Daley et al. (2010) and Van Elswyk and McNeill [94,95] reviewed research comparing the fatty acid composition of beef from forage- and grain-based production systems. In general, beef from forage-fed cattle is higher in saturated fatty acids, lower in monounsaturated fatty acids (mainly oleic acid), and higher in polyunsaturated fatty acids. The total amount of fat and fatty acid composition has been shown to affect the sensory characteristics of forage- and grain-fed beef [38,96–112]. In addition to changes in lipid content and fatty acid levels in forage- versus grain-fed beef, beef from forage-based systems has been shown to have higher levels of off-flavors compared to beef from grain-fed systems [94,96,97,113]. Elmore et al. [96] reported that grass-fed beef contained higher levels of diterpenoids—derivatives of chlorophyll called phyt-1-ene and phyt2-ene—that contributed to the flavor differences between grass- and grain-fed beef. The green odor found in grass-fed beef has also been associated with higher levels of hexanals derived from oleic and -linoleic acids [114]. In the United States, Canada, and Australia, beef consumers tend to like beef derived from grain-based production systems [38,78,91,101,103]. With increased exposure to beef from forage-based systems, larger consumer segments may develop preferences for beef from forage-based systems in the US. Japanese and South Korean consumers like highly marbled beef that is derived from grain-based beef production systems [72,74,75,108]. In other countries, varying results have been reported, and the drivers of liking differed across consumer segments. Beef from forage-based systems tends to be higher in lean-type flavors, such as beef identity, bloody/serumy, metallic, and liver-like, and lower in lipid-derived flavors, such as fat-like and cardboardy [38,114]. Oliver and others [115] showed that consumers in Germany, Spain, and the United Kingdom had different overall liking ratings when evaluating beef from forage-based systems in Uruguay versus locally grown beef from their respective countries. They identified three main clusters of consumers: consumers that preferred locally grown beef, consumers that preferred beef from Uruguay, and consumers that did not differentiate. In Chili, Morales et al. [59] reported that, in blind consumer tests, consumers in Chili liked beef from feedlot finished steers that had higher levels of marbling compared to both beef from pasture-fed low and high marbled beef and grain-fed low marbled beef. Garcia-Torres et al. [77] used beef from three production systems (organic beef fed on grass, organic beef fed on concentrate, and conventional concentrate feeding) in Spain. Spanish consumers segmented into two clusters. Cluster 1 mainly consisted of young and mature women from low to middle income levels with educational levels from secondary school to university studies. Consumers in Cluster 2 were mainly mature men with middle to high incomes with university education. Interestingly, consumers from Clusters 1 and 2, in an overall assessment, rated cooked meat lower for organic beef fed on grass compared to beef from either organic or conventional concentrate-fed treatments. Results from most of these studies are presented in Table 1. In summary, diets prior to slaughter for beef, either forage-, forage- and grain-, or grain-based diets, impact the sensory characteristics and consumer liking of beef; however, consumer segments for drivers of liking differ, and preferences are apparent across countries.

Extensive research has been conducted to examine other pre- and post-harvest factors that impact consumer liking or beef sensory properties. These papers will not be discussed, as extensive reviews have been published, though this research is important, as it provides factors that impact beef tenderness, juiciness, and flavor. It should be noted that, based on the data presented, factors affecting the eating quality of beef most likely will impact overall consumer liking.

While much of the meat science consumer research examines consumers who consume beef, pork, lamb, chicken, or other protein sources, the objectives of the research is usually to examine consumers of specific species. It should always be recognized that consumers who are beef consumers most likely are consumers of other protein sources. Therefore, the previous discussion of drivers of liking for beef most likely extend to pork and lamb consumers. However, pork and lamb consumers may have additional drivers of liking, and other factors may influence their overall liking. The following discussion will incorporate these additional drivers of consumer liking in addition to the aforementioned tenderness, juiciness, and flavor aspects.

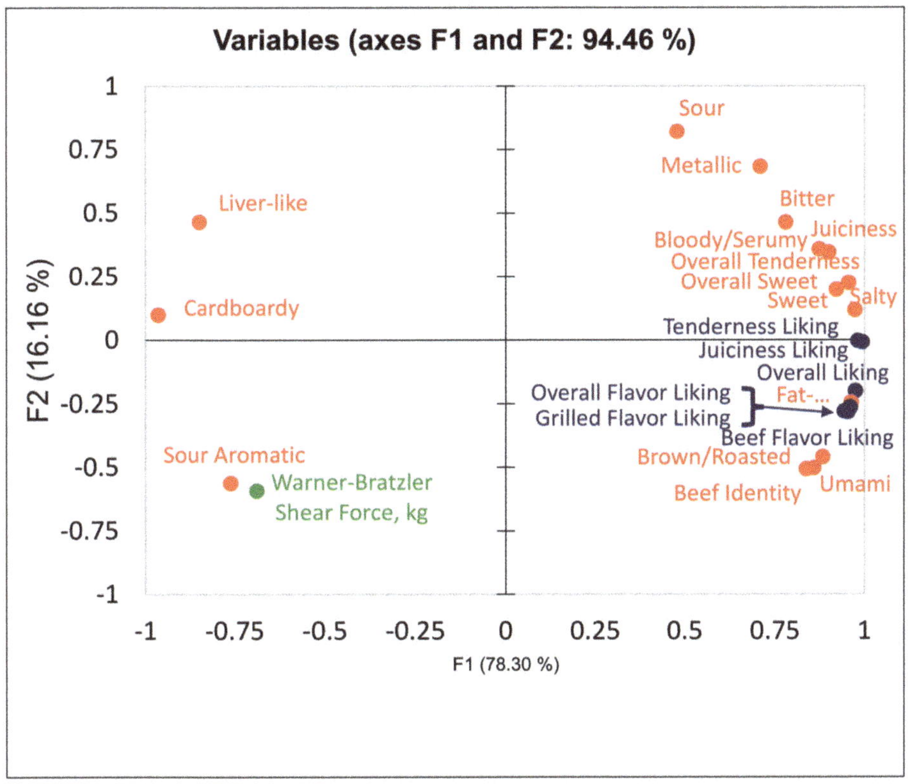

Figure 1. Principal component biplot of data adapted from Laird (2015) [90] and Luckemeyer (2015) [82] where descriptive flavor and texture attributes are in red, Warner-Bratzler shear force (WBSF) values are in green, and consumer sensory attributes are in blue.

3. Pork Consumers

Quality drivers of pork are generally recognized to be pH, water holding capacity, color, and marbling. Tenderness is considered a quality component for pork, but it has not always been evaluated. As pork is more variable in pH, water holding capacity, and color, how these factors influence consumer liking has been extensively examined across multiple countries and regions. In summarizing this research, pork that is high in pH is generally darker in color, has higher water holding capacity, is more intense in some flavor attributes, is juicier, and tends to be more tender than pork with a normal pH. Pork with a lower than normal pH has been shown to be lighter in color, have lower water holding capacity, be less intense in flavor, be drier, and be tougher than normal pH pork. Additionally, pork is traditionally cooked to higher degrees of doneness, and as low pH pork has lower water holding capacity, pork cooked to higher degrees of doneness tend to be tougher and drier. Extensive research globally has been conducted to examine these relationships.

In the United States, the last large comprehensive consumer study was to examine the effects of the aforementioned quality attributes on consumer perception [116]. Pork loins ($n = 679$) were selected to differ in pH, intramuscular fat, Minolta L* color values, and WBSF. Prior to serving to consumers, within the aforementioned parameters, pork was cooked to four endpoint cooking temperatures. The intent was to examine the main and interactive effects of the treatments. Consumers ($n = 2280$) were selected in three cities and were regular pork consumers. It should be noted that pork consumers are usually beef consumers as well. Data were analyzed using ordered logistical regression, and predicted mean consumer pork loin liking responses were presented. Consumer liking increased slightly as intramuscular fat levels increased, but increased levels of pH and decreased levels of WBSF influenced consumer liking ratings to a greater extent. Additionally, they found that, as internal cooking temperature increased, overall, tenderness and juiciness liking decreased [116]. They concluded that pork that had lower WBSF values, a higher pH, and intramuscular fat and that was cooked to lower internal temperature endpoints were liked to a greater extent by consumers.

Other consumer studies have examined factors driving consumer liking for pork. While the magnitude of the results may not be similar to those discussed above [116], consumer research has generally shown that pH, water holding capacity, tenderness as defined by either mechanical methods or trained sensory panel methods, intramuscular fat, and endpoint cooking temperature affect consumer liking [117–125]. However, pork production systems vary across some countries, and pork diets do differ in some areas of the world. In the United States, commodity pork is derived from animals fed soy bean- and corn-based diets. Additionally, live weight to slaughter varies by production system, and some counties castrate their males, while other countries traditionally slaughter intact males. These effects from production system/nutrition, live weight endpoint, and sex, mainly castrated versus non-castrated males, have been shown to affect pork sensory characteristics and consumer perception.

As pigs are monogastrics, flavor components from their diet can be absorbed and stored in the water portion or cytoplasm of adipose cells. Some of these compounds may be volatile aromatics and affect the flavor of the subsequent meat. The questions are whether consumers can detect these flavor aromatic compounds and, though they may not be able to describe the flavor, whether this change in flavor affects their overall liking. Research that examined changes in the energy level, grain source, supplementation sources, fat sources, and minor dietary constituents for pork has been presented [126]. While some dietary components influenced the pork's flavor and lipid oxidation level, within the realm of standard high concentrate diets, little effect was found by adjusting small components of the diet. The adage that pork tastes like what the pig eats has some credence. Pigs fed specialty diets result in pork that generally has flavor attributes associated with that diet. The classic example is the production of Iberian pork meat in Spain, where pigs are fed acorns as a component of their diet. It has been well documented that fatty acid levels are affected by the diet and feeding system. For example, flavor and texture differences were apparent in meat from Iberian pigs fed on acorn and pasture versus confinement feeding [127]. Extensive research has also been conducted to examine breed type in pigs and the subsequent effect on consumer sensory perception. In general, breed or genetic types that produce pork that is higher in marbling and/or pH, such as Berkshire pigs for example, result in improved consumer perception.

Pork meat containing boar taint and the subsequent effect on consumer acceptability has been extensively examined [128]. Two main compounds have been associated with boar taint: androsterone (specifically androst-16-en-3-one) [129] and skatole (3-methylindole) [130,131]. Androsterone is a steroid hormone synthesized in the Leydig cells of the testis, and levels are affected by animal age, weight, sexual maturity, feeding and rearing conditions, herd factors, and genetics [132–141]. Skatole has a fecal odor and is the product of the anaerobic degradation of tryptophan in the gut. Its levels depend on the diet, rearing conditions, and handling of the pigs as well as sex, age, and genetics [134,142–147]. These compounds are mainly, but not solely, found in intact male pigs. With increased concerns over animal welfare, the practice of castration has been discontinued in some countries. It is generally accepted that as skatole and androsterone increase, consumers can detect off-flavors in pork, and consumer liking subsequently decreases [128]. The cooking procedure, the type of meat, and the age of the animal (intact males that have not shown signs of the onset of puberty) can affect consumer detection and levels of boar taint. Interestingly, consumers can vary in their detection of boar taint. The cooking method's effect on masking boar taint in pork has been reported [148]. The most common thresholds for bore taint in pork are 0.5 and 1.0 µg/g for androstenone in fat tissue [134,149,150] and 0.10 and 0.20 µg/g for skatole in fat tissue [134,149,151–153]. However, approximately 40%–50% of consumers are anosmics or cannot detect androstenone [149,154–156]. Skatole on the other hand is perceived by 99% of consumers [157]. Research showed that consumer acceptability for pork with boar taint containing different levels of androstenone resulted in three consumer segments: pork lovers, boar meat lovers, and reject boar taint meat lovers [158]. Some studies have also shown that women are more sensitive than men to androstenone [154,155,159–161]. Immunocastration of pig has been proposed as a method to decrease the incidence of boar taint in pork with the subsequent strategy that immuocastration would decrease androstenone in intact males and increase consumer acceptance [162].

4. Lamb Consumers

Lamb meat across the globe has more variation in diet/nutrition and production parameters than pork and beef, resulting in more variability in lamb eating quality. Lamb is generally produced on either native forage, native forage with supplementation, or concentrate diets. However, more lamb is finished on forage-based diets than on concentrate diets globally. Feeding lambs concentrate diets is a common practice in the United States, but it is not a common practice in major lamb production counties of Australia, New Zealand, some European countries, and South America. Lamb flavor differs from beef and pork flavor and has mainly been reported based on differences in fatty acid content. When comparing lamb from animals fed on pastures versus lambs fed on high concentrate diets, differences in the fatty acid level and the subsequent differences in flavor have been reported [163–167]. In general, lambs fed on forage or pasture have more intense lamb, rancid, and liver flavors, as well as more intense levels of off-odors and off-flavors than lambs fed on concentrate diets. The type of forage also impacts lamb flavor [164–166]. The major fatty acid change in forage versus concentrate feeding is that forage-fed lambs have been shown to have higher levels of linolenic acid [168,169]. Linolenic acid is susceptible to lipid oxidation and off-flavors associated with lipid oxidation [170,171]. Caneque et al. [168] showed that pasture-fed lamb has been shown to have lower levels of linoleic acid [168]; linoleic acid is a component of lamb flavor identity [172,173]. While skatole has been associated with boar taint, meat from pasture-fed castrate and non-castrate lambs had higher skatole levels and may have a similar development of undesirable flavors in lamb as reported for pork [171]. It has been reported that increased skatole levels in lamb resulted in more intense sheep meat odor [163]. As in beef and pork, feeding diets with higher levels of grain or high concentrate diets result in lamb carcasses with higher fat levels and meat with higher levels of marbling [162,166]. Other research has shown that tenderness in lamb was related to marbling levels [174,175].

The question is whether differences in lamb eating quality induced by the feeding system affect consumer acceptability. Acceptance of Spanish, German, and French consumers for Uruguayan lamb fed a concentrate diet, a pasture-only diet, or two diets that consisted of a combination of pasture and concentrate diet was examined [167]. Consumers across countries preferred lamb fed the concentrate diet for overall, tenderness, and flavor acceptability. Similarly, Spanish consumers preferred the lamb fed concentrate diets [176]; whereas German consumers, even though they ate lamb less than once per month, preferred lamb from animals that had been fed a high concentrate diet [177]. Consumers in the UK preferred lamb that was from younger animals, and they indicated that they disliked intense mutton odor and flavor [178]. Mutton flavor has been associated with lamb from older animals and from pasture-fed animals. However, consumer clusters within country have been reported, and it was found that some consumers did prefer pasture-fed lamb, especially if the lambs had some supplementation of grain [167].

Lamb from Australia is predominantly derived from animals that are grass-fed and may have some grain supplementation [179–181]. Australian consumers did not distinguish lamb from animals fed pasture- or grain-based diets [182]. However, predominantly pasture-fed lamb was found to have a "pastoral" flavor [183]. "Mutton" flavor is traditionally associated with meat from older animals and has been identified in Australian lamb. Australian lambs are commonly harvested from 4 to 24 months of age, and older animals, defined as mutton, may also be harvested. Meat from older sheep or mutton has been shown to have detectable and higher levels of mutton flavor [184], and in general meat from older animals were less acceptable to consumers for tenderness, juiciness, flavor, and overall liking than meat from younger lambs [185]. Flavors defined as "pastoral" and "mutton" have been associated with increased levels of branched chain fatty acids. It has been reported that branched chained fatty acids most likely do not affect lamb flavor until lambs are over one year of age [186]. Levels of branched chain fatty acids were lower in animals less than one year of age, and the relationship between the level of branched chain fatty acids and lamb flavor was not found in younger animals. The relationship between the level of branched chain fatty acids and the intensity of pastoral and mutton flavors in lamb and subsequent consumer acceptance was examined [187]. They found that branched chain fatty acids of 4-methyloctanoic and 4-ethyloctanoic were related to consumer odor and overall liking attributes, but not to consumer flavor liking [187]. While other consumer studies have been conducted with Australian consumers and have reported a slight preference for grain-fed lamb [182,188], other drivers may influence consumer acceptance, such as frequency of consumption and environmental and health concerns [189].

Australian, Chinese, and United States consumers rated Australian lamb [57,189]. In general, consumers rated lamb loin and topside cuts similarly across countries, except for Chinese consumers, who rated topside cuts slightly lower in tenderness liking than Australian and United States consumers. Twelve different types of lamb produced in Greece, Italy, Spain, France, Iceland, and the United Kingdom were evaluated in a Home Use Test by consumers in these same countries [170]. Lamb was from animals that were either intact males, females or castrated males from 10 breeds and crossbreeds, fed pre-slaughter diets composed of either milk, concentrates, and various forages as main ingredients, of slaughter ages from 1 and 12 months, and with carcass weights from 5.5 to 30.4 kg. Family groups were identified that differed in their lamb preferences [190]. Those families with Mediterranean origin preferred lamb fed either milk or mainly concentrate diets, and families of mainly a northern origin preferred lamb from grass or with grass included in the diet. The remainder of consumers had a wider taste preference. Sanudo et al. [191] concluded that there are two categories of European lamb consumers of lamb: those who prefer a "milk or concentrate taste" and those who prefer a "grass taste." While other consumer research has been conducted for lamb consumer liking, these studies provide some assessment of variation in taste preference for consumers in different markets.

In general, tenderness and juiciness are important to lamb consumers, as similarly identified for pork and beef consumers; however, flavor differences in lamb can impact flavor and overall acceptability. Consumer preferences for flavor are most likely affected by past experience in lamb consumption and cooking methods.

5. Consumer Segments

Consumer segments are extensively discussed in marketing and consumer insight research. It is well known that consumers today differ from consumers in the 1980s. The consumer segment defined as millennials, individuals born between 1977 and 1995, represent about 25% of the buying power in the US economy [192]. Millennials have been described as special, sheltered, confident, team-oriented, achieving, pressured, and conventional, and they are more numerous, more affluent, better educated, and more ethnically diverse than the previous major consumer segment, the baby boomers [193]. The US census identified millennials as the most diverse generation in American history, with 44.2% of millennials defined from a minority group [194]. Since millennials have had constant internet and social media access, they are more likely to be open to change and are more self-expressive than older generations [195]. Millennials use technology much more than previous generations did. Based on the changes in the environment for consumer segments, the question that Laird [83] addressed was as follows: Do millennials have the same drivers of overall liking as the combination of generation X and baby boomer consumer segments? It should be noted that generation X and baby boomer segments were combined. She recruited consumers in four cities that were either light (eat beef 2 or 3 times per month) or heavy (eat beef 3 or more times per week) beef eaters that were millennials or were older than millennials, which included generation X and baby boomer consumer segments. Consumers evaluated two different beef, chicken, and pork cuts in a Central Location Test (CLT) and in a Home Use Test (HUT). While millennials tended to cook beef steak, chicken breasts, and pork chops using pan-frying cooking methods more frequently than grilling, differences in consumer sensory attributes did not differ, except light beef eaters in both age groups had lower scores than heavy beef eaters (Table 2). Consumers from the four segments indicated that price was the first driver of meat purchase intent and that overall flavor liking was most closely related to overall liking for beef consumers, as similarly reported in Figure 1. Additionally, juiciness and tenderness liking were closely associated with each other and somewhat closely related to overall liking. Interestingly, consumers across the four segments rated the beef, chicken, and pork similarly, regardless of whether they were evaluated in the CLT or HUT portion of the study. These data demonstrated that, while there may be other drivers of purchase intent and of potential repeat purchases, demographics had little impact on consumer liking. It should be noted that consumer segments based on emotional or environmental concerns were not included. Therefore, questions on the drivers of liking for consumers who want, for example, locally grown, grass-fed beef without hormone applications was not addressed.

Table 2. Least square means for consumer sensory attributes across consumer groups adapted from Laird (2015) [83].

Treatment	Overall Liking	Overall Flavor Liking	Beef/Pork/ Chicken Liking	Grilled Flavor Liking	Juiciness Liking	Tender-ness Liking
p-value	*0.01*	*0.07*	*0.08*	*0.36*	*0.34*	*0.44*
Millennial, light beef eater	5.9[a]	5.8	6.0	5.4	6.3	6.2
Millennial, heavy beef eater	6.2[bc]	6.0	6.1	5.6	6.2	6.3
Non-millennial, light beef eater	5.9[ab]	5.9	5.9	5.5	6.2	6.3
Non-millennial, heavy beef eater	6.3[c]	6.1	6.3	5.6	6.5	6.5
Root Mean Square Error	2.21	2.23	2.22	2.34	2.28	2.27

[abc] Mean values within a column followed by the same or no letter are not significantly different ($p > 0.05$).

If consumer segments or demographics have changed, but are not related to consumer sensory perception, are there other consumer clusters or segments that may affect consumer perception? Consumer clustering techniques have evolved and are commonly used in marketing or consumer insight research. Statistical clustering techniques, such as k-means, agglomerative hierarchical clustering, Gaussian mixture models, and univariate clustering, have been developed to segment consumers into groups based on similarities defined by the model. These techniques have had minimal use in the meat science literature. In examining data where beef top loin, top sirloin, tenderloin, and beef bottom round roasts were evaluated in four cities with 3541 consumer observations, six consumer segments were identified using agglomerative hierarchical cluster analysis, where consumer overall liking, flavor liking, and beef flavor liking were used to develop clusters (Figure 2) [82,83]. The six clusters differed in overall, flavor, and beef flavor liking. The objective of this analysis was to see if consumers who rated beef samples differently across beef products had different drivers of liking.

Using principal component analysis, segmentations of clusters were reported (Figure 2). It is apparent that consumers in Cluster 1 were more closely related to muscle fiber tenderness, sweet, overall sweet, bitter, and umami and negatively related to WBSF. These consumers appear to be more driven by tenderness than by flavor attributes. Consumers in Clusters 3, 4, 5, and 6 rated beef with more emphasis on beef flavor attributes. The segmentation of consumers in Clusters 4 and 6 indicates that flavor attributes affected their liking ratings differently. Consumers in Cluster 6 rated beef samples with more intense amounts of beef identity, sour, fat-like, and warmed over flavor aromatics higher for overall liking and flavor liking. However, consumers in Cluster 4 rated beef that had more intense levels of bloody/serumy, cardboardy, and liver-like flavor attributes higher for overall and flavor liking, and a higher WBSF value did not affect their overall liking ratings. While these are general trends, this analysis begins to segment consumers, and it is apparent that consumers have different drivers for overall liking. Oliver et al. [115] reported three clusters when examining consumer liking for Uruguayan meat and locally sourced meat in Germany, Spain, and the United Kingdom. They were able to detect differences in consumer age and beef product usage across consumers segments within each country. They concluded that Uruguayan beef was acceptable in Germany and to a lesser extent in Britain and Spain. Borgogno et al. [13] used k-mean cluster analysis to segment consumers based on familiarity with different types of fresh meat. The reported two clusters for familiarity ratings where Cluster 1 was defined as the low familiarity segment and Cluster 2 was the high familiarity group. They examined the effect of label information, the liking of appearance, the liking of taste, and the expectations of fresh meat quality. Consumer clustered differed for some of these attributes. Garcia-Toirres et al. [77] also used cluster analysis to differentiate consumers based on their knowledge of organic production, the frequency of consumption of organic food, and sociodemographic variables. While there were only slight differences in consumer overall liking for three types of beef, there were differences in the relative importance of attributes associated with color, origin, production system, and price for beef. Morales et al. [59] calculated three clusters for consumers in Chile when evaluating beef from four production systems (pasture-low marbling; pasture-high marbling; feedlot-low marbling; feedlot-high marbling). Cluster 1 included consumers that typically eat lean beef and was 75% women. Cluster 2 was composed mainly of men over 40 and was characterized as high expectation consumers, and Cluster 3 was composed of high education level consumers. Bernues et al. [14] segmented Spanish consumers based on a survey into four clusters. Cluster 1 was defined as the traditional or conservative cluster and was characterized as containing consumers who liked to cook, did not like foreign meals, and considered planning meals important for family nutrition and health. Cluster 2 consumers did not like or spend time cooking and did not like changes in meals. The consumers in Cluster 3 had open attitudes toward innovation and were deemed adventurous. They liked to cook, tried foreign recipes, and frequently changed their meals. Cluster 4 contained the lowest number of consumers and was categorized as careless. These consumers had simple eating and cooking habits, did not like going to restaurants, ordered meat when eating out, and were mainly young men who lived in cities and had low incomes and low levels of education. While consumer liking attributes were not evaluated, these data showed that consumer segmentation affects consumer attributes about meat. Nie and Zepeda [196] segmented US consumers into four clusters: rational, adventurous, careless, and a fourth segment that had some characteristics of both conservative and uninvolved consumers. The segments exhibited significant differences in organic and local food consumption. It was apparent that consumers across the four clusters differed in lifestyle and food purchasing habits.

Font i Furnols et al. [197] used cluster analysis to segment Spanish, French, and British consumers into clusters based on country of origin, price, and feeding system for lamb meat. They showed that consumers in different clusters rated lamb differently for overall acceptability depending on the lamb production system of pasture, pasture + 0.6% live weight of concentrate, pasture + 1.2% live weight of concentrate, and concentrate and alfalfa. These studies show that consumer clustering can be a valuable tool in understanding drivers of overall liking. As consumers have their own unique attitudes, experiences, expectations, and lifestyles, additional information on drivers of overall liking can be obtained by the addition of these type of analyses.

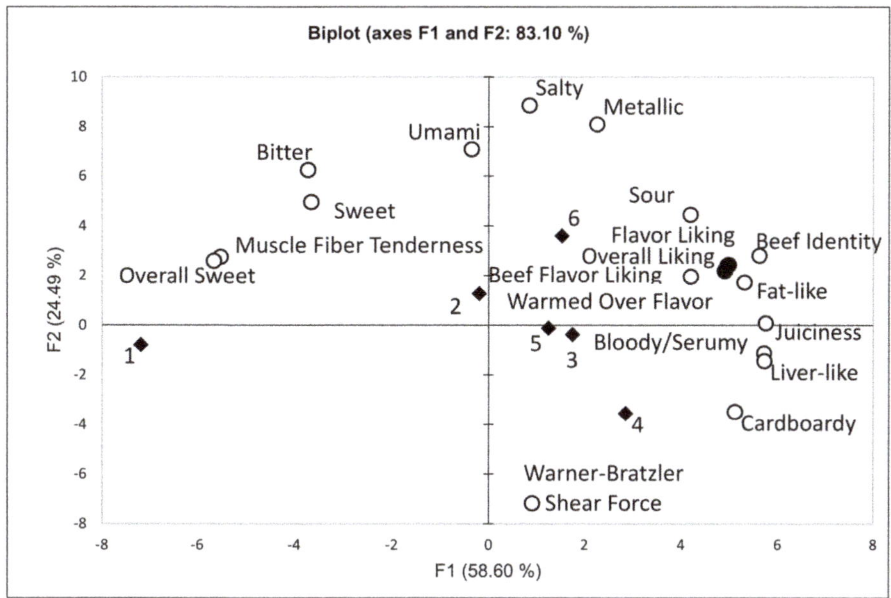

Figure 2. Principal component biplot of selected data from Laird (2015) [83] and Luckemeyer (2015) [82] using the six consumer liking clusters from Table 2. (♦), descriptive flavor and texture attributes and WBSF values (○), and consumer sensory attributes (●).

6. Other Consumer Sensory Methods

As a community, meat scientists have traditionally used either Central Location or Home Use consumer testing methods to address consumer perception. Some researchers have branched out to work with either agricultural economists or consumer insight/marketing experts to understand purchase intent. While these methods have application and provide additional information on consumer perception, other methods used by psychologists, neural psychologists, and human behaviorists can assist meat scientists in further understanding consumer acceptance. It is generally accepted that 95% of consumer decisions are influenced by the subconscious. Regardless of the percentage, it is greatly recognized that the subconscious plays a major role in consumer perception. However, how do we as meat scientists measure subconscious effects and its impact on consumer acceptance? Research tools are available for assessing and further understanding consumer perception and what drives consumer liking from the human behavior aspect [198]. Human behavior tools assess different aspects of consumer cognitive processes and can be used to further understand consumer perception. They include eye tracking [199–201], electrophysical responses such as heart rate, blood pressure, and skin conductance, electrodermal activity or galvanic skin response, brain activity through electroencephalography [202–205], fMRI [206–210], and facial expression [211]. These methods measure emotional and physiological responses and provide additional information on how consumers perceive a product. Use of these tools may help meat scientists fully understand consumer segments and the environmental and social factors that drive consumer liking within segments.

One of the greatest challenges consumer sensory scientists face is the impact of environments on consumer preferences [11]. Consumers respond differently in Central Location tests where every aspect of the environment and product presentation is controlled versus consumer responses in Home Use tests where the consumer has the opportunity to handle the product and prepare it as it would normally be used in the home, and where the family environment may influence overall liking. It has been well documented that an environment impacts sensory responses. Augmented and virtual reality systems have been used to emulate an environment similar to the environment where consumers would consume the product [212,213]. They reported differences in consumer liking based on the environment. Incorporation and understanding of environmental factors may assist researchers in more accurately understanding consumer perception and provide additional research tools.

7. Conclusions

Consumers can detect differences in eating quality of meat products. Consumer preference and acceptance of meat products across whole muscle beef, pork, and lamb showed that consumers perceive differences in overall liking in meat products mainly using assessments of juiciness, tenderness, and flavor. In general, as long as tenderness and juiciness are at acceptable levels, flavor is the main driver of overall liking. It should always be understood that all three components influence overall liking. However, there are different drivers of overall liking across consumer segments. Consumers usually assess tenderness and juiciness together or similarly. For flavor, consumers differ in drivers of liking. Product usage or exposure, environmental factors, socioeconomic factors, and cooking methods, for example, can influence consumer liking and consumer flavor liking. Consumer research has been extensively used as a research tool since the 1990s, and Australia has based their MSA system on consumer research, indicating that consumer evaluation of meat products is a viable, repeatable research tool. New advances in statistical tools to understand consumer segments are being incorporated into meat science research, but further use of these tools may assist scientists in understanding drivers of consumer liking. Additionally, the use of human behavior research tools may assist meat scientists in further understanding consumer drivers of liking.

Funding: This research received no external funding.

Conflicts of Interest: The author declares no conflict of interest.

References

1. Watkins, R.M. Finish in beef cattle from the standpoint of the consumer. *J. Anim. Sci.* **1936**, *1936b*, 67–70. [CrossRef]
2. Scott, A.L. How much fat do consumers want in beef? *J. Anim. Sci.* **1939**, *1939*, 103–107. [CrossRef]
3. Macintosh, D.L.; Hall, J.L.; Vail, G.E. Some observations pertaining to tenderness of meat. *J. Anim. Sci.* **1936**, *1936b*, 285–289. [CrossRef]
4. Brady, D.E. Results of consumer preference studies. *J. Anim. Sci.* **1957**, *16*, 233–240. [CrossRef]
5. Francis, J.J.; Romans, J.R.; Norton, H.W. Consumer rating of two beef marbling levels. *J. Anim. Sci.* **1977**, *45*, 67–70. [CrossRef]
6. USDA. *United States Standards for Grades of Carcass Cattle*; Liwest. Seed Program, Agricultural Market Service: Washington, DC, USA, 2016.
7. ASTM STP434. *Manual on Sensory Testing Methods*; Committee E18; ASTM International: West Conshohocken, PA, USA, 1968.
8. AMSA. *Guidelines for Cookery and Sensory Evaluation of Meat*; American Meat Science Association: Chicago, IL, USA, 1978.
9. AMSA. *Research Guidelines for Cookery, Sensory Evaluation and Instrumental Tenderness Measurements of Fresh Meat*; American Meat Science Association: Chicago, IL, USA, 1995.
10. IFT. Guidelines for the preparation and review of papers reporting sensory evaluation data. *J. Food Sci.* **1995**, *60*, 211.
11. Meilgaard, M.; Civille, C.V.; Carr, B.T. *Sensory Evaluation Techniques*, 5th ed.; CRC Press/Taylor & Francis Group: Boca Raton, FL, USA, 2016.
12. Savell, J.W.; Branson, R.E.; Cross, H.R.; Stiffler, D.M.; Wise, J.W.; Griffin, D.B.; Smith, G.C. National consumer retail beef study: Palatability evaluations of beef loin steaks that differed in marbling. *J. Food Sci.* **1987**, *52*, 517–519, 532. [CrossRef]
13. Borgogono, M.; Favotto, S.; Corazzin, M.; Cardello, A.V.; Piasentier, E. The role of product familiarity and consumer involvement on liking and perception of fresh meat. *Food Qual. Pref.* **2015**, *44*, 139–147. [CrossRef]

14. Bernues, A.; Ripoll, G.; Panea, B. Consumer segmentation based on convenience orientation and attitudes towards quality attributes of lamb meat. *Food Qual. Pref.* **2012**, *26*, 211–220. [CrossRef]

15. Verbeke, W.; Van Wezemael, L.; de Barcellos, M.D.; Kugler, J.O.; Hocquette, J.F.; Ueland, O.; Grunert, K.G. European beef consumers' interest in a beef eating-quality guarantee. Insights from a qualitative study in four EU countries. *Appetite* **2010**, *54*, 289–296. [CrossRef]

16. Font-i-Furols, M.; Guerrero, L. Consumer preference, behavior and perception about meat and meat products: An overview. *Meat Sci.* **2014**, *98*, 361–371. [CrossRef] [PubMed]

17. Savell, J.W.; Cross, H.R.; Francis, J.J.; Wise, J.W.; Hale, D.S.; Wilkes, D.L.; Smith, G.C. National consumer retail beef study: Interaction of trim level, price and grade on consumer acceptance of beef steaks and roasts. *J. Food Qual.* **1989**, *12*, 251–274. [CrossRef]

18. Neely, T.; Lorenzen, C.; Miller, R.; Tatum, J.; Wise, J.; Taylor, J.; Buyck, M.; Reagan, J.; Savell, J. Beef customer satisfaction: Role of cut, USDA quality grade, and city on in-home consumer ratings. *J. Anim. Sci.* **1998**, *76*, 1027–1033. [CrossRef] [PubMed]

19. Lorenzen, C.L.; Miller, R.K.; Taylor, J.F.; Neely, T.R.; Tatum, J.D.; Wise, J.W.; Buyck, M.J.; Reagan, J.O.; Savell, J.W. Beef customer satisfaction: Cooking method and degree of doneness effects on the top loin steak. *J. Anim. Sci.* **1999**, *77*, 637–644. [CrossRef]

20. Neely, T.; Lorenzen, C.; Miller, R.; Tatum, J.; Wise, J.; Taylor, J.; Buyck, M.; Reagan, J.; Savell, J. Beef customer satisfaction: Cooking method and degree of doneness effects on the top round steak. *J. Anim. Sci.* **1999**, *77*, 653–660. [CrossRef]

21. Savell, J.W.; Lorenzen, C.L.; Neely, T.R.; Miller, R.K.; Tatum, J.D.; Wise, J.W.; Taylor, J.F.; Buyck, M.J.; Reagan, J.O. Beef customer satisfaction: Cooking method and degree of doneness effects on the top sirloin steak. *J. Anim. Sci.* **1999**, *77*, 645–652. [CrossRef]

22. Lorenzen, C.L.; Miller, R.K.; Taylor, J.F.; Neely, T.R.; Tatum, J.D.; Wise, J.W.; Buyck, M.J.; Reagan, J.O.; Savell, J.W. Beef Customer Satisfaction: Trained sensory panel ratings and Warner-Bratzler shear force values. *J. Anim. Sci.* **2003**, *81*, 143–149. [CrossRef]

23. Shackelford, S.D.; Koohmaraie, M.; Miller, M.F.; Crouse, J.D.; Reagan, J.O. An evaluation of tenderness of the longissimus muscle of Angus by Hereford versus Brahman crossbred heifers. *J. Anim. Sci.* **1991**, *69*, 171–177. [CrossRef]

24. Hoover, L.C.; Cook, K.D.; Miller, M.F.; Huffman, K.L.; Wu, C.K.; Lansdell, J.L.; Ramsey, C.B. Restaurant consumer acceptance of beef loin strip steaks tenderized with calcium chloride. *J. Anim. Sci.* **1995**, *73*, 3633–3638. [CrossRef]

25. Miller, M.F.; Hoover, L.C.; Cook, K.D.; Guerra, A.L.; Huffman, K.L.; Tinney, K.S.; Ramsey, C.B.; Brittin, H.C.; Huffman, L.M. Consumer acceptability of beef steak tenderness in the home and restaurant. *J. Food Sci.* **1995**, *60*, 963–965. [CrossRef]

26. Boleman, S.J.; Boleman, S.L.; Miller, R.K.; Taylor, J.F.; Cross, H.R.; Wheeler, T.L.; Koohmaraie, M.; Shackelford, S.D.; Miller, M.F.; West, R.L.; et al. Consumer evaluation of beef of known categories of tenderness. *J. Anim. Sci.* **1997**, *75*, 1521–1524. [CrossRef] [PubMed]

27. Roeber, D.L.; Cannell, R.C.; Belk, K.E.; Miller, R.K.; Tatum, J.D.; Smith, G.C. Implant strategies during feeding: Impact on carcass grades and consumer acceptability. *J. Anim. Sci.* **2000**, *78*, 1867–1874. [CrossRef] [PubMed]

28. Miller, M.F.; Carr, M.A.; Ramsey, C.B.; Crockett, K.L.; Hoover, L.C. Consumer thresholds for establishing value of beef tenderness. *J. Anim. Sci.* **2001**, *79*, 3062–3068. [CrossRef] [PubMed]

29. Shackelford, S.D.; Wheeler, T.L.; Meade, M.K.; Reagan, J.O.; Byrnes, B.L.; Koohmaraie, M. Consumer impressions of Tender Select beef. *J. Anim. Sci.* **2001**, *79*, 2605–2614. [CrossRef]

30. Barham, B.L.; Brooks, J.C., Jr.; Blanton, J.R.; Herring, A.D.; Carr, M.A.; Kerth, C.R.; Miller, M.F. Effects of growth implants on consumer perceptions of meat tenderness in beef steers. *J. Anim. Sci.* **2003**, *81*, 3052–3056. [CrossRef]

31. Platter, W.J.; Tatum, J.D.; Belk, K.E.; Chapman, P.L.; Scanga, J.A.; Smith, G.C. Relationships of consumer sensory ratings, marbling score, and shear force value to consumer acceptance of beef strip loin steaks. *J. Anim. Sci.* **2003**, *81*, 2741–2750. [CrossRef]

32. Platter, W.J.; Tatum, J.D.; Belk, K.E.; Scanga, J.A.; Smith, G.C. Effects of repetitive use of hormonal implants on beef carcass quality, tenderness, and consumer ratings of beef palatability. *J. Anim. Sci.* **2003**, *81*, 984–996. [CrossRef]

33. Carr, M.A.; Crockett, K.L.; Ramsey, C.B.; Miller, M.F. Consumer acceptance of calcium chloride-marinated top loin steaks. *J. Anim. Sci.* **2004**, *82*, 1471–1474. [CrossRef]

34. Killinger, K.M.; Calkins, C.R.; Umberger, W.J.; Feuz, D.M.; Eskridge, K.M. A comparison of consumer sensory acceptance and value of domestic beef steaks and steaks from a branded, Argentine beef program. *J. Anim. Sci.* **2004**, *82*, 3302–3307. [CrossRef]

35. Killinger, K.M.; Calkins, C.R.; Umberger, W.J.; Feuz, D.M.; Eskridge, K.M. Consumer sensory acceptance and value for beef steaks of similar tenderness, but differing in marbling level. *J. Anim. Sci.* **2004**, *82*, 3294–3301. [CrossRef]

36. Killinger, K.M.; Calkins, C.R.; Umberger, W.J.; Feuz, D.M.; Eskridge, K.M. Consumer visual preference and value for beef steaks differing in marbling level and color. *J. Anim. Sci.* **2004**, *82*, 3288–3292. [CrossRef] [PubMed]

37. McKenna, D.; Mies, P.; Baird, B.; Peiffer, K.; Ellebracht, J.; Savell, J. Biochemical and physical factors affecting discoloration characteristics of 19 bovine muscles. *Meat Sci.* **2005**, *70*, 665–682. [CrossRef] [PubMed]

38. Sitz, B.M.; Calkins, C.R.; Feuz, D.M.; Umberger, W.J.; Eskridge, K.M. Consumer sensory acceptance and value of domestic, Canadian, and Australian grass-fed beef steaks. *J. Anim. Sci.* **2005**, *83*, 2863–2868. [CrossRef] [PubMed]

39. Cox, R.J.; Kerth, C.R.; Gentry, J.G.; Prevatt, J.W.; Braden, K.W.; Jones, W.R. Determining acceptance of domestic forage- or grain-finished beef by consumers from three southeastern U.S. states. *J. Food Sci.* **2006**, *71*, S542–S546. [CrossRef]

40. Kerth, C.R.; Braden, K.W.; Cox, R.; Kerth, L.K.; Rankins, D.L., Jr. Carcass sensory, fat color, and consumer acceptance characteristics of Angus-cross steers finished on ryegrass (Lolium multiflorum) forage or on a high-concentrate diet. *Meat Sci.* **2007**, *75*, 324–331. [CrossRef] [PubMed]

41. Wismer, W.V.; Okine, E.K.; Stein, A.; Seibel, M.R.; Goonewardene, L.A. Physical and sensory characterization and consumer preference of corn and barley-fed beef. *Meat Sci.* **2008**, *80*, 857–863. [CrossRef]

42. Mehaffey, J.M.; Brooks, J.C.; Rathmann, R.J.; Alsup, E.M.; Hutcheson, J.P.; Nichols, W.T.; Streeter, M.N.; Yates, D.A.; Johnson, B.J.; Miller, M.F. Effect of feeding zilpaterol hydrochloride to beef and calf-fed Holstein cattle on consumer palatability ratings. *J. Anim. Sci.* **2009**, *87*, 3712–3721. [CrossRef]

43. Jackman, P.; Sun, D.W.; Allen, P.; Brandon, K.; White, A.M. Correlation of consumer assessment of longissimus dorsi beef palatability with image colour, marbling and surface texture features. *Meat Sci.* **2010**, *84*, 564–568. [CrossRef]

44. Yancey, J.W.S.; Apple, J.K.; Muellenet, J.-F.; Sawyer, J.T. Consumer responses for tenderness and overall impression can be predicted by visible and near-infrared spectroscopy, Meullenet-Owens razo shear, and Warner-Bratzler shear force. *Meat Sci.* **2010**, *85*, 487–492. [CrossRef]

45. Igo, J.L.; Brooks, J.C.; Johnson, B.J.; Starkey, J.; Rathmann, R.J.; Garmyn, A.J.; Nichols, W.T.; Hutcheson, J.P.; Miller, M.F. Characterization of estrogen-trenbolone acetate implants on tenderness and consumer acceptability of beef under the effect of two aging times. *J. Anim. Sci.* **2011**, *89*, 792–797. [CrossRef]

46. Maughan, C.; Tansawat, R.; Cornforth, D.; Ward, R.; Martini, S. Development of a beef flavor lexicon and its application to compare the flavor profile and consumer acceptance of rib steaks from grass- or grain-fed cattle. *Meat Sci.* **2011**, *90*, 116–121. [CrossRef] [PubMed]

47. O'Quinn, T.G.; Brooks, J.C.; Polkinghorne, R.J.; Garmyn, A.F.; Johnson, B.J.; Starkey, J.D.; Rathmann, R.J.; Miller, M.F. Consumer assessment of beef strip loin steaks of varying fat levels. *J. Anim. Sci.* **2012**, *90*, 626–634. [CrossRef] [PubMed]

48. Robinson, D.L.; Café, L.M.; McIntyre, B.L.; Geesink, G.H.; Barendse, W.; Pethick, D.W.; Thompson, J.M.; Polkinghorne, R.; Greenwood, P.L. Production and processing studies on calpain-system gene markers for beef tenderness: Consumer assessments of eating quality. *J. Anim. Sci.* **2012**, *90*, 2850–2860. [CrossRef] [PubMed]

49. Garmyn, A.J.; Miller, M.F. Meat science and muscle biology symposium: Implant and beta agonist impacts on beef palatability. *J. Anim. Sci.* **2014**, *92*, 10–20. [CrossRef] [PubMed]

50. Garmyn, A.J.; Brooks, J.C.; Hodgen, J.M.; Nichols, W.T.; Hutcheson, J.P.; Rathmann, R.J.; Miller, M.F. Comparative effects of supplementing beef steers with zilpaterol hydrochloride, ractopamine hydrochloride, or no beta-agonist on strip loin composition, raw and cooked color properties, shear force, and consumer assessment of steaks aged for fourteen or twenty-one days postmortem. *J. Anim. Sci.* **2014**, *92*, 3670–3864. [CrossRef]

51. Hunt, M.R.; Garmyn, A.J.; O'Quinn, T.G.; Corbin, C.H.; Legako, J.F.; Rathmann, R.J.; Brooks, J.C.; Miller, M.F. Consumer assessment of beef palatability from four beef muscles from USDA Choice and Select graded carcasses. *Meat Sci.* **2014**, *98*, 1–8. [CrossRef] [PubMed]

52. Tedford, J.L.; Rodas-Gonzolez, A.; Garmyn, A.J.; Brooks, J.C.; Johnson, B.J.; Starkey, J.D.; Clark, G.O.; Collins, J.A.; Miller, M.F. U.S. consumer perception of U.S. and Canadian beef quality grades. *J. Anim. Sci.* **2014**, *92*, 3685–3692. [CrossRef]

53. Egan, A.F.; Ferguson, D.M.; Thompson, J.M. Consumer sensory requirements for beef and their implication for the Australian beef industry. *Aust. J. Exper. Agric.* **2001**, *41*, 855–859. [CrossRef]

54. Polkinghorne, R.; Thompson, J.M.; Watson, R.; Gee, A.; Porter, M. Evaluation of the Meat Standards Australia (MSA) Beef Grading System. *Aust. J. Exper. Agric.* **2008**, *48*, 1351–1359. [CrossRef]

55. Watson, R.; Gee, A.; Polkinghorne, R.; Porter, M. Consumer assessment of eating–quality–Development of protocols for Meat Standards Australia (MSA) testing. *Anim. Prod. Sci.* **2008**, *48*, 1360–1367. [CrossRef]

56. Lyford, C.; Thompson, J.; Polkinghorne, R.; Miller, M.; Nishimura, T.; Neath, K.; Allen, P.; Belasco, E. Is willingness to pay (WTP) for beef quality grades affected by consumer demographics and meat consumption preferences? *Aust. Agribus. Rev.* **2010**, *18*, 1–17.

57. Bonny, S.P.F.; O'Reilly, R.A.; Pethick, D.W.; Gardner, G.E.; Hocquette, J.-F.; Pannier, L. Update of Meat Standards Australia and the cuts based grading scheme for beef and sheepmeat. *J. Integr. Agric.* **2018**, *17*, 1641–1654. [CrossRef]

58. Legrand, I.; Hocquette, J.F.; Polkinghorne, R.J.; Pethick, D.W. Prediction of beef eating quality in France using the Meat Standards Australia system. *Animal* **2013**, *7*, 524–529. [CrossRef] [PubMed]

59. Morales, R.; Aguiar, A.P.S.; Subiabre, I.; Realini, C.E. Beef acceptability and consumer expectations associated with production systems and marbling. *Food Qual. Prefer.* **2013**, *29*, 166–173. [CrossRef]

60. Hocquette, J.-F.; Van Wezemael, L.; Chriki, S.; Legrand, I.; Verbeke, W.; Farmer, L.; Scollan, N.D.; Polkinghorne, R.; Rødbotten, R.; Allen, P.; et al. Modelling of beef sensory quality for a better prediction of palatability. *Meat Sci.* **2014**, *97*, 316–322. [CrossRef]

61. Van Wezemail, L.; De Smet, S.; Ueland, O.; Verbeke, W. Relationships between sensory evaluations of beef tenderness, shear force measurements and consumer characteristics. *Meat Sci.* **2014**, *97*, 310–315. [CrossRef]

62. Allen, P. Testing the MSA palatability grading scheme on Irish beef. Viandes et Produits Carnés. 2015, VPC-2015-31-1-5. Available online: https://www.viandesetproduitscarnes.fr/index.%20php/fr/processtechnologies/624-test-du-systeme-msa-pour-predire-la-qualite-de-la-viande-bovine-irlandaise (accessed on 28 April 2018).

63. Guzek, D.; Glabska, D.; Gutkowska, K.; Wierzbicki, J.; Woznicak, A.; Wierbicka, A. Influence of cut and thermal treatment on consumer perception of beef in Polish trials. *Pak. J. Agric. Sci.* **2015**, *52*, 521–526.

64. Bonny, S.P.F.; Hocquette, J.F.; Pethick, D.W.; Farmer, L.J.; Legrand, I.; Wierzbicki, J.; Allen, P.; Polkinghorne, R.J.; Gardner, G.E. The variation in the eating quality of beef from different sexes and breed classes cannot be completely explained by carcass measurements. *Animal* **2016**, *10*, 987–995. [CrossRef]

65. Bonny, S.P.F.; Pethick, D.W.; Legrand, I.; Wierzbicki, J.; Allen, P.; Farmer, L.J.; Polkinghorne, R.J.; Hocquette, J.F.; Gardner, G.E. European conformation and fat scores have no relationship with eating quality. *Animal* **2016**, *10*, 996–1006. [CrossRef]

66. Bonny, S.P.F.; Hocquette, J.F.; Pethick, D.W.; Legrand, I.; Wierzbicki, J.; Allen, P.; Farmer, L.J.; Polkinghorne, R.J.; Gardner, G.E. Untrained consumer assessment of the eating quality of beef: 1. A single composite score can predict beef quality grades. *Animal* **2017**, *11*, 1389–1398. [CrossRef]

67. Devlin, D.J.; Gault, N.F.S.; Moss, B.W.; Tolland, E.; Tollerton, J.; Farmer, L.J.; Gordon, A.W. Factors affecting eating quality of beef. *Adv. Anim. Biosci.* **2017**, *8*, s2–s5. [CrossRef]

68. McCarthy, S.N.; Henchion, M.; White, A.; Brandon, K.; Allen, P. Evaluation of beef eating quality by Irish consumers. *Meat Sci.* **2017**, *132*, 118–124. [CrossRef] [PubMed]

69. Chong, F.S.; Farmer, L.J.; Hagan, T.D.J.; Speers, J.S.; Sanderson, D.W.; Devlin, D.J.; Tollerton, I.J.; Gordon, A.W.; Methven, L.; Moloney, A.P.; et al. Regional, socioeconomic and behavioural- impacts on consumer acceptability of beef in Northern Ireland, Republic of Ireland and Great Britain. *Meat Sci.* **2019**, *154*, 86–95. [CrossRef] [PubMed]

70. Crownover, R.D.; Garmyn, A.J.; Polkinghorne, R.J.; Rathmann, R.J.; Bernhard, B.C.; Miller, M.F. The effects of hot vs. cold boning on eating quality of New Zealand grass fed beef. *Meat Muscle Biol.* **2017**, *1*, 207–217. [CrossRef]

71. Cho, S.H.; Kim, J.H.; Kim, J.H.; Seong, P.N.; Park, B.Y.; Kim, K.E.; Seo, G.; Lee, J.M.; Kim, D.H. Effect of socio-demographic factors on sensory properties of Korean Hanwoo bull beef by different cut and cooking methods. *J. Anim. Sci. Technol.* **2007**, *49*, 857–870.

72. Hwang, I.H.; Polkinghorne, R.; Lee, J.M.; Thompson, J.M. Demographic and Design Effects on Beef Sensory Scores Given by Korean and Australian Consumers. *Aust. J. Exper. Agric.* **2008**, *48*, 1387–1395. [CrossRef]

73. Park, B.Y.; Hwang, I.H.; Cho, S.H.; Yoo, Y.M.; Kim, J.H.; Lee, J.M.; Polkinghorne, R.; Thompson, J.M. Effect of carcass suspension and cooking method on the palatability of three beef muscle as assessed by Korean and Australian consumers. *Aust. J. Exper. Agric.* **2008**, *48*, 1396–1404. [CrossRef]

74. Thompson, J.M.; Polkinghorne, R.; Hwang, I.H.; Gee, A.M.; Cho, S.H.; Park, B.Y.; Lee, J.M. Beef quality grades as determined by Korean and Australian consumers. *Aust. J. Exper. Agric* **2008**, *48*, 1380–1386. [CrossRef]

75. Polkinghorne, R.J.; Nishimura, T.; Neath, K.E.; Watson, R. Japanese consumer categorisation of beef into quality grades, based on Meat Standards Australia methodology. *Anim. Sci. J.* **2011**, *82*, 325–333. [CrossRef]

76. Polkinghorne, R.J.; Nishimura, T.; Neath, K.E.; Watson, R. A comparison of Japanese and Australian consumers' sensory perceptions of beef. *Anim. Sci. J.* **2014**, *85*, 69–74. [CrossRef]

77. Thompson, J.; Polkinghorne, R.; Gee, A.; Motiang, D.; Strydom, P.; Mashau, M.; Ng'ambi, J.; deKock, R.; Burrow, H. *Beef Palatability in the Republic of South Africa: Implications for Niche-Marketing Strategies*; ACIAR Technical Reports; Australian Centre for International Agricultural Research; ACIAR: Canberra, ACT, Australia, 2010; Volume 72, pp. 1–56.

78. Corbin, C.H.; O'Quinn, T.G.; Garmyn, A.J.; Legako, J.F.; Hunt, M.R.; Dinh, T.T.N.; Rathmann, R.J.; Brooks, J.C.; Miller, M.F. Sensory evaluation of tender beef strip loin steaks of varying marbling levels and quality treatments. *Meat Sci.* **2015**, *100*, 24–31. [CrossRef] [PubMed]

79. Legako, J.F.; Dinh, T.T.N.; Miller, M.F.; Adhikari, K.; Brooks, J.C. Consumer palatability scores, sensory descriptive attributes, and volatile compounds of grilled beef steaks from three USDA Quality Grades. *Meat Sci.* **2016**, *112*, 77–85. [CrossRef] [PubMed]

80. Garcia-Torres, S.; Lopez-Gajardo, A.; Mesia, F.J. Intensive vs. free-range organic beef. A preference study through consumer liking and conjoint analysis. *Meat Sci.* **2016**, *114*, 114–120. [CrossRef]

81. Glascock, R. Beef Flavor Attributes and Consumer Perception. Master's Thesis, Texas A&M University, College Station, TX, USA, 2014.

82. Luckemeyer, T.J. Beef Flavor Attributes and Consumer Perception of Light Beef Eaters. Master's Thesis, Texas A&M University, College Station, TX, USA, 2015.

83. Laird, H.L. Millennial's Perception of Beef Flavor. Master's Thesis, Texas A&M University, College Station, TX, USA, 2015.

84. Miller, R.K.; Kerth, C.R.; Berto, M.C.; Laird, H.L.; Savell, J.W. Steak Thickness, Cook Surface Temperature and Quality Grade Affected Top Loin Steak Consumer and Descriptive Sensory Attributes. *Meat Muscle Biol.* **2019**, *3*, 467–478. [CrossRef]

85. Brooks, J.C.; Belew, J.B.; Griffin, D.B.; Gwartney, B.L.; Hale, D.S.; Henning, W.R.; Johnson, D.D.; Morgan, J.B.; Parrish, F.C., Jr.; Reagan, J.O.; et al. National Beef Tenderness Survey–1998. *J. Anim. Sci.* **2000**, *78*, 1852–1860. [CrossRef]

86. Voges, K.L.; Mason, C.L.; Brooks, J.C.; Delmore, R.J., Jr.; Griffin, D.B.; Hale, D.S.; Henning, W.R.; Johnson, D.D.; Lorenzen, C.L.; Maddock, R.J.; et al. National beef tenderness survey–2006: Assessment of Warner-Bratzler shear and sensory panel ratings for beef from US retail and foodservice establishments. *Meat Sci.* **2007**, *77*, 357–364. [CrossRef]

87. Guelker, M.R.; Haneklaus, A.N.; Brooks, J.C.; Carr, C.C.; Delmore, R.J., Jr.; Griffin, D.B.; Hale, D.S.; Harris, K.B.; Mafi, G.G.; Johnson, D.D.; et al. National Beef Tenderness Survey–2010: Warner-Bratzler shear force values and sensory panel ratings for beef steaks from United States retail and food service establishments. *J. Anim. Sci.* **2013**, *91*, 1005–1014. [CrossRef] [PubMed]

88. Igo, M.W.; Arnold, A.N.; Miller, R.K.; Gehring, K.B.; Mehall, L.N.; Lorenzen, C.L.; Delmore, R.J., Jr.; Woerner, D.R.; Wasser, B.E.; Savell, J.W. Tenderness assessments of top loin steaks from retail markets in four U.S. cities. *J. Anim. Sci.* **2015**, *93*, 4610–4616. [CrossRef] [PubMed]

89. Martinez, H.A.; Arnold, A.N.; Brooks, J.C.; Carr, C.C.; Gehring, K.B.; Griffin, D.B.; Hale, D.S.; Hafi, G.G.; Johnson, D.D.; Lorenzen, C.L.; et al. National Beef Tenderness Survey–2015: Palatability and Shear Force Assessments of Retail and Foodservice Beef. *Meat Muscle Biol.* **2017**, *1*, 138–148. [CrossRef]

90. Realini, C.E.; Font i Furnols, M.; Sanudo, C.; Montossi, F.; Oliver, M.A.; Guerrero, L. Spanish, French and British consumers' acceptability of Uruguayan beef, and consumers' beef choice associated with country of origin, finishing diet and meat price. *Meat Sci.* **2013**, *96*, 14–21. [CrossRef]

91. Sepulveda, C.A.; Garmyn, A.J.; Legako, J.F.; Miller, M.F. Cooking method and USDA Quality grade affect consumer palatability and flavor of beef strip loin steaks. *Meat Muscle Biol.* **2019**, *3*, 375–388. [CrossRef]

92. Bueso, M.E.; Garmyn, A.J.; O'Quinn, T.G.; Brooks, J.C.; Brashears, M.M.; Miller, M.F. Comparing Honduran and United States consumers' sensory perceptions of Honduran and U.S. beef loin steaks. *Meat Muscle Biol.* **2018**, *2*, 233–241. [CrossRef]

93. Mathews, K.H.; Johnson, R.J. Alternative beef production systems: Issues and implications. *Econ. Res. Serv.* **2013**, LDPM-128-01. Available online: https://pdfs.semanticscholar.org/181b/7a105e4d874bc7478afec7895dfd9cf3f0f7.pdf (accessed on 4 February 2020).

94. Daley, C.A.; Abbott, A.; Doyle, P.S.; Nader, G.A.; Larson, S. A review of fatty acid profiles and antioxidant content in grass-fed and grain-fed beef. *Nutr. J.* **2010**, *9*, 1–12. [CrossRef] [PubMed]

95. Van Elswyk, M.E.; McNeill, S.H. Impact of grass/forage feeding versus grain finishing on beef nutrients and sensory quality: The, U.S. experience. *Meat Sci.* **2014**, *96*, 535–540. [CrossRef]

96. Elmore, J.S.; Warren, H.E.; Mottram, D.S.; Scollan, N.D.; Enser, M.; Richardson, R.I. A comparison of the aroma volatiles and fatty acid compositions of grilled beef muscle from Aberdeen angus and Holstein-Friesian steers fed diets based on silage or concentrates. *Meat Sci.* **2004**, *68*, 27–33. [CrossRef]

97. Larick, D.K.; Hedrick, H.B.; Bailey, M.E.; Williams, J.E.; Hancock, D.L.; Garner, G.B. Flavor constituents of beef as influenced by forage- and grain-feeding. *J. Food Sci.* **1987**, *52*, 245–251. [CrossRef]

98. Lorenz, S.; Buettner, A.; Ender, K.; Nuernberg, G.; Papstein, H.J.; Schieberle, P. Influence of keeping system on the fatty acid composition in the longissimus muscle of bulls and odorants formed after pressure-cooking. *Eur. Food Res. Technol.* **2002**, *214*, 112–118. [CrossRef]

99. Scolan, N.; Hocquette, J.-F.; Nuernberg, K.; Dannenberger, D.; Richardson, I.; Moloney, A. Innovations in beef production systems that enhance the nutritional and health value of beef lipids and their relationship with meat quality. *Meat Sci.* **2006**, *74*, 17–33. [CrossRef]

100. Raes, K.; De Smet, S.; Balcaen, A.; Claeys, E.; Demeyer, D. Effect of diets rich in n 3 polunsaturated fatty acids on muscle lipids and fatty acids in Belgian Blue double-muscled young bulls. *Rep. Nutri. Develop.* **2003**, *43*, 331–345. [CrossRef]

101. Duckett, S.K.; Neel, J.P.S.; Fontenot, J.P.; Clapham, W.M. Effects of winter stocker growth rate and finishing system on: III. Tissue proximate, fatty acid, vitamin and cholesterol content. *J. Anim. Sci.* **2009**, *87*, 2961–2970. [CrossRef] [PubMed]

102. Leheska, J.M.; Thompson, L.D.; Howe, J.C.; Hentges, E.; Boyce, J.; Brooks, J.C.; Shriver, B.; Hoover, L.; Miller, M.F. Effects of conventional and grass-feeding systems on the nutrient composition of beef. *J. Anim. Sci.* **2008**, *8*, 3575–3685. [CrossRef] [PubMed]

103. Lorenzen, C.L.; Golden, J.W.; Martz, F.A.; Grun, I.U.; Ellersieck, M.R.; Gerrish, J.R.; Moore, K.C. Conjugated linoleic acid content of beef differs by feeding regime and muscle. *Meat Sci.* **2007**, *75*, 159–167. [CrossRef] [PubMed]

104. Schroeder, J.W.; Cramer, D.A.; Bowling, R.A.; Cook, C.W. Palatability, shelflife and chemical differences between forage- and grain-finished beef. *J. Anim. Sci.* **1980**, *50*, 852–859. [CrossRef]

105. Realini, C.E.; Duckett, S.K.; Hill, N.S.; Hoveland, C.S.; Lyon, B.G.; Sackmann, J.R.; Gillis, M.H. Effect of endophyte type on carcass traits, meat quality, and fatty acid composition of beef cattle grazing tall fescue. *J. Anim. Sci.* **2005**, *83*, 430–439. [CrossRef]

106. Reagan, J.O.; Carpenter, J.A.; Bauer, F.T.; Lowrey, R.S. Packaging and palatability characteristics of grass and grass-grain fed beef. *J. Anim. Sci.* **1977**, *45*, 716–721. [CrossRef]

107. Hedrick, H.B.; Paterson, J.A.; Matches, A.G.; Thomas, J.D.; Morrow, R.E.; Stringer, W.G.; Lipsey, R.J. Carcass and palatability characteristics of beef produced on pasture, corn silage and corn grain. *J. Anim. Sci.* **1983**, *57*, 791–801. [CrossRef]

108. Iida, F.; Saitou, K.; Kawamura, T.; Yamaguchi, S.; Nishimura, T. Effect of fat content on sensory characteristics of marbled beef from Japanese black steers. *Anim. Sci. J.* **2015**, *86*, 707–715. [CrossRef]

109. Larick, D.K.; Turner, B.E. Influence of Finishing Diet on the Phospholipid Composition and Fatty Acid Profile of Individual Phospholipids in Lean Muscle of Beef Cattle. *J. Anim. Sci.* **1989**, *67*, 2282–2293. [CrossRef]

110. Dransfield, E.; Zamora, F.; Bayle, M.C. Consumer selection of steaks as influenced by information and price index. *Food Qual. Pref.* **1998**, *9*, 321–326. [CrossRef]

111. Wood, J.D.; Richardson, R.I.; Nute, G.R.; Fisher, A.V.; Campo, M.M.; Kasapidou, E. Effects of fatty acids on meat quality: A review. *Meat Sci.* **2003**, *66*, 21–32. [CrossRef]

112. Nuernberg, K.; Wood, J.D.; Scollan, N.D.; Richardson, R.I.; Nute, G.R.; Nuernberg, G. Effect of a grass-based and a concentrate feeding system on meat quality characteristics and fatty acid composition of longissimus muscle in different cattle breeds. *Livest. Prod. Sci.* **2005**, *94*, 137–147. [CrossRef]

113. Resconi, V.C.; Campo, M.M.; Montossi, F.; Ferreira, V.; Sanudo, C.; Escudero, A. Relationship between odour-active compounds and flavour perception in meat from lambs fed different diet. *Meat Sci.* **2010**, *85*, 700–706. [CrossRef] [PubMed]

114. Calkins, C.R.; Hodgens, J.M. A fresh look at meat flavor. *Meat Sci.* **2007**, *77*, 63–80. [CrossRef] [PubMed]

115. Oliver, M.A.; Nute, G.R.; Font i Furnols, M.; San Julian, R.; Campo, M.M.; Sanudo, C.; Caneque, V.; Guerrero, L.; Alvarez, I.; Diaz, M.T.; et al. Eating quality of beef, from different production systems, assessed by German, Spanish and British consumers. *Meat Sci.* **2006**, *74*, 435–442. [CrossRef]

116. Moeller, S.J.; Miller, R.K.; Edwards, K.K.; Zerby, H.N.; Logan, K.E.; Aldredge, T.L.; Stahl, C.A.; Boggess, M.; Box-Steffensmeier, J.M. Consumer perceptions of pork eating quality as affected by pork quality attributes and end-point cooked temperature. *Meat Sci.* **2010**, *84*, 14–22. [CrossRef]

117. Brewer, M.S.; McKeith, F.K. Consumer-rated quality characteristics as related to purchase intent of fresh pork. *J. Food Sci.* **1999**, *64*, 171–174. [CrossRef]

118. Verbeke, W.; Van Oeckel, M.J.; Warnants, N.; Viaene, J.; Boucque, C.V. Consumer perception, facts and possibilities to improve acceptability of health and sensory characteristics of pork. *Meat Sci.* **1999**, *53*, 77–99. [CrossRef]

119. Brewer, M.S.; Zhu, L.G.; McKeith, F.K. Marbling effects on quality characteristics of pork loin chops: Consumer purchase intent, visual and sensory characteristics. *Meat Sci.* **2001**, *59*, 153–163. [CrossRef]

120. Bryhni, E.A.; Byrne, D.V.; Rodbotten, M.; Moller, S.; Claudi-Magnussen, C.; Karlsson, A.; Agerhem, H.; Johansson, M.; Martens, M. Consumer and sensory investigations in relation to physical/ chemical aspects of cooked pork in Scandinavia. *Meat Sci.* **2003**, *65*, 737–748. [CrossRef]

121. Lonergan, S.M.; Stalder, K.J.; Huff-Lonergan, E.; Knight, T.J.; Goodwin, R.N.; Prusa, K.J.; Beitz, D.C. Influence of lipid content on pork sensory quality within pH classification. *J. Anim. Sci.* **2007**, *85*, 1074–1079. [CrossRef] [PubMed]

122. Norman, J.L.; Berg, E.P.; Heymann, H.; Lorenzen, C.L. Pork loin color relative to sensory and instrumental tenderness and consumer acceptance. *Meat Sci.* **2003**, *65*, 927–933. [CrossRef]

123. Rincker, P.J.; Killefer, J.; Ellis, M.; Brewer, M.S.; McKeith, F.K. Intramuscular fat content has little influence on the eating quality of fresh pork loin chops. *J. Anim. Sci.* **2008**, *86*, 730–737. [CrossRef] [PubMed]

124. Rosenvold, K.; Andersen, H.J. Factors of significance for pork quality—A review. *Meat Sci.* **2003**, *64*, 219–237. [CrossRef]

125. Van Laack, R.L.; Stevens, S.G.; Stalder, K.J. The influence of ultimate pH and intramuscular fat content on pork tenderness and tenderization. *J. Anim. Sci.* **2001**, *79*, 392–397. [CrossRef] [PubMed]

126. Melton, S.L. Effects of feeds on flavor of red meat: A review. *J. Anim. Sci.* **1990**, *68*, 4421–4435. [CrossRef] [PubMed]

127. Cava, R.; Ventanas, J.; Ruiz, J.; Andres, A.I.; Antequera, T. Sensory characteristics of Iberian ham: Influence of rearing system and muscle location. *Food Sci. Technol. Int.* **2000**, *6*, 235–242. [CrossRef]

128. Font-i-Furnols, M.; Tous, N.; Esteve-Garcia, E.; Gispert, M. Do all the consumers accept the marbling in the same way? The relation between visual and sensory acceptability of pork. *Meat Sci.* **2012**, *91*, 448–453. [CrossRef]

129. Patterson, R.L.S. 5α-androst-16-ene-3-one: Compound responsible for taint in boar fat. *J. Sci. Food Agric.* **1968**, *19*, 31–37. [CrossRef]

130. Vold, E. Fleishproduktionseigenschaften bei Ebern and Kastraten IV. Organoleptische und gaschromatografische Untersuchungen wassed- ampfflüchtiger stooffe des rückenspeckes von ebern. *Meld. Nord.* **1970**, *49*, 1–25.

131. Bonneau, M. Use of entire males for pig meat in the European Union. *Meat Sci.* **1998**, *49* (Suppl. 1), S257–S272. [CrossRef]

132. Aldal, I.; Andresen, Ø.; Egeli, A.K.; Haugen, J.E.; GrØdum, A.O.; Fjetland, O.; Eikaas, J.L. Levels of androstenone and skatole and the occurrence of boar taint in fat from young boars. *Live. Prod. Sci.* **2005**, *95*, 121–129. [CrossRef]

133. Bonneau, M.; Desmoulin, B.; Dumont, B.L. Qualités organoleptiques des viandes de porcs mâles entiers ou castrés: Composition des graisses et odeurs sexuelles chez les races hypermusclées. *Ann. Zootech.* **1979**, *28*, 53–72. [CrossRef]

134. Claus, R.; Weiler, U.; Herzog, A. Physiological aspects of androstenone and skatole formation in the boar: A review with experimental data. *Meat Sci.* **1994**, *38*, 239–305. [CrossRef]

135. Doran, E.; Whittington, F.M.; Wood, J.D.; McGivan, D. Characterisation of androstenone metabolism in pig liver microsomes. *Chem. Biol. Interac.* **2004**, *147*, 141–149. [CrossRef]

136. FØrland, D.M.; Lundström, K.; Andresen, Ø. Relationship between androstenone content in fat, intensity of boar taint and size of accessory sex glands in boars. *Nord. Veterinaemed.* **1980**, *32*, 201–206. [PubMed]

137. Nicolau-Solano, S.I.; McGivan, J.D.; Whittington, F.M.; Nieuwhof, G.J.; Wood, J.D.; Doran, O. Relationship between the expression of hepatic but not testicular 3beta-hydroxysteroid dehydrogenase with androstenone deposition in pig adipose tissue. *J. Anim. Sci.* **2006**, *84*, 2809–2817. [CrossRef]

138. Rydhmer, L.; Zamarastkaia, G.; Andersson, K.; Algers, B.; Lundström, K. Aggressive and sexual behaviour of entire male pigs. In Proceedings of the European association for Animal Production Working Group, Rome, Italy, 31 August–3 September 2003.

139. Salmon, E.L.R.; Edwards, S.A. Effects of gender contact on the behaviour and performance of entire boars and gilts from 60–130 kg. *Proc. Br. Soc. Anim. Sci.* **2006**, *2006*, 72.

140. Squires, E.J.; Lou, Y.; Gibson, J.P. *Boar Taint: How Much Is too Much?* Gibson, J.P., Aker, C.A., Ball, R.O., Eds.; Ontario Pork Carcass Appraisal Project Symposium: Guelph, ON, Canada, 1996.

141. Walstra, P.; Garssen, G.J. *Influence of Quality of the Pigs and Season on Androstenone Level*; European Association for Animal Production Working Group: Milton Keynes, UK, 1995.

142. Babol, J.; Zamaratskaia, G.; Juneja, R.K.; Lundström, K. The effect of age on distribution of skatole and indole levels in entire male pigs in four breeds: Yorkshire, Landrace, Hampshire and Duroc. *Meat Sci.* **2004**, *67*, 351–358. [CrossRef]

143. Friis, C. *Is Boar Taint Aelated to Sex Differences or Polymorphism of Skatole Metabolism?* European Association for Animal Production Working Group: Milton Keynes, UK, 1995.

144. Hansen, L.L.; Mikkelsen, L.L.; Agerhem, H.; Laue, A.; Jensen, M.T.; Jensen, B.B. *Effect of Fermented Liquid Feed and Zinc Bacitracin on Microbial Metabolism in the Gut and Sensoric Profile of m. longissimus Dorsi from Entire Male and Female Pigs. Boar Taint in Entire Male Pigs*; Bonneau, M., Lundström, K., Malmfors, B., Eds.; EAAP Publication: Stockholm, Sweden, 1997; Volume 92, pp. 92–96.

145. Hansen, L.L.; Larsen, A.E.; Hansen-Moller, J. Influence of keeping pigs heavily fouled with faeces plus urine on skatole and indole concentration (boar taint) in subcutaneous fat. 1995, Acta Agriculturae Scandinavica. *Sec. A Anim. Sci.* **1995**, *45*, 178–185. [CrossRef]

146. Lin, Z.; Lou, Y.; Squires, E.J. Functional polymorphism in porcine CYP2E1 gene: Its association with skatole levels. *J. Steroid Biochem. Mol. Biol.* **2006**, *99*, 231–237. [CrossRef] [PubMed]

147. Lundström, K.; Malmfors, B.; Fjelkner-Modig, S.; Szatek, A. Consumer testing of boar meat in Sweden. *Swed. J. Agric. Res.* **1982**, *13*, 39–46.

148. Borrisser-Pairo, F.; Panella-Riera, N.; Gil, M.; Kallas, Z.; LOinares, M.B.; Egea, M.; Garrido, M.D.; Oliver, M.A. Consumers' sensitivity to androstenone and the evaluation of different cooking methods to mask boar taint. *Meat Sci.* **2017**, *123*, 198–204. [CrossRef] [PubMed]

149. Font i Furnols, M.; Gispert, M.; Diestre, A.; Oliver, M.A. Acceptability of boar meat by consumers depending on their age, gender, culinary habits, sensitivity and appreciation of androstenone smell. *Meat Sci.* **2003**, *64*, 433–440. [CrossRef]

150. Rhodes, D.N. Consumer testing of bacon from boar and gilt pigs. *J. Sci. Food Agric.* **1971**, *22*, 485–490. [CrossRef]

151. Bonneau, M.; Le Denmat, M.; Vaudelet, J.C.; Veloso Nunes, J.R.; Mortensen, A.B.; Mortensen, H.P. Contributions of fat androstenone and skatole to boar taint: I. Sensory attributes of fat and pork meat. *Livest. Prod. Sci.* **1992**, *32*, 63–80. [CrossRef]

152. Desmoulin, B.; Bonneau, M.; Frouin, A.; Bidard, J.F. Consumer testing of pork and processed meat from boars: The influence of fat androstenone level. *Livest. Prod. Sci.* **1982**, *9*, 707–715. [CrossRef]

153. Walstra, P.; Claudi-Magnussen, C.; Chevillon, P.; von Seth, G.; Diestre, A.; Matthews, K.; Homer, D.B.; Bonneau, M. An international study on the importance of androstenone and skatole for boar taint: Levels of androstenone and skatole by country and season. *Livest. Prod. Sci.* **1999**, *62*, 15–28. [CrossRef]
154. Weiler, U.; Font i Furnols, M.; Fischer, K.; Kemmer, H.; Oliver, M.A.; Gispert, M.; Dobrowolski, A.; Claus, R. Influence of differences in sensitivity of Spanish and German consumers to perceive androstenone on the acceptance of boar meat differing in skatole and androstenone concentrations. *Meat Sci.* **2000**, *54*, 297–304. [CrossRef]
155. Blanch, M.; Panella-Riera, N.; Chevillon, P.; Font-i-Furnols, M.; Gil, M.; Gil, J.M.; Kallas, Z.; Oliver, M.A. Impact of consumer's sensitivity to androstenone on acceptability of meat from entire male pigs in three European countries: France, Spain and United Kingdom. *Meat Sci.* **2012**, *90*, 572–578. [CrossRef]
156. Razafindrazaka, H.; Monnereau, A.; Razafindrazaka, D.; Tonasso, L.; Schiavinato, S.; Rakotoarisoa, J.-A.; Radimilahy, C.; Letellier, T.; Pierron, D. Genetic admixture and flavor preferences: Androstenone sensitivity in Malagasy populations. *Hum. Biol.* **2015**, *87*, 59–70. [CrossRef] [PubMed]
157. Weiler, U.; Fischer, K.; Kemmer, H.; Dobrowolski, A.; Claus, R. *Influence of androstenone sensitivity on consumer reactions to boar meat. Boar taint in entire male pigs*; Bonneau, M., Lundström, K., Malmfors, B., Eds.; EAAP Publication No. 92; European Association for Animal Production Working Group: Stockholm, Sweden, 1997; pp. 147–151.
158. Panella-Riera, N.; Blanch, M.; Kallas, Z.; Chevillon, P.; Garavaldi, A.M.; Gil, M.; Gil, J.M.; Font-i-Furnols, M.; Oliver, M.A. Consumers' segmentation based on the acceptability of meat from entire male pigs with different boar taint levels in four European countries: France, Italy, Spain and United Kingdom. *Meat Sci.* **2016**, *114*, 137–145. [CrossRef]
159. Bremner, E.A.; Mainland, J.D.; Khan, R.M.; Sobel, N. The prevalence of androstenone anosmia. *Chem. Senses* **2003**, *28*, 423–432. [CrossRef] [PubMed]
160. Lunde, K.; Skuterud, E.; Nilsen, A.; Egelandsdal, B. A new method for differentiat- ing the androstenone sensitivity among consumers. *Food Qual. Prefer.* **2009**, *20*, 304–311. [CrossRef]
161. Wysocki, C.J.; Beauchamp, G.K. Ability to smell androstenone is genetically determined. *Proc. Nat. Acad. Sci. USA* **1984**, *81*, 4899–4902. [CrossRef] [PubMed]
162. Font i Furnols, M.; Gispert, M.; Guerrero, L.; Velarde, A.; Tibau, J.; Soler, J.; Hortos, M.; Garcia-Regueiro, J.A.; Perez, J.; Saurez, P.; et al. Consumers' sensory acceptability of pork from immunocastrated male pigs. *Meat Sci.* **2008**, *80*, 1013–1018. [CrossRef]
163. Rousset-Akrim, S.; Young, O.A.; Berdagué, J.L. Diet and growth effects in panel assessment of sheep meat odour and flavor. *Meat Sci.* **1997**, *45*, 169–181. [CrossRef]
164. Borton, R.J.; McClure, K.E.; Wulf, D.M. Sensory evaluation of loin chops from lambs fed concentrate or grazed on ryegrass to traditional or heavy weights. *J. Anim. Sci.* **1999**, *77* (Suppl. 1), 168.
165. Kemp, J.D.; Mahyuddin, M.; Ely, D.G.; Fox, J.D.; Moody, W.G. Effect of feeding system, slaughter weight and sex on organoleptic properties, and fatty acid composition of lamb. *J. Anim. Sci.* **1981**, *51*, 321–330. [CrossRef]
166. Priolo, D.M.; Agabiel, J.; Prache, S.; Dransfield, E. Effect of grass or concentrate feeding systems on lamb carcass and meat quality. *Meat Sci.* **2002**, *62*, 179–185. [CrossRef]
167. Font i Furnols, M.; Realini, C.E.; Guerrero, L.; Oliver, M.A.; Sañudo, C.; Campo, M.M.; Nute, G.R.; Caneque, V.; Alvarez, I.; San Julian, R.; et al. Acceptability of lamb fed on pasture, concentrate or combination of both systems by Eropean consumers. *Meat Sci.* **2009**, *81*, 196–202. [CrossRef] [PubMed]
168. Cañeque, V.; De la Fuente, J.; Díaz, M.T.; Álvarez, I. Composición en ácidos grasos y vitamina E de la carne de corderos alimentados con niveles diferentes de concentrado. *INIA Uruguay Série Técnica* **2007**, *168*, 97–102.
169. Díaz, M.T.; Álvarez, I.; De la Fuente, J.; Sañudo, C.; Campo, M.M.; Oliver, M.A.; Font, I.; Furnols, M.; Montossi, F.; Sam Julian, R.; et al. Fatty acid composition of lamb meat from typical lamb production systems in Spain, the United Kingdom, Germany and Uruguay. *Meat Sci.* **2005**, *71*, 256–263. [CrossRef]
170. Young, O.A.; Baumeister, B.M.B. The effect of diet on the flavour of cooked beef and the odour compounds in beef fat. *N. Z. J. Agric. Res.* **1999**, *42*, 297–304. [CrossRef]
171. Young, O.A.; Lane, G.A.; Priolo, A.; Fraser, K. Pastoral and species flavour in lambs raised on pasture, lucerne or maize. *J. Sci. Food Agric.* **2002**, *83*, 93–104. [CrossRef]

172. Fisher, A.; Enser, M.; Richardson, R.I.; Wood, J.D.; Nute, G.R.; Kurt, E.; Sinclair, L.A.; Wilkinson, R.G. Fatty acid composition and eating quality of lamb types derived from four diverse breeds production systems. *Meat Sci.* **2000**, *55*, 141–147. [CrossRef]

173. Nute, G.R.; Richardson, R.I.; Wood, J.D.; Hughes, S.I.; Wilkinson, R.G.; Cooper, S.L.; Sinclair, L.A. Effect of dietary oil source on the flavour and the colour and lipid stability of lamb meat. *Meat Sci.* **2007**, *77*, 547–555. [CrossRef]

174. Schönfeldt, H.C.; Naudé, R.T.; Bok, W.; Van Heerden, S.M.; Sowden, L. Cooking- and juiciness-related quality characteristics of goat and sheep meat. *Meat Sci.* **1993**, *34*, 381–394. [CrossRef]

175. Smith, G.C.; Dutson, T.R.; Hostetler, R.L.; Carpenter, Z.L. Fatness, rate of chilling and tenderness of lamb. *J. Food Sci.* **1976**, *41*, 748–756. [CrossRef]

176. Sañudo, C.; Enser, M.; Campo, M.M.; Nute, G.R.; María, G.; Sierra, I.; Wood, J.D. Fatty acid composition and fatty acid characteristics of lamb carcasses from Britain and Spain. *Meat Sci.* **2000**, *54*, 339–346. [CrossRef]

177. Fonti Furnols, M.; San Julián, R.; Guerrero, L.; Sañudo, C.; Campo, M.M.; Olleta, J.L.; Oliver, M.A.; Cañeque, V.; Álvarez, I.; Díaz, M.T.; et al. Acceptability of lamb meat from different producing systems and ageing time to German, Spanish and British consumers. *Meat Sci.* **2006**, *72*, 545–554. [CrossRef] [PubMed]

178. Wood, J.D.; Fisher, A.V. *Improving the Quality of Lamb Meat-Taste, Fatness and Consumer Appeal*; Slade, C.F.R., Lawrence, T.L.J., Eds.; New Developments in Sheep Production, Occasional Publication (No. 14); British Society of Animal Production: London, UK, 1990; pp. 88–108.

179. Rowe, J.B. Supplementary feeds for sheep. *J. Agric. West. Aust.* **1986**, *27*, 100–102.

180. Wales, W.J.; Doyle, P.T.; Pearce, G.R. The feeding value of cereal straws for sheep. I. Wheat straws. *Anim. Feed Sci. Technol.* **1990**, *29*, 1–14. [CrossRef]

181. McFarland, I.; Curnow, M.; Hyder, M.; Ashton, B.; Roberts, D. Feeding and managing sheep in dry times. *Dep. Agric. Food* **2006**, *4697*. Available online: https://researchlibrary.agric.wa.gov.au/bulletins/112/ (accessed on 4 February 2020).

182. Pethick, D.W.; Fergusson, D.M.; Gardner, G.E.; Hocquette, J.F.; Thompson, J.M.; Warner, R. *Muscle Metabolism in Relation to Genotypic and Environmental Influences on Consumer Defined Quality of Red Meat*; Hocquette, J.F., Gigli, S., Eds.; EAAP Publication: Rome, Italy, 2015; Volume 112, pp. 95–110. Available online: https://www.researchgate.net/profile/David_Pethick/publication/228628675_Muscle_metabolism_in_relation_to_genotypic_and_environmental_influences_on_consumer_defined_quality_of_red_meat/links/00b4951834481d9e69000000.pdf.

183. Young, O.A.; Berdague, J.L.; Viallon, C.; Rousset-Akrim, S.; Theriez, M. Fat- borne volatiles and sheepmeat odour. *Meat Sci.* **1997**, *45*, 183–200. [CrossRef]

184. Watkins, P.J.; Rose, G.; Salvatore, L.; Allen, D.; Tucman, D.; Warner, R.D.; Dunshea, F.R.; Petick, D.W. Age and nutri9tion influence the concentration sof three branched chain fatty acids in sheep fat from Australian abattoirs. *Meat Sci.* **2010**, *86*, 594–599. [CrossRef] [PubMed]

185. Hopkins, D.L.; Hegarty, R.S.; Walker, P.J.; Pethick, D.W. Relationship between animal age, intramuscular fat, cooking loss, pH, shear force and eating quality of aged meat from sheep. *Aust. J. Exper. Agric.* **2006**, *46*, 879–884. [CrossRef]

186. Watkins, P.J.; Kearney, G.; Rose, G.; Allen, D.; Ball, A.J.; Pethick, D.W.; Warner, R.D. Effect of branched-chain fatty acids, 3-methylindole and 4-mehtylphenol on consumer sensory scores of grilled lamb meat. *Meat Sci.* **2014**, *96*, 1088–1094. [CrossRef]

187. Flakemore, A.R.; Malau-Aduli, B.S.; Nichols, P.D.; Malau-Aduli, A.E.O. Omega-3 fatty acids, nutrient retention values, and sensory meat eating quality in cooked and raw Australian lamb. *Meat Sci.* **2014**, *123*, 79–87. [CrossRef]

188. Pannier, L.; Gardner, G.E.; Pearce, K.L.; McDonagh, M.; Ball, A.J.; Jacob, R.H.; Pethick, D.W. Associations of sire estimated breeding values and objective meat quality measurements with sensory scores in Australian lamb. *Meat Sci.* **2014**, *96*, 1076–1087. [CrossRef]

189. Pannier, L.; Gardner, G.E.; O'Reilly, R.A.; Pethick, D.W. Factors affecting lamb eating quality and the potential for their interaction into an MSA sheepmeat grading model. *Meat Sci.* **2018**, *144*, 43–52. [CrossRef] [PubMed]

190. O'Reilly, R.A.; Pannier, L.; Gardner, G.E.; Pethick, D.W. Sensory evaluation of Australian lamb and yearling, a comparison of Australian, American and Chinese consumers. In Proceedings of the 7th European Conference on Sensory and Consumer Research, Dijon, France, 11–14 September 2016.

191. Sañudo, C.; Alfonso, M.; San Julián, R.; Thorkelsson, G.; Valdimarsdottir, T.; Zygoyiannins, D.; Stamataris, C.; Piasentier, E.; Mills, C.; Berge, P.; et al. Regional variation in the hedonic evaluation of lamb meat from diverse production systems by consumers in six European countries. *Meat Sci.* **2007**, *75*, 610–621. [CrossRef] [PubMed]

192. Stegelin, F. Food and the millennial generation. *J. Food Distrib. Res.* **2002**, *33*, 182–184.

193. Howe, N.; Strauss, W. *Millennials Rising: The Next Great Generation*; Vintage Books by Random House: New York, NY, USA, 2000.

194. US Census Bureau. Bureau, U.S.C. *Millennials Outnumber Baby Aoomers and Are Far More Diverse*; Census Bureau Reports; US Census Bureau: Suitland, MA, USA, 2015.

195. Taylor, P.; Keeter, S. *Millennials: A Portrait of Generation Next: Confident, Connected Open to Change*; Pew Research Center: Washington, DC, USA, 2010.

196. Nie, C.; Zepeda, Z. Lifestyle segmentation of US food shoppers to examine organic and local food consumption. *Appetite* **2018**, *57*, 28–37. [CrossRef] [PubMed]

197. Font i Furnols, M.; Realini, C.; Montossi, F.; Sanudo, C.; Campo, M.M.; Oliver, M.A.; Nute, G.R.; Guerrer, L. Consumer's purchasing intention for lamb meat affected by country of origin, feeding system and meat price: A conjoint study in Spain, France and United Kingdom. *Food Qual. Pref.* **2011**, *22*, 443–451. [CrossRef]

198. Huseynov, S.; Kassas, B.; Segovia, M.S.; Palma, M.A. Incorporating Biometric Data in Models of Consumer Choice. *Appl. Econ.* **2018**, *51*, 1514–1531. [CrossRef]

199. Duchowski, A.T. *Eye Tracking Methodology: Theory and Practice*, 3rd ed.; Springer: Berlin/Heidelberg, Germany, 2007.

200. Farnsworth, B. How Eye Tracking Technology Is Changing the World (5 examples). 2018. Available online: https://imotions.com/blog/future-eye-tracking-technology/ (accessed on 11 May 2019).

201. Walker, C.; Federici, F.M. *Eye Tracking and Multidisciplinary Studies on Translation*; John Benjamins Publishing Company: Amsterdam, The Netherlands; Philadelphia, PA, USA, 2018.

202. Bento, V.; Paula, L.; Ferreira, A.; Figueiredo, N.; Tome, A.; Silva, F.; Paulo Cunha, J.; Georgieva, P. Advances in eeg-based brain-computer interfaces for control and biometry. In Proceedings of the Int. Symposium on Computational Intelligence for Engineering System, Nashville, TN, USA, 30 March–2 April 2009.

203. Brown, C.; Randolph, A.B.; Burkhalter, J.N. The story of taste: Using EEGs and self-reports to understand consumer choice. *Kennesaw J. Undergrad. Res.* **2012**, *2*, 5.

204. Campisi, P.; La Rocca, D. Brain waves for automatic biometric—Based user recognition. Information Forensics and Security. *IEEE Trans. Inf. Forensics Secur.* **2014**, *9*, 782–800. [CrossRef]

205. Gonzalez Viejo, C.; Fuentes, S.; Howell, K.; Torrico, D.D.; Dunshea, F.R. Integration of non-invasive biometrics with sensory analysis techniques to assess acceptability of beer by consumers. *Physiol. Behav.* **2019**, *200*, 139–147. [CrossRef]

206. Van der Laan, L.N.; de Ridder, D.T.D.; Viergever, M.A.; Smeets, P.A.M. The first taste is always with the eyes: A meta-analysis on the neural correlates of processing visual food cues. *NeuroImage* **2011**, *55*, 296–303. [CrossRef]

207. Tang, D.W.; Fellows, L.K.; Small, D.M.; Dagher, A. Food and drug cues activate similar brain regions: A meta-analysis of functional MRI studies. *Physiol. Behav.* **2012**, *106*, 317–324. [CrossRef] [PubMed]

208. Garcia-Garcia, I.; Narberhaus, A.; Marques-Iturria, I.; Garolera, M.; Radoi, A.; Serura, B.; Pueyo, R.; Ariza, M.; Jurado, M.A. Neural responses to visual food cues: Insights from functional magnetic resonance imaging. *Eur. Eat. Disord. Rev.* **2013**, *21*, 89–98. [CrossRef] [PubMed]

209. Chen, J.; Papies, E.K.; Barsalou, L.W. A core eating network and its modulations underlie diverse eating phenomena. *Brain Cogn.* **2016**, *110*, 20–42. [CrossRef] [PubMed]

210. Buodo, G.; Rumiati, R.; Lotto, L.; Sarlo, M. Does food-drink parings affect appetitive processing of food cues with different rewarding properties? Evidence from subjective, behavioral, and neural measures. *Food Qual. Prefer.* **2019**, *75*, 124–132. [CrossRef]

211. Brouwer, A.M.; Hogervorst, M.A.; van Erp, J.B.F.; Grootjen, M.; van Dam, E.; Zandstra, E.H. Measuring cooking experience implicitly and explicitly: Physiology, facial expression and subjective ratings. *Food Qual. Prefer.* **2019**, *78*, 103726. [CrossRef]

212. Crofton, E.C.; Botinestean, C.; Fenelon, M.; Gallaher, E. Potential applications for virtual and augmented reality technologies in sensory science. *Innov. Food Sci. Emerg. Technol.* **2019**, *56*, 102178. [CrossRef]
213. Sinesio, F.; Moneta, E.; Porcherot, C.; Abba, S.; Dryfuss, L.; Guillamet, K.; Bruyninckx, S.; Laporte, C.; Henneberg, S.; McEwa, J.A. Do immersive techniques help to capture consumer reality? *Food Qual. Prefer.* **2019**, *77*, 123–134. [CrossRef]

Article

Influence of Kiwifruit Extract Infusion on Consumer Sensory Outcomes of Striploin (*M. longissimus lumborum*) and Outside Flat (*M. biceps femoris*) from Beef Carcasses

Angela Lees [1], Małgorzata Konarska [2], Garth Tarr [3], Rod Polkinghorne [4] and Peter McGilchrist [1,5,*]

1 School of Environmental and Rural Science, University of New England, Armidale 2350, Australia
2 Division of Engineering in Nutrition, Warsaw University of Life Sciences, 02-787 Warsaw, Poland
3 School of Mathematics and Statistics, The University of Sydney, Sydney 2006, Australia
4 Birkenwood Pty. Ltd., Murrurundi 2338, Australia
5 School of Veterinary & Life Sciences, Murdoch University, Perth 6150, Australia
* Correspondence: peter.mcgilchrist@une.edu.au; Tel.: +61-2-6773-1845

Received: 30 June 2019; Accepted: 5 August 2019; Published: 8 August 2019

Abstract: Actinidin is a cysteine protease enzyme which occurs in kiwifruit and has been associated with improved tenderness in red meat. This study evaluated the impact of actinidin, derived from kiwifruit, on consumer sensory outcomes for striploin (*M. longissimus lumborum*) and outside flat (*M. biceps femoris*). Striploins and outside flats were collected from 87 grass-fed steers. Carcasses were graded to the Meat Standards Australia (MSA) protocols. Striploins and outside flats were then dissected in half and allocated to one of the following two treatments: (1) not infused (control) and (2) infused with a kiwifruit extract (enhanced), and then prepared as grill and roast samples. Grill and roast samples were then aged for 10 or 28 days. Consumer evaluations for tenderness, juiciness, flavor, and overall liking were conducted using untrained consumer sensory panels consisting of 2080 individual consumers, in accordance with the MSA protocols. These scores were then used to calculate an overall eating quality (MQ4) score. Consumer sensory scores for tenderness, juiciness, flavor, overall liking, and MQ4 score were analyzed using a linear mixed-effects model. Kiwifruit extract improved consumer scores for tenderness, juiciness, flavor, overall liking, and MQ4 scores for striploins and outside flat ($p < 0.05$). These results suggest that kiwifruit extract provides an opportunity to improve eating experiences for consumers.

Keywords: actinidin; consumer acceptance; consumer sensory testing; eating quality; grass-fed beef; Meat Standards Australia (MSA); proteolysis; sensory testing

---

1. Introduction

A challenge to the beef industry has been to provide consumers with a consistent and enjoyable eating experience. Lyford et al. [1] established that consumers are willing to pay higher premiums for higher quality beef. In Australia, the eating quality of beef is underpinned by the Meat Standards Australia (MSA) cut-based grading program [2]. The development of eating quality predictors was conducted through numerous consumer sensory panels evaluating the tenderness, juiciness, flavor, and overall liking of 39 cuts from beef carcasses. These cuts can be prepared using up to eight different methods [3], providing eating quality predictions for 135 cut × cook method combinations [2]. Traditionally, eating quality predictions have been used to assign one of four quality grades consisting of: (1) unsatisfactory, (2) "3 star" good everyday quality, (3) "4 star" better than everyday quality, and (4) "5 star" premium quality [4,5]. However, it is well established that eating quality is highly variable across different cuts and cut × cook method combinations [6]. Eating quality is impacted by numerous animal and management factors including, but not limited to the following: *Bos indicus* content, carcass hanging method, intramuscular fat, ultimate pH (pHu), and post-mortem aging of the cut and cooking method [5]. Thus, eating quality of beef can be influenced by many facets of the supply chain [2,4]. However, to ensure consumer enjoyment and satisfaction of all beef purchases, improving the potential consumer experience of unsatisfactory and lower quality cuts requires value adding, which also has the potential to increase industry profitability.

Techniques for value adding, preparing, and cooking beef are continually growing, increasing the value proposition of all muscles in the carcass. Kiwifruit (*Actinidia deliciosa*) contains a cysteine protease enzyme, actinidin [7], and has been identified to improve the tenderness in beef [8–10], lamb [11] and pork [12]. Toohey et al. [8] concluded that kiwifruit infusion of beef topsides (*M. semimembranosus*) reduced shear force values. The findings of Toohey et al. [8] suggest that kiwifruit extract infusions may improve consumer tenderness scores. Therefore, a kiwifruit extract infusion has the potential to increase the quality grades of low-quality cuts, improving the potential value of the cut, while improving consumer satisfaction. Eating quality, as defined by the MSA model, evaluates consumer satisfaction by four attributes including: (1) tenderness, (2) juiciness, (3) flavor, and (4) overall liking. These attributes are then used to determine an overall eating quality score (MQ4) for each cut × cook method combination. The influence of actinidin derived from kiwifruit on consumer sensory outcomes has not yet been elucidated for any cook method. Therefore, the hypothesis of this study was that consumer sensory scores will improve for the striploin (*M. longissimus lumborum*) and outside flat (*M. biceps femoris*) after infusion with a kiwifruit extract solution cooked as a grill or roast.

2. Materials and Methods

2.1. Animals

A total of 87 grass-fed steers were utilized in the current study. All cattle were sourced from a single cross-breeding herd comprising of *Bos indicus* crossbred maternal lines and were sired by Red Poll, Wagyu, and Brahman bulls. Hormone growth promotants (HGPs) were not used within this study.

2.2. Slaughter Procedure, Carcass Grading, and Muscle Collection

After 8 h in lairage, cattle were slaughtered at a commercial abattoir (Queensland, Australia). Post-slaughter carcasses were marshalled into a spray chiller. Carcass temperature and pH rate of decline were recorded at hourly intervals from chiller entry until a muscle pH of 6 was obtained [2,5]. This was done to ensure that all carcasses passed through pH 6 between 15 °C and 35 °C, ensuring conformance to MSA pH and temperature decline requirements [5,13]. Post chilling, carcasses were evaluated by a single accredited MSA grader 20 h post slaughter [13]. Hump heights were measured during carcass grading using a 5 mm graduated metal ruler [6]. Striploins (*M. Longissimus lumborum*) and outside flats (*M. biceps femoris*) from the left side of each carcass were collected at boning. Striploins and outside flats were vacuum packed and chilled for 24 h prior to collection from the abattoir. These muscles were then transported by refrigerated transport at 1 °C for further processing.

2.3. Muscle Preparation

On day 6, striploins and outside flats were dissected from any secondary muscles and denuded to remove all external fat and epimysium. Striploins and outside flats were then dissected in half and then allocated to one of two treatments: (1) not infused (control) and (2) infused with a kiwifruit extract (enhanced). The kiwifruit infusion solution was prepared according to the specifications, where 10 kg kiwifruit extract was completely dissolved in 72 L of water (Earlee Products Pty Ltd.; Wunda Brine CFD 5000, Code: 044-224M, Batch No: 170727, Brisbane, Australia). Enhanced muscle portions were injected at a rate of approximately 10% initial weight using a Fomaco Machine equipped with 4 mm needles (Copenhagen, Denmark). Post-enhancement muscle samples were reweighed to determine change in muscle weights associated with the enhancement process.

Enhanced and control striploins and outside flat samples were then prepared as grill (GRL) and roast (RST) samples, as per the MSA protocols as described by Watson et al. [14,15]. Briefly, GRL and RST samples were portioned into $75 \times 25 \times 150$ mm and $75 \times 75 \times 150$ mm portions, respectively. Grill samples were then individually wrapped in freezer film prior to packing into vacuum sealed bags [15]. Roast samples were individually packed into labelled vacuum-sealed bags [15]. Samples were then aged at 1 °C until 10 or 28 days after slaughter (4 and 22 days post-infusion treatment) at which point they were frozen and stored at −20 °C until being thawed for consumer sensory testing.

2.4. Consumer Sensory Testing

Consumer sensory testing sessions were conducted using the MSA protocols, as described by Watson et al. [14,15]. Briefly, GRL samples were cooked on a Silex grill (Silex S-Tronic Single Grill, Piotis Pty Ltd., Marrickville, Australia) heated to approximately 200 °C. Grill samples were cooked on the Silex in a predetermined order for 5 min to ensure samples were cooked to a medium doneness, samples were then rested for 2 min as per the MSA protocols, as described by Watson et al. [14,15]. Samples were then halved and served to two consumers.

Roast samples were presented to untrained consumers in three presentations (i) as a hot roast (RST) sample as per standard MSA protocol, portioned into a 10 mm slice [15]; (ii) cold roast sample portioned as a 10 mm slice (RST_{10}); and then (iii) cold roast sample portioned as a 2 mm slice (RST_2). Roast samples were cooked in a commercial fan-forced gas oven at 160 °C until samples reached an internal temperature of 65 °C. Samples were stored in a Bain Marie for a minimum of 5 min until preparation for serving. Samples were then trimmed to a standard size of $65 \times 65 \times 110$ mm, before being returned to a Bain Marie maintained at 48 °C. Prior to service, the RST samples were portioned into 10 mm slices. The remaining roast samples were placed in a specifically designed roast holder and chilled overnight, and then served cold to a different consumer group (the following day). This was identical to the hot service, where RST_{10} samples were prepared into 10 mm slices and offered to a second group of consumers. Similarly, at the conclusion of this service the final remaining section of the roast was prepared as a RST_2 on a deli slicer, set at 2 mm, then served to a third group of consumers, two days after the initial hot roast sensory testing. Excluding these exceptions, other procedures followed the MSA protocols as described by Watson et al. [14,15].

Grill and RST samples were presented in a controlled 6×6 latin square design ensuring that each sample was presented an equal number of times in the serving order from two to seven and an equal number of times before and after each of the other products, effectively balancing for potential order or halo effects [15]. Samples were prepared over 35 consumer sensory sessions consisting of 60 untrained consumers in each session, suggesting 2100 consumer scores. However, there were 20 untrained consumers that did not consume all sensory samples, thus a total of 2080 individual consumers were used within the current study. Consumers ($n = 2080$) evaluated each sample served for tenderness, juiciness, flavor, and overall liking by scoring a line on a 100 mm visual analogue scale ranging from 0–100. The visual analogue scale was anchored by descriptions which were not tender/very tender, not juicy/very juicy, and dislike extremely/like extremely for both liking of flavor and overall liking scores, i.e., 0 indicated a not tender sample and 100 was used to describe a very tender sample.

2.5. Statistical Analysis

The overall eating quality score (MQ4) was determined from the consumer tenderness, juiciness, flavor, and overall liking [6]. Tenderness, juiciness, flavor, and overall liking scores were weighted by 0.4, 0.1, 0.2, and 0.3, respectively, providing a MQ4 score between 0 and 100 [6]. The raw means of each sensory trait were calculated together with clipped means calculated by removing the highest and lowest 2 scores for each trait [6].

All data exploration and statistical analyses were conducted in R [16]. Data merging and manipulation, data visualizations, and summary data were conducted using the "dplyr" [17], "ggplot2" [18], and "table1" [19] packages, respectively.

Initially, correlations between raw and clipped consumer sensory scores for meat tenderness, juiciness, flavor, overall liking, and MQ4 score were conducted. Raw and clipped consumer sensory scores for meat tenderness, juiciness, flavor, overall liking, and MQ4 score were analyzed using a linear mixed-effects model in the "lme4" package [20] and estimated marginal means were generated using the "emmeans" package [21].

Models incorporated muscle, number of days aged, treatment and cooking method, and their interactions as fixed effects. Models were refined to remove relevant insignificant interactions in a step-wise manner. The final models for tenderness, juiciness, flavor, overall liking, and MQ4 included muscle, number of days aged, treatment, cooking method, muscle × treatment, number of days aged × treatment, muscle × treatment, muscle × cooking method, number of days aged × cooking method, treatment × cooking method, and number of days aged × treatment × cooking method. Additionally, an individual animal/carcass identification was incorporated as a random effect in all models, to account for animal factors. The term cooking method was used to describe GRL, RST, RST_2, and RST_{10}.

3. Results

Strong correlations between raw and clipped consumer sensory scores for tenderness ($R^2 = 0.99$, $p \leq 0.0001$), juiciness ($R^2 = 0.99$, $p \leq 0.0001$), flavor ($R^2 = 0.99$, $p \leq 0.0001$), overall liking ($R^2 = 0.99$, $p \leq 0.0001$) and MQ4 ($R^2 = 0.99$, $p \leq 0.0001$) were identified. Furthermore, there were no differences in the significant terms between models conducted on raw and clipped data, therefore, data herein pertains to analysis conducted on the raw consumer sensory data.

3.1. Carcass Traits

All 87 carcasses were graded as per the MSA grading specifications [13]. Carcass characteristics as evaluated by the MSA carcass grading specifications are summarized in Table 1.

Table 1. Mean (± SEM), minimum and maximum carcass characteristics as determined by the Meat Standards Australia (MSA) beef carcass grading specifications.

Carcass Trait	Mean	Minimum	Maximum
HSCW, kg	301.6 ± 2.1	262	357.5
Hump height, mm	86.2 ± 2.4	40	150
Eye muscle area, cm^2	79.3 ± 0.94	65	100
Rib fat, mm	7.5 ± 0.32	3	19
Ossification	155.4 ± 2.1	130	230
MSA marbling score	327.5 ± 7.8	220	510
pHu	5.5 ± 0.01	5.43	6.12

3.2. Kiwifruit Extract Infusion

The average weight increase post kiwifruit extract enhancement was 11.2% ± 0.29%, however, there was some variability in the enhancement rate between striploin and outside flat muscles (Table 2).

Table 2. Mean (± SEM), minimum, and maximum increase in muscle weights for striploins (*M. longissimus lumborum*) and outside flats (*M. biceps femoris*) enhanced with a kiwifruit extract.

Item	Mean	Minimum	Maximum
Striploin			
Initial weight, kg	2.18 ± 0.04	1.39	3.2
Final weight, kg	2.41 ± 0.05	1.54	3.5
Percent increase, %	10.75 ± 0.04	4.54	16.2
Outside flat			
Initial weight, kg	1.41 ± 0.03	0.82	2.31
Final weight, kg	1.57 ± 0.03	0.87	2.56
Percent increase, %	11.64 ± 0.35	5.83	25

3.3. Consumer Sensory Outcomes

Overall, on average striploins had greater tenderness, juiciness, flavor, overall liking, and MQ4 scores as compared with outside flats ($p < 0.001$, Table 3). Consumers exhibited an increased preference for RST_2 samples ($p < 0.001$, Table 4).

Table 3. Estimated marginal means (± 95% confidence interval) and the difference between striploins (*M. longissimus lumborum*) and outside flats (*M. biceps femoris*) scores for tenderness, juiciness, flavor, overall liking, and eating quality (MQ4) [1,2].

Consumer Scores	Striploin	Outside Flat	Difference
Tenderness	58.4 ± 0.86 [a]	42 ± 0.86 [b]	16.4
Juiciness	54.9 ± 0.84 [a]	46.7 ± 0.84 [b]	8.2
Flavor	57.4 ± 0.72 [a]	48.4 ± 0.71 [b]	9
Overall liking	57.7 ± 0.76 [a]	45.5 ± 0.76 [b]	12.2
MQ4	57.5 ± 0.76 [a]	45.4 ± 0.75 [b]	12.1

[1] Tenderness, juiciness, flavor, and overall liking were scored by consumers on a 100 mm visual analogue scale ranging from 0 to 100. [2] Eating quality scores (MQ4) were calculated by weighting tenderness (0.4), juiciness (0.1), flavor (0.2), and overall liking (0.3) providing a MQ4 score between 0 and 100 [6]. [3] Data presented represents consumer sensory scores from 2080 individual untrained consumers. [a,b] Within a row, means without a common superscript differ ($p < 0.0001$).

Table 4. Estimated marginal means (± 95% confidence interval), and difference between striploins (*M. longissimus lumborum*) and outside flats (*M. biceps femoris*) scores for tenderness, juiciness, flavor, overall liking, and eating quality (MQ4) served as grill (GRL), hot roast (RST), cold roast as a 2 mm slice (RST_2) and cold roast as a 10 mm slice (RST_{10}) for tenderness, juiciness, flavor, overall liking, and eating quality (MQ4) [1,2,3].

Consumer Scores	Striploin				Outside Flat				Difference [4]			
	GRL	RST	RST_2	RST_{10}	GRL	RST	RST_2	RST_{10}	GRL	RST	RST_2	RST_{10}
Tenderness	60.8 ± 0.93	60.9 ± 1.11	55.8 ± 1.89	56 ± 1.89	39.8 ± 0.91	39 ± 1.11	54.5 ± 1.87	34.9 ± 1.87	21 *	21.9 *	1.3	21.1 *
Juiciness	59.1 ± 0.91	54 ± 1.1	55.1 ± 1.91	51.3 ± 1.91	49.7 ± 0.9	44.5 ± 1.11	54.6 ± 1.88	37.9 ± 1.88	9.4 *	9.5 *	0.5	13.4 *
Flavor	60.7 ± 0.77	58.6 ± 0.93	56.4 ± 1.6	54 ± 1.6	49.1 ± 0.76	45.6 ± 0.93	57.1 ± 1.57	41.7 ± 1.57	11.6 *	13 *	−0.7	12.3 *
Overall liking	60.8 ± 0.82	58.6 ± 1	56.9 ± 1.72	54.4 ± 1.72	44.5 ± 0.81	42.5 ± 1	56 ± 1.69	38.9 ± 1.69	16.3 *	16.1*	0.9	15.5 *
MQ4	60.6 ± 0.81	58.8 ± 0.98	56.2 ± 1.67	54.4 ± 1.67	45 ± 0.8	42.6 ± 0.98	55.7 ± 1.65	38.4 ± 1.65	15.6 *	16.2*	0.5	16 *

[1] Tenderness, juiciness, flavor, and overall liking were scored by consumers on a 100 mm visual analogue scale ranging from 0 to 100. [2] Eating quality scores (MQ4) were calculated by weighting tenderness (0.4), juiciness (0.1), flavor (0.2), and overall liking (0.3) providing a MQ4 score between 0 and 100 [6]. [3] Data presented represents consumer sensory scores from 2080 individual untrained consumers. [4],* Within column indicates significant differences ($p < 0.0001$) between consumer sensory traits between striploins (*M. longissimus lumborum*) and outside flats (*M. biceps femoris*).

The number of days muscles were aged did not influence consumer evaluations of tenderness ($p = 0.11$), juiciness ($p = 0.59$), flavor ($p = 0.2$), overall liking ($p = 0.7$) or MQ4 ($p = 0.99$) scores. Furthermore, there were no number of days aged × cooking method influences on consumer perceptions of tenderness ($p = 0.65$), juiciness ($p = 0.23$), flavor ($p = 0.55$), overall liking ($p = 0.7$) or MQ4 ($p = 0.67$) scores.

Kiwifruit extract infusion improved consumer evaluations of tenderness, juiciness, flavor, overall liking, and MQ4 by 11.4, 13.2, 12.2, 9.9, and 10.8 points, respectively, ($p < 0.001$), when compared to the control samples. Similarly, there was a treatment × cooking method interaction as the GRL, RST, RST_2, and RST_{10} tenderness scores increased by 16.4, 14.1, 10.1, and 12.5 points ($p = 0.04$, Figure 1); overall liking scores increased by 14, 10.3, 8.4, and 10.4 points ($p = 0.01$, Figure 1); and MQ4 scores increased by 14, 11.4, 9, and 11.1 points ($p = 0.05$, Figure 1), respectively, when compared with the control samples. However, there was no effect on consumer juiciness ($p = 0.36$, Figure 1) or flavor ($p = 0.08$, Figure 1) scores.

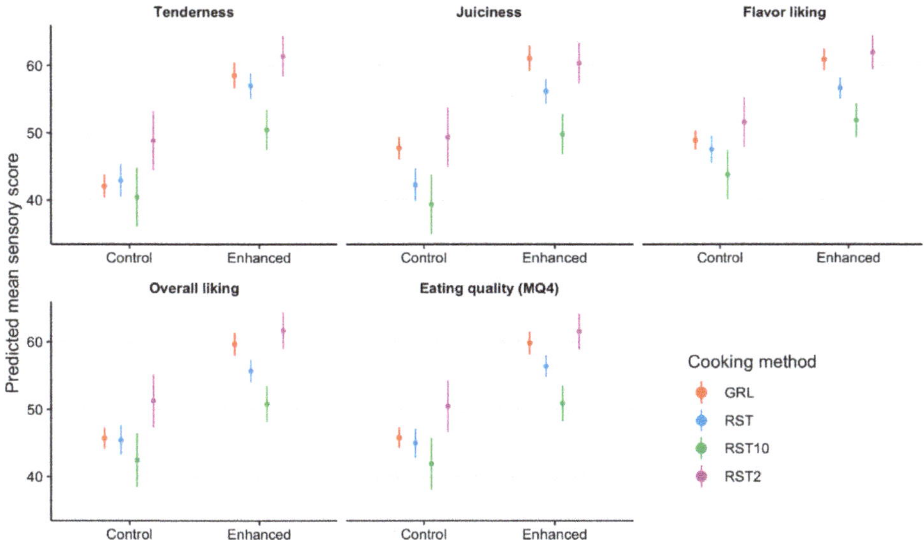

Figure 1. Estimated marginal means with 95% confidence intervals for tenderness, juiciness, flavor, overall liking, and eating quality (MQ4) scores of striploins (*M. longissimus lumborum*) and outside flats (*M. biceps femoris*) not infused (control) and infused with a kiwifruit extract (enhanced) presented to consumers ($n = 2080$) as grill (GRL), 10 mm hot roast (RST), 2 mm cold roast (RST_2) and 10 mm cold roast (RST_{10}).

Kiwifruit infusion improved consumer scores for tenderness, juiciness, flavor, overall liking, and MQ4 of striploins (Figure 2) by between 11.7 and 15.7 points and outside flats (Figure 3) by between 8 and 10.9 points ($p < 0.02$; Figures 2–4), relative to the control samples. Furthermore, there was a number of days aged × treatment interactions with consumer perceptions of tenderness ($p = 0.02$), juiciness ($p = 0.05$), flavor ($p = 0.0001$), overall liking ($p = 0.0006$), and MQ4 ($p = 0.0009$), however, there was a reduction in consumer acceptance of the enhanced treatment at 28 days aging (Figures 2 and 3).

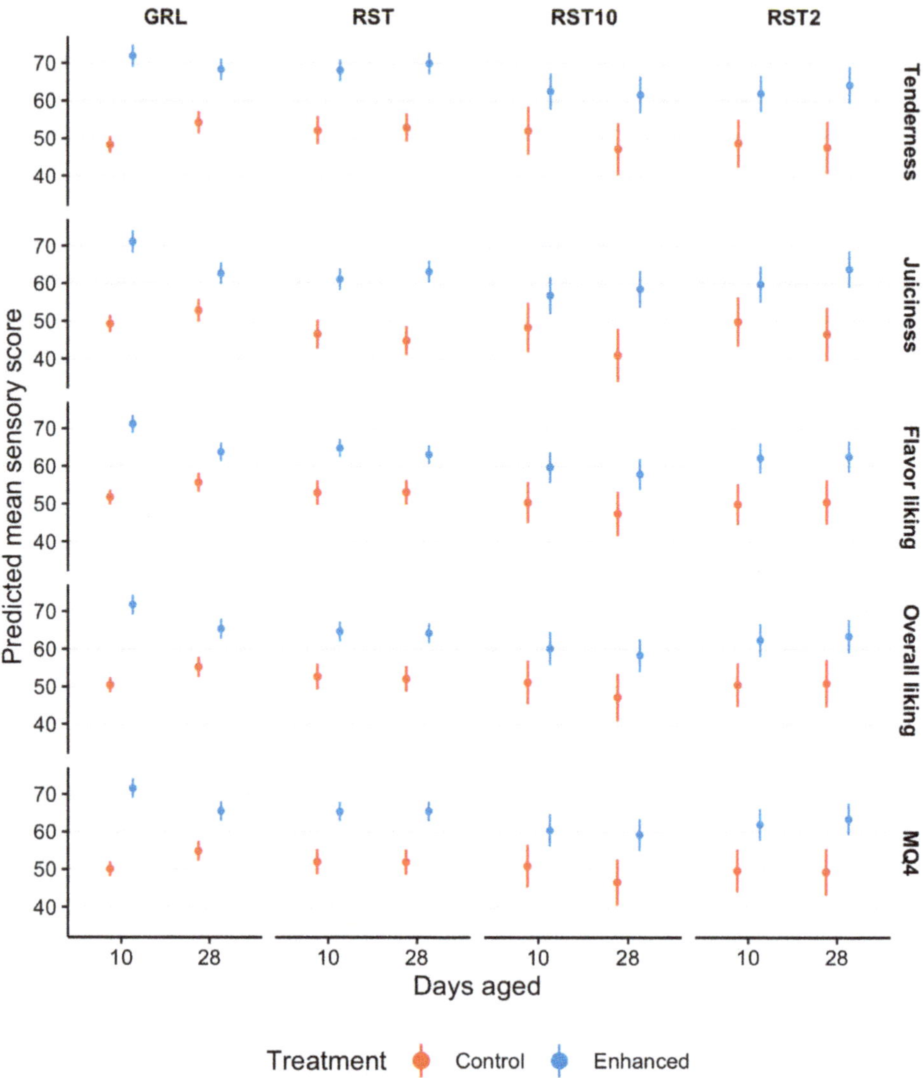

Figure 2. Estimated marginal means with 95% confidence intervals for tenderness, juiciness, flavor liking, overall liking, and eating quality (MQ4) scores of striploins (STR045; *M. longissimus lumborum*) presented to consumers (n = 2080) as grill (GRL), 10 mm hot roast (RST), 2 mm cold roast (RST$_2$), and 10 mm cold roast (RST$_{10}$) that were not infused (control) and infused with a kiwifruit extract (enhanced), and aged for 10 or 28 days.

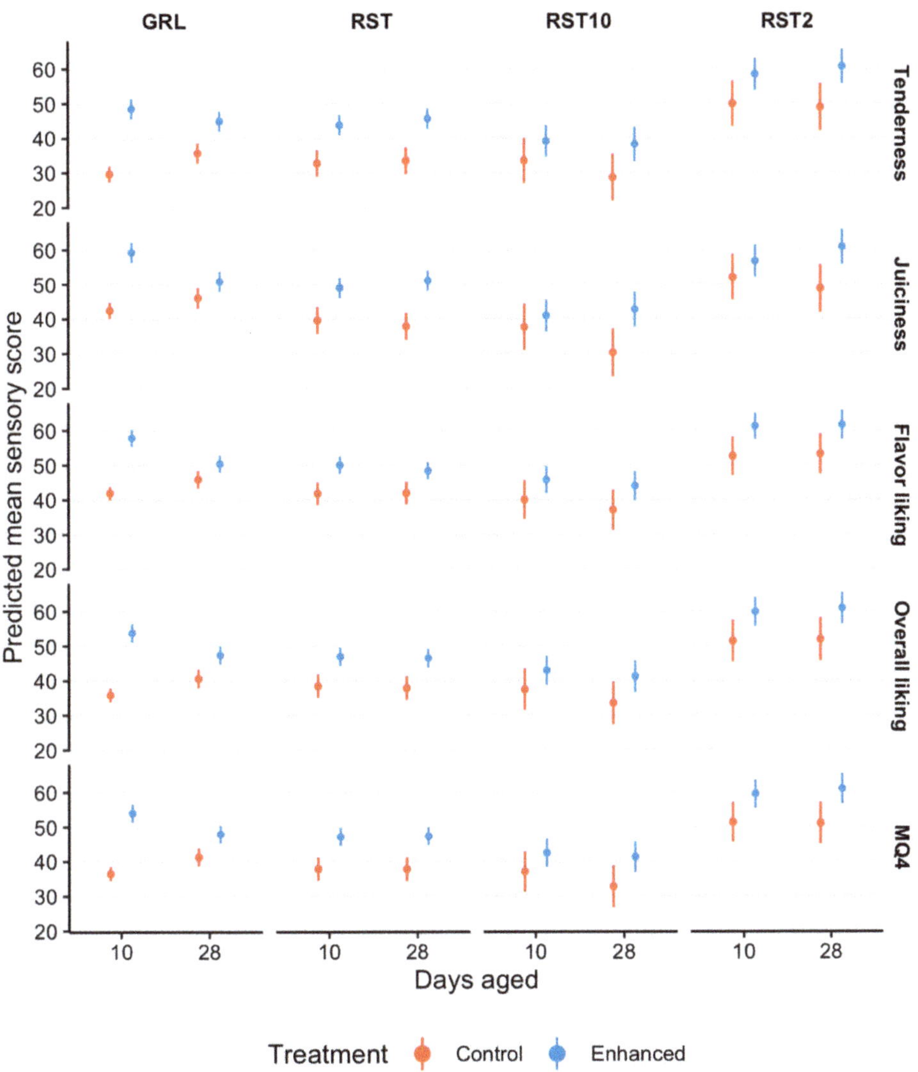

Figure 3. Estimated marginal means with 95% confidence intervals for tenderness, juiciness, flavor liking, overall liking, and eating quality (MQ4) scores of outside flats (OUT005, *M. biceps femoris*) presented to consumers ($n = 2080$) as grill (GRL), 10 mm hot roast (RST), 2 mm cold roast (RST$_2$) and 10 mm cold roast (RST$_{10}$) that were not infused (control) and infused with a kiwifruit extract (enhanced), and aged for 10 or 28 days.

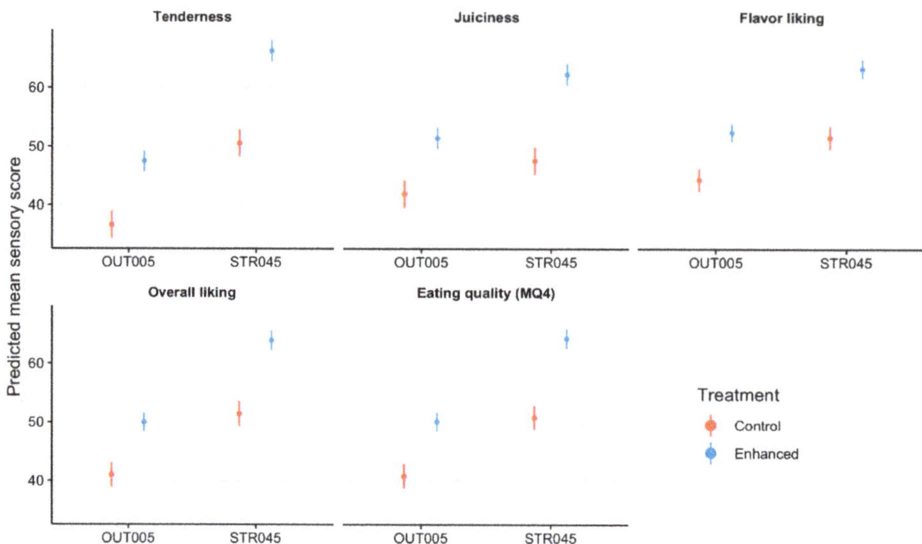

Figure 4. Estimated marginal means with 95% confidence intervals for tenderness, juiciness, flavor, overall liking, and eating quality (MQ4) scores of striploins (STR045, *M. longissimus lumborum*) and outside flats (OUT005, *M. biceps femoris*) presented to consumers (*n* = 2080) that were not infused (control) and infused with a kiwifruit extract (enhanced).

There were a number of days aged × treatment × cooking method influences (Figures 2 and 3) on consumer evaluations of tenderness (*p* = 0.005), juiciness (*p* < 0.001), flavor (*p* = 0.003), overall liking (*p* = 0.001), and MQ4 (*p* = 0.0007). However, consumer tenderness evaluations had similar increases for RST, RST$_2$, and RST$_{10}$ samples aged for 10 and 28 days. At 10 days aging there was an increase of 13.6, 10.8, and 8 points for enhanced RST, RST$_2$, and RST$_{10}$ as compared with the control samples. Comparably, at 28 days aging, consumer evaluations for tenderness increased 14.6, 14.2, and 12 pointsfor the enhanced RST, RST$_2$ and RST$_{10}$.

Consumer sensory scores indicated that the juiciness of enhanced GRL samples that were aged for 10 days, were 19.3 points higher than control samples. Whereas juiciness scores for enhanced GRL that were aged for 28 days, increased by 7.4 points. The kiwifruit extract increased the juiciness scores of RST, RST$_2$, and RST$_{10}$ that were aged for 10 days by 11.9, 7.2, and 5.8 points, respectively, when compared with the control samples. Although the consumer evaluations of juiciness exhibited an increase of 15.9, 14.7, and 12 points in tenderness scores of enhanced RST, RST$_2$, and RST$_{10}$ that were aged for 28 days, relative to the control samples.

Flavor scores highlight that enhanced GRL samples that were aged for 10 days, were 17.6 points higher than control samples, however, 28 days aging only observed an increase of 6.3 points. Similar increases in consumer scores were observed for enhanced RST (10 days, 10 points and 28 days, 8.2 points), RST$_2$ (10 days, 10.5 points and 28 days, 10.2 points), and RST$_{10}$ (10 days, 7.5 points and 28 days, 8.7 points) samples aged for 10 and 28 days.

At 10 days aging, there was an increase of 19.6 points in consumer overall liking for enhanced GRL samples, relative to the control samples. Although, increasing the number of days aged to 28 days only increased the overall liking of enhanced GRL samples by 8.4 points. Consumer overall liking scores had similar increases for enhanced RST (10 days, 10.2 points and 28 days, 10.4 points), RST$_2$ (10 days, 10.2 points and 28 days, 10.7 points) and RST$_{10}$ (10 days, 7.3 points and 28 days, 9.5 points) samples aged for 10 and 28 days, when compared with the control samples.

There was an increase of 19.5 points in MQ4 scores of enhanced GRL samples aged for 10 days, relative to the control samples. However, this did not persist to 28 days aging where the increase in MQ4 was 8.6 points. Additionally, there were increases in MQ4 scores of enhanced RST (10 days, 11.4 points and 28 days, 11.5 points), RST_2 (10 days, 10.2 points and 28 days, 12 points) and RST_{10} (10 days, 7.4 points and 28 days, 10.5 points) samples aged for 10 and 28 days, in comparison to the control samples.

4. Discussion

This study is the first to evaluate the influence of infusing beef cuts with actinidin derived from kiwifruit on untrained consumer sensory panels. The results from this study show that infusing striploins and outside flats with a kiwifruit extract improved consumer scores for tenderness, juiciness, flavor, overall liking, and MQ4 scores, which supported the initial hypothesis. Specifically, there was an increase of 13.4 points in the MQ4 score of enhanced striploins, which related to a quality grade increase from a 3-star good everyday product to a 4-star better than everyday quality product [4,5]. In comparison, outside flats had an increase of 9.3 points in the MQ4 score, which indicated a quality grade increase from unsatisfactory to a 3-star good everyday product [4,5]. Therefore, it is clearly evident that the kiwifruit enhancement was capable of improving consumer acceptance of beef striploins and outside flats, providing an opportunity for the beef industry to generate greater revenue from these primals.

Similarly, for GRL, RST, RST_2, and RST_{10} infusion with kiwifruit extract increased the MQ4 scores from 45.8 ± 0.78, 45.0 ± 1.10, 50.4 ± 1.97, and 41.9 ± 1.97 to 59.8 ± 0.85, 56.4 ± 0.85, 61.5 ± 1.36, and 50.9 ± 1.36. This equates to an increase in the eating quality predictions of RST and RST_{10} from an unsatisfactory product to a 3-star good everyday grade [4,5]. However, for GRL and RST_2, kiwifruit extract improved the eating quality from a low grade 3-star good everyday quality to a higher good everyday quality [4,5]. This suggests that the kiwifruit extract improved the overall eating experience of consumers while not impacting on the flavor of striploins or outside flats. Similarly, Christensen et al. [12] highlighted similar findings in pork *M. biceps femoris*. The authors reported that an improvement in textural attributes, while juiciness, flavor, and taste were not influenced by actinidin infusions [12]. Overall, this suggests that kiwifruit extract can be used to effectively tenderize meat without adversely affecting other sensory attributes.

At 10 and 28 days aging, enhanced GRL, RST, RST_2 and RST_{10} exhibited increases in consumer evaluations of tenderness, juiciness, flavor, overall liking, and MQ4 scores, as compared to the control samples. However, tenderness, juiciness, flavor, overall liking, and MQ4 scores for enhanced samples aged for 28 days generally decreased in comparison to enhanced samples that were aged for 10 days, or four days post infusion. Regardless, beef infused with the kiwifruit solution and aged for 28 days still had higher consumer sensory scores as compared with the control samples. The mechanisms that resulted in the decreased consumer scores for enhanced samples aged for 28 days as compared with 10 days are yet to be clarified.

Lewis and Luh [10] indicated that actinidin may hydrolyze muscle proteins that are associated with toughness in beef cuts, rather than having a hydrolyzing action on all of the proteins in beef muscle. The tenderization of beef is a complex process that involves numerous structural changes in myofibrillar components associated with proteolytic enzymes [9]. This suggests that the improvement of eating quality from kiwifruit extracts may be due to an increased proteolytic activity on muscle proteins [11], possibly specifically myofibrillar proteins [8], by actinidin.

The results presented here suggest that proteolysis is sustained in beef cuts infused with kiwifruit extracts up to 28 days and possibly longer. This suggests that there may be a risk of over tenderization occurring due to the continued proteolytic activity [22]. Hence, aging beyond 10, or 14 days [8], may not provide the optimum eating experience for consumers. As such, these results suggest that the optimum number of days for aging beef infused with kiwifruit extracts warrants further investigations. Furthermore, it is important to consider that the optimum number of days for aging may vary across different cuts of meat, as some beef cuts may benefit from a longer period of aging [14,23]. Understanding the optimum number of days for aging post infusion process would ensure that a consistent and enjoyable eating experience for consumers is achieved.

This study has highlighted that using a kiwifruit infusion improved the quality grades of striploin and outside flats within this study, suggesting that kiwifruit extract provides an opportunity to improve the value of low- and poor-quality beef. Lyford et al. [1] indicated that consumer willingness to pay increased by approximately 50% for increased quality of (1) unsatisfactory to 3-star good everyday quality and (2) 3-star good to 4-star better everyday quality beef. Therefore, developing value adding methods for improving the quality grades of low-quality cuts is important for the long-term financial sustainability of the beef industry. On the basis of these findings, it could be reasoned that the kiwifruit infusion of low-quality striploins and outside flats has the potential to "double" the market value within Australia. Nevertheless, these results suggest that cuts infused with the kiwifruit solution should be cooked or frozen within a short period after the infusion process is conducted to reduce the activity of actinidin and ensure an optimum consumer experience.

5. Conclusions

Enhancing striploins and outside flats improved the eating quality scores of GRL, RST, RST_2, and RST_{10} as evaluated by untrained consumers. In both muscles, at 10 and 28 days post-mortem aging, consumed as GRL or RST, the tenderness, juiciness, flavor, and overall liking of the kiwifruit enhanced product was increased, improving the intrinsic eating quality. As such, this study suggests that kiwifruit extract has the potential to increase consumer acceptance while increasing profitability of processors by improving unsatisfactory and low-quality beef cuts into acceptable and high-quality cuts.

Author Contributions: R.P., M.K., and P.M. contributed to the conceptualization and experimental design of the study; M.K., P.M., and R.P., undertook the study; A.L. and G.T. conducted the formal data analysis and presentation of the data; A.L. wrote the original manuscript draft; reviews and editing of the completed manuscript was conducted by G.T., M.K., R.P., and P.M.; project administration and funding acquisition was the responsibility of R.P. and P.M.

Funding: This research was funded by the Meat and Livestock Australia P/L., North Sydney, NSW, Australia; grant number: NBP.0722, "Transport Duration Effects on MSA Eating Quality".

Acknowledgments: The authors would like to acknowledge Earlee Products Pty Ltd. (Australia) for the use of their product and facilities and the Meat Standards Australia for grading the carcasses and collecting the products used within this study. The authors also wish to acknowledge and offer thanks to Sensory Solutions for conducting the consumer sensory panels. The authors would also like to thank Jessira Perovic (MSA), Janine Lau (MSA), Sarah Strachan (MSA), and Jarrod Lees (UNE) for their feedback on this manuscript prior to submission.

Conflicts of Interest: The authors declare no conflict of interest.

References

1. Lyford, C.P.; Thompson, J.M.; Polkinghorne, R.; Miller, M.F.; Nishimura, T.; Neath, K.; Allen, P.; Belasco, E.J. Is willingness to pay (WTP) for beef quality grades affected by consumer demographics and meat consumption preferences? *Australas. Agribus. Rev.* **2010**, *18*, 1.
2. Polkinghorne, R.; Thompson, J.M.; Watson, R.; Gee, A.; Porter, M. Evolution of the Meat Standards Australia (MSA) beef grading system. *Aust. J. Exp. Agric.* **2008**, *48*, 1351–1359. [CrossRef]
3. McGilchrist, P.; Polkinghorne, R.J.; Ball, A.J.; Thompson, J.M. The Meat Standards Australia Index indicates beef carcass quality. *Animal* **2019**, *13*, 1750–1751. [CrossRef] [PubMed]
4. Polkinghorne, R.; Watson, R.; Thompson, J.M.; Pethick, D.W. Current usage and future development of the Meat Standards Australia (MSA) grading system. *Aust. J. Exp. Agric.* **2008**, *48*, 1459–1464. [CrossRef]
5. Thompson, J. Managing meat tenderness. *Meat Sci.* **2002**, *62*, 295–308. [CrossRef]
6. Watson, R.; Polkinghorne, R.; Thompson, J.M. Development of the Meat Standards Australia (MSA) prediction model for beef palatability. *Aust. J. Exp. Agric.* **2008**, *48*, 1368–1379. [CrossRef]
7. Wada, M.; Suzuki, T.; Yaguti, Y.; Hasegawa, T. The effects of pressure treatments with kiwi fruit protease on adult cattle semitendinosus muscle. *Food Chem.* **2002**, *78*, 167–171. [CrossRef]

8. Toohey, E.S.; Kerr, M.J.; van de Ven, R.; Hopkins, D.L. The effect of a kiwi fruit based solution on meat traits in beef m. semimembranosus (topside). *Meat Sci.* **2011**, *88*, 468–471. [CrossRef] [PubMed]

9. Aminlari, M.; Shekarforoush, S.S.; Gheisari, H.R.; Golestan, L. Effect of Actinidin on the Protein Solubility, Water Holding Capacity, Texture, Electrophoretic Pattern of Beef, and on the Quality Attributes of a Sausage Product. *J. Food Sci.* **2009**, *74*, C221–C226. [CrossRef] [PubMed]

10. Lewis, D.A.; Luh, B.S. Application of Actinidin from kiwifruit to meat tenderization and characterization of beef muscle protein hydrolysis. *J. Food Biochem.* **1988**, *12*, 147–158. [CrossRef]

11. Han, J.; Morton, J.D.; Bekhit, A.E.D.; Sedcole, J.R. Pre-rigor infusion with kiwifruit juice improves lamb tenderness. *Meat Sci.* **2009**, *82*, 324–330. [CrossRef] [PubMed]

12. Christensen, M.; Tørngren, M.A.; Gunvig, A.; Rozlosnik, N.; Lametsch, R.; Karlsson, A.H.; Ertbjerg, P. Injection of marinade with actinidin increases tenderness of porcine *M. biceps femoris* and affects myofibrils and connective tissue. *J. Sci. Food Agric.* **2009**, *89*, 1607–1614. [CrossRef]

13. Meat Standards Australia. *MSA Standards Manual for Beef Grading*; Meat & Livestock Australia: Sydney, Australia, 2007.

14. Watson, R.; Gee, A.; Polkinghorne, R.; Porter, M. Consumer assessment of eating quality development of protocols for Meat Standards Australia (MSA) testing. *Aust. J. Exp. Agric.* **2008**, *48*, 1360–1367. [CrossRef]

15. Watson, R.; Gee, A.; Polkinghorne, R.; Porter, M. Consumer assessment of eating quality development of protocols for Meat Standards Australia (MSA) testing. Accessory Publication: MSA sensory testing protocols. *Aust. J. Exp. Agric.* **2008**, *48*, 1360–1367. [CrossRef]

16. R Core Team. *R: A Language and Environment for Statistical Computing*; R Foundation for Statistical Computing: Vienna, Austria, 2018; Available online: https://www.R-project.org/ (accessed on 8 May 2019).

17. Wickham, H.; François, R.; Henry, L.; Müller, K. Dplyr: A Grammar of Data Manipulation. 2019. Available online: https://CRAN.R-project.org/package=dplyr (accessed on 8 May 2019).

18. Wickham, H.; Chang, W.; Henry, L.; Pedersen, T.L.; Takahashi, K.; Wilke, C.; Woo, K. Ggplot2: Create Elegant Data Visualisations Using the Grammar of Graphics. 2019. Available online: https://CRAN.R-project.org/package=ggplot2 (accessed on 8 May 2019).

19. Rich, B. Table1: Tables of Descriptive Statistics in Html. 2018. Available online: https://CRAN.R-project.org/package=table1 (accessed on 30 June 2019).

20. Bates, D.; Maechler, M.; Bolker, B.; Walker, S. Lme4: Linear Mixed-Effects Models Using 'Eigen' and S4. 2019. Available online: https://CRAN.R-project.org/package=lme4 (accessed on 30 June 2019).

21. Length, R. emmeans: Estimated Marginal Means, aka Least-Squares Means. 2019. Available online: https://Cran.R-Proj.Org/Package=Emmeans (accessed on 30 June 2019).

22. Zhu, X.; Kaur, L.; Staincliffe, M.; Boland, M. Actinidin pretreatment and sous vide cooking of beef brisket: Effects on meat microstructure, texture and in vitro protein digestibility. *Meat Sci.* **2018**, *145*, 256–265. [CrossRef] [PubMed]

23. Watson, R.; Polkinghorne, R.; Gee, A.; Porter, M.; Thompson, J.M.; Ferguson, D.; Pethick, D.; McIntyre, B. Effect of hormonal growth promotants on palatability and carcass traits of various muscles from steer and heifer carcasses from a *Bos indicus-Bos taurus* composite cross. *Aust. J. Exp. Agric.* **2008**, *48*, 1415–1424. [CrossRef]

 *foods*

Article

High Oxygen Modified Atmosphere Packaging Negatively Influences Consumer Acceptability Traits of Pork

Yunling Peng [1,†]**, Karunia Adhiputra** [1,†]**, Anneline Padayachee** [1]**, Heather Channon** [2]**, Minh Ha** [1] **and Robyn Dorothy Warner** [1,*]

[1] Faculty of Veterinary and Agricultural Sciences, The University of Melbourne, Parkville 3010, Australia; yunlingp@yahoo.com (Y.P.); k.adhiputra@gmail.com (K.A.); hello@drannelinepadayachee.com (A.P.); minh.ha@unimelb.edu.au (M.H.)
[2] Australian Pork Ltd, P.O. Box 4746, Kingston 2604, Australia; heather.channon@australianpork.com.au
* Correspondence: robyn.warner@unimelb.edu.au; Tel.: +61-407-316760 or +61-390-356663
† These authors contributed equally to the work.

Received: 18 October 2019; Accepted: 11 November 2019; Published: 13 November 2019

Abstract: Current trends in meat packaging have seen a shift from conventional overwrap to vacuum packing (VAC) and modified atmosphere packaging (MAP). The aim of this experiment was to investigate the effects of high oxygen MAP (HiOxMAP) of pork loins compared with vacuum packed (VAC) on eating quality and colour, after storage in simulated illuminated retail display conditions. Pork loins ($n = 40$) were cut and stored under two packaging methods (HiOxMAP, 80% O_2, 20% CO_2; VAC) for up to 14 days, with samples taken at various times for measurements. After 7 days of storage, HiOxMAP samples exhibited inferior consumer acceptability for tenderness, flavor, overall liking, quality and re-purchase intention as well as higher shear force and hardness, relative to VAC samples ($p < 0.05$ for all). Loins stored in HiOxMAP had higher lightness (L*), redness (a*) and yellowness (b*) values at 3 and 7 days, but lower ratio of oxymoglobin to metmyoglobin (oxy/met) values in the meat surface at 14 days of display, relative to VAC samples ($p < 0.05$ for all). The oxy/met ratio declined from 2.3 to 1.7 between days 3 and 14 of display in HiOxMAP samples ($p < 0.05$), whereas the ratio was similar and stayed relatively high for VAC samples. VAC samples produced consistently higher colour values (a*, b*, oxy/met) when left to bloom 30 min after removal from packaging ($p < 0.05$). Lipid oxidation values, measured using thiobarbituric acid reactive substances, in HiOxMAP pork loins, were higher at all time points compared to VAC during the 14 day storage period ($p < 0.05$). The use of vacuum packing for retail shelves, should be considered as the preferred option, over HiOxMAP.

Keywords: quality; tenderness; oxidation; colour; sensory

1. Introduction

Pork is the one of the most consumed meats and is the most common protein source in the human diet in the world. Meeting the consumers' demand for high quality pork is a main focus for the pork industry [1] and it has been shown that consumers are increasingly expecting high quality meat [2]. The pork consumers' purchase intention is negatively affected by inconsistent eating quality [3–5]. However, factors such as the supply chain pathways from packaging to retailers to consumers affecting eating quality of pork need to be determined. Understanding the importance of these factors for pork eating quality is essential for the pork industry to improve pork quality and complete an integrated assurance system of pork eating quality [6].

Porks' presentation is a fundamental gateway between consumers and producers. It has been shown that the consumers' decision to purchase fresh meat is initially based on the visual qualities, such as colour, marbling and purge loss [7], while sensory attributes including tenderness, juiciness and flavour of meat have been regarded as critical factors for re-purchase intention by consumers [8]. Packaging of pork has been shown to be a factor which could affect pork quality [9,10]. Packaging is not only important to extend meat shelf-life [9], but has also been found to influence eating qualities of meat [9,11]. A number of studies have investigated the impacts that packaging systems have on beef eating quality [12–14]. However, the effect of packaging systems on pork eating quality has not been well studied in comparison. MAP with a high oxygen content (70–80%; HiOxMAP) has been shown to improve beef colour, while it negatively affects eating quality of beef [13,15]. In contrast, vacuum packaging, although less desirable in appearance, has been shown to lead to an increase in tenderness and juiciness of beef [13]. A few studies [9,10] have been conducted to examine differential effects of vacuum packaging and HiOxMAP on pork quality. These studies, however, lacked integration of physical measurements with sensory attributes.

The surface colour of meat is predominantly dependent on the pigment myoglobin (Mb), an iron and oxygen binding protein. Mb incorporates an oxygen atom into its structure when exposed to an oxygen-rich environment. Oxygen-bound Mb (oxymyoglobin) is primarily responsible for the bright cherry-red colour in meat, particularly in the surface of meat in HiOxMAP [16]. In an oxygen-deprived environment, Mb takes the form of an oxygen-free protein deoxymyoglobin and is the prevalent form of myoglobin in vacuum packed meat [17]. When myoglobin or oxymyoglobin, are oxidized and lose an electron, metmyoglobin is formed, which has a brown colour [17]. When metmyoglobin starts to appear in the surface of meat, the meat is taken off the retail shelves, hence measurements of metmyoglobin accumulation in the meat surface over time are used as an indicator of shelf-life [18].

As discussed above, although the effects of HiOxMAP and vacuum packaging (VAC) on colour stability and lipid oxidation have been extensively researched for beef [19–21], little has been conducted on pork loins, particularly in relation to consumer sensory attributes. As it is well-known that pork is quite different metabolically and biochemically to beef, data on pork is also required. As pork loins are commonly displayed inside HiOxMAP in retail, it is important to determine the full effects of HiOxMAP on the eating quality and shelf quality of pork loins. This study therefore proposed to investigate the effects of pork loin ageing in HiOxMAP or VAC on the texture, colour stability, lipid oxidation, and consumer acceptability and eating quality.

2. Materials and Methods

2.1. Experiment Design and Pork Loins

Sensory procedures used in this study were approved by the Human Research Ethics Committee of the University of Melbourne. The experiment was split into two replicates, being two collection days which were two weeks apart. Forty loins from the left and right side of 20 pig carcasses were collected at 1–2 days post-slaughter, from carcasses weighing about 65 kg. Loin samples were placed in a refrigerated trailer at 4 °C and transferred to the meat laboratory in the University of Melbourne.

2.2. Sample Cutting and Packaging

Before pork loins arrived at the sensory lab, all the tools and equipment were sterilised with 80% ethanol. Sequentially, two paired loins from the one carcass were transferred from the chiller, 2–3 cm was removed from the posterior end of each loin and discarded. Then, from the anterior end and middle of both loins, three × 12.5 cm sections and five × 4 cm sections were removed and allocated to treatments as shown in Table 1. The allocation allowed for randomization of position between sides and location within the anterior end of the loin. From the posterior end, seven × 2.5 cm sections were removed from both sides and these were allocated to treatments, as shown in Table 1, with randomization of the allocation of samples to treatments according to the position as occurred for the samples for the anterior end. The 4 cm sections were used for objective assessment of Warner-Bratzler shear force (WBSF), compression and cooking loss. These samples were processed on the day of collection (day 0, control) or after the prescribed packaging and ageing period. The 12.5 cm sections were sliced further into 5×2.5 cm chops, within each section, and used for sensory assessment. These samples were packaged, subjected to prescribed packaging and ageing and frozen at -20 °C, prior to sensory assessment, with all samples being processed within one month of freezing. The 2.5 cm sections were subjected to treatments shown in Table 1. Colour was measured on samples on day of collection or removal from ageing/packaging treatment. Thiobarbituric acid reactive substances (TBARS) were measured after packaging and freezing of samples at -20 °C, with all samples being processed within one month of freezing.

Table 1. Allocation of samples within the length of the striploin, to measurements and treatments across 0, 3, 7, and 14 days of ageing.

Position in Both Loins	0 Days	3 Days VAC [1] & HiOxMAP [1]	7 Days VAC & HiOxMAP	14 Days VAC & HiOxMAP
Anterior section 5×4 cm	Warner-Bratzler, Compression, cooking loss	Warner-Bratzler, Compression, cooking loss	Warner-Bratzler, Compression, cooking loss	
Middle section 3×12.5 cm	Consumer sensory		Consumer sensory	
Posterior 7×2.5 cm	Colour, TBARS	Colour, TBARS	Colour, TBARS	Colour, TBARS

[1] VAC, vacuum packaging; MAP, high oxygen modified atmosphere packaging (80% O_2, 20% CO_2).

All the chops (2.5 cm and 4 cm) were first weighed, then packed according to treatment, shown in Table 1. Vacuum packaging was performed on a Multivac C200 (Sepp Haggenmüller GmbH & Co., Wolferschwenden, Germany) using polyamide and polyetheylene vacuum pouches PA/PE 70 (Multivac) with an oxygen permeability less than 65 cc/m^2 (24 h) and water transmission less than 5 g/m^2/24 h. HiOxMAP packaging was conducted on a Multivac T200 (Sepp Haggenmüller GmbH & Co., Wolferschwenden, Germany). Packing trays were clear Cryovac (T0D0901C 170 mm $\times$ 223 mm, Sealed Air, Australia), a cello soaker pad (130 $\times$ 90mm; CBS, Carrum Downs, Australia) was used in all MAP trays and a Biaxially Oriented PolyAmide/Polyethylene/Ethylene vinyl alcohol based film with less than 10 cc/m^2/24 h and less than 3 g/m^2/24 h was also used. The gas composition in the MAP was 80% O_2 and 20% CO_2. The gas ratio of two packaging packs after ageing treatments were checked by a gas analyser and was 80% O_2, 20% CO_2 $\pm$ 0.1%. Thiobarbituric acid reactive substances (TBARS) are formed as a by-product of lipid peroxidation (as degradation products of fats), which can be detected by the TBARS assay using thiobarbituric acid as a reagent. The TBARS assay measures malondialdehyde (MDA) present in the sample, as well as malondialdehyde generated from lipid hydroperoxides by the hydrolytic conditions of the reaction [22].

2.3. Display and Storage

Control samples for sensory assessment and for objective and chemical measurement were trimmed and packed in vacuum bags on day 0 and stored in a -20 °C freezer. Vacuum packed and HiOxMAP samples were displayed in simulated retail display in a double flat glass door upright refrigerator with LED light (Bromic Refrigeration, Ingleburn, New South Wales, Australia) at 2 ± 1 °C for 3, 7 or 14 days. The samples were rotated between shelves and position daily to minimise the effects of any variation in illumination between shelves. After display, samples for sensory assessment were packed in vacuum bags and stored at -20 °C. Samples for objective measurements were trimmed, then packed in vacuum bags for storage at -20 °C.

2.4. Colour Measurement

Instrumental colour analysis on the surface of meat samples was conducted using a Hunterlab Miniscan EZ (Hunter Assoc. Labs Inc., Virginia, VA, USA) that had been calibrated against white and black reference tiles. The settings used for colour measurement were D65 illuminant and 10° observer angle. CIE L* (lightness), a* (redness) and b* (yellowness) values [23] were obtained from the average value of two readings on the surface of loin samples. Control samples were measured prior to packaging and after a 30 min bloom at 4 °C. HiOxMAP treated samples were measured as soon as the pack was opened, at the completion of ageing/display. VAC packaged meat was measured as soon as the meat was opened ('unbloomed') and also after 30 min blooming ('bloomed') at a low temperature (4 °C) in the refrigerator. The Hunterlab Miniscan also gives percentage reflectance across the wavelengths of light from 400 nm to 700 nm. The ratio of oxymyoglobin:metmyoglobin (oxy/met) was calculated by dividing the percentage of light reflectance at wavelength 630 nm by the percentage of light reflectance at wavelength 580 nm as recommended by Hunt and King [17] and used in previous studies [18,24–26].

2.5. Objective Measurements of Texture and Water Loss

For purge loss, samples from 3 and 7 days in VAC or HiOxMAP were weighed before and after packaging. The purge loss was calculated as the initial weight minus the final weight, divided by the initial weight, multiplied by 100 to obtain percentage as described previously [27].

Cooking of pork samples was performed according to procedures outlined in [28], with modifications. Pork samples were prepared into 2 × 70 g ± 5 g blocks to be used for either Warner-Bratzler Shear Force (WBSF) or compression assessment. For WBSF and compression, meat samples were placed in individual plastic bags and suspended in a preheated 70 °C water bath (F38-ME, Julabo, 77,960 Seelbach, Germany) and cooked until an internal temperature of 70 °C was reached. Internal temperature was monitored using temperature probes inserted into the core of the sample (Grant Instruments, Cambridge, United Kingdom). After cooking, samples were cooled in iced water for 30 min. Samples were then dried with paper towel and weighed. After weighing, samples were placed on a tray and covered by plastic wrap to minimise moisture loss, and stored at 4 °C overnight in preparation for WBSF and compression assessments. Cooking loss is calculated as the initial weight minus the final weight, divided by the initial weight, multiplied by 100 to obtain percentage.

Objective texture measurements were conducted, using the Warner-Bratzler shear force (WBSF) method of [29] with modifications [28], on 0, 3 and 7 day samples after completion of treatments. For each sample, six 1 cm^2 × 4–5 cm rectangular strips were cut parallel to the direction of muscle fibres, and the WBSF was measured using a Warner-Bratzler shear force blade (V-shaped) adapted to a Lloyd machine (Lloyd Instruments Ltd, Largo, FL, USA) with a 500 N load cell, and the shearing speed was set at 300 mm/min. The peak of the shear force was recorded and the mean was calculated from the 6 sub-samples.

A modified compression method was used to analyse the effects of connective tissue on tenderness of meat using methods of [30] with modifications described by [28]. A 0.63 cm diameter flat-ended probe was attached to the Lloyd Machine and a 500 N load cell was used. The procedure included 2 penetrations. Firstly, the probe was driven vertically down at a speed of 50 mm/min into 80% of the thickness of a 1 cm thick sample. The muscle fibres were horizontal to the placement of the sample on the plate and perpendicular to the probe. The force work was measured, which was called initial penetration. Secondly, the probe was raised and then lowered, to penetrate the sample a second time. The force work was recorded as second penetration. The parameters measured from compression were:

- Hardness: the force work required for the initial penetration,
- Cohesiveness (or called ease of break down): the reduced force work required for second penetration from work needed for the initial penetration,
- Chewiness: the force required to achieve hardness and cohesiveness.

2.6. TBARS Analysis

TBARS analysis was conducted using a modified extraction method of Witte, et al. [31]. 20 g duplicate samples of pork were homogenized with 50 mL 20% trichloroacetic acid in 2 M phosphoric acid in a Nutribullet Pro 900 (Nutriliving, Pacoima, CA, USA) for 50 s. The homogenate was then diluted with 50 mL deionised water and blended for another 15 s. 50 mL of the homogenate was filtered through Whatman no.1 filter paper. 4mL of the filtrate was transferred to a test tube in duplicate followed by the addition of 4 mL of 2-thiobarbituric acid (5 mM). The test tube was stoppered and kept in the dark for 15 h at room temperature with appropriate standard solutions using 1, 1, 3, 3 tetraethoxypropane as standard. The resulting colour was measured at 532 nm in a spectrophotometer, and absorbance converted to malonaldehyde mg/kg of meat.

2.7. Sensory Assessment

The design of this sensory assessment was in accordance with previous sensory studies on pork [28] but with higher numbers of consumers and samples. These procedures used, including allocation, cooking and serving, have previously been used in research in beef, pork and lamb [27,28,32,33] and are described in Watson, et al. [34] with modifications to the cooking protocol for pork described in Channon, Taverner, D'Souza and Warner [28]. Consumer panelists aged between 18- and 60-years-old were recruited from students and staff of the University of Melbourne. Consumers were asked to participate in a one-hour sensory panel consisting of 10 panelists. Five consumer panel sessions were conducted with samples for cooking and consumption from each carcass and treatment within a carcass, allocated evenly across the five sessions. The total number of consumers was 50, and the total number of samples was 300, from 20 carcasses. Each consumer consumed six pieces of meat, 60 samples were served and consumed in a session, and no consumers were used in more than one session.

The allocation of samples within one sensory session was randomized based on 10 consumers consuming pieces of meat from 10 carcasses, and was designed by using Latin–square design. Every consumer tasted the three treatments twice, enabling a high degree of comparison between treatments and between carcasses. Serving orders can be a factor in influencing sensory results and therefore random serving orders were used in the sensory evaluation. The randomisation of samples presented to consumers was also designed based on Latin-square design.

Allocation of frozen samples into sensory sessions was achieved by sorting samples in the sensory lab three days before sensory sessions. Samples were taken out from freezer and laid out on tables. Using an allocation sheet, samples were sorted into different sessions with 30 steaks in each session, then placed back in the freezer. Before thawing, carton numbers and samples in cartons were checked to avoid errors and also samples were placed in a single layer to ensure consistent rates of sample thawing. Samples were thawed in a refrigerator at 2 ± 1 °C for 24 h before cooking. Prior to cooking, the temperature of steaks were recorded and ranged between 5 and 8 °C. After thawing, samples were taken out from vacuum bags ready for cooking.

Cooking was conducted on two Silex grillers (S-161 Silex Elektrogerate GmbH, 22143 Hamburg, Germany). Five steaks were cooked at a time on each Silex griller, thus 10 samples were served simultaneously to 10 consumers. As outlined by Channon, Taverner, D'Souza and Warner [28], pork samples were cooked at 160 °C for 3 min 10 s until a medium level of doneness was achieved. After two min resting, the internal temperature measured was 70 °C, consistent with a pilot reference study conducted. After cooking, samples were placed on a cutting board and the 4 edges were removed from samples with the middle part used for consumption. Each consumer was presented with 6 cooked samples over a 35 min period.

In each session, consumers were seated in individual sensory booths. Before evaluation, consumers were given instruction and explanations and were asked to fill out a questionnaire to record the demographic information including name, age group, and pork consumption habit.

Each consumer was asked to assess each sample for aroma (dislike extremely to like extremely), tenderness (not tender to very tender), juiciness (not juicy to very juicy), flavour (dislike extremely to like extremely), overall liking (dislike extremely to like extremely) and quality grade (unsatisfactory, average, premium) and purchase intention (not buy it, might buy it, will buy it) on a 100 mm line scale with 0 representing the minimum and 100 representing the maximum. In the case of quality grade and purchase intention, they were anchored with words at either end of the scale, as for the other traits, but also had the words 'average', and 'might buy it' in the middle of the scale. The 10 tastings for each sample were averaged to give the final eating quality score for each. The methodology for sample collection, allocation to treatments, allocation to consumer sessions, cooking, running of consumer sessions, analysis of the line scales, and averaging were first described in Watson, Gee, Polkinghorne and Porter [34], and have been used in consumer research on beef [32,33,35], pork [28] and lamb [27] in Australia and in many countries including Japan [36], South Africa [37] and Korea [38].

2.8. Statistical Analysis

The GENSTAT program (Version 16, release 16.1.0.10916, 64-bit) was used to do an analysis of variance to determine the effects of packaging and ageing on the texture and the sensory attributes of the pork loin samples. Block factors of carcass and replicate were used in ANOVA for purge loss, cooking loss, WBSF, hardness, chewiness, cohesiveness, TBARS and colour measurements. The analysis of variance for sensory attributes included blocking factors of carcass, replicate and session in the ANOVA.

3. Results

3.1. Sensory Assessment

The impact of packaging and ageing on sensory attributes of pork are shown in Figure 1. Samples vacuum packed for 7 days had higher sensory scores, and hence were preferred, for tenderness, flavour, overall liking, and quality relative to control and 7 day HiOxMAP samples ($p < 0.05$ for all except flavour, $p < 0.10$). In addition, the 7 day VAC samples had higher scores for re-purchase intention, relative to the 7 day HiOxMAP samples ($p < 0.05$). In terms of objectively measured tenderness, vacuum packaging of pork resulted in an average WBSF value of 54.1N, which were approximately 5 units higher than the control and HiOxMAP treated samples (see below for further results for WBSF). Generally, meat must have a PSF of <40 N (4.1 kg) in order to ensure high levels of consumer acceptability [39].

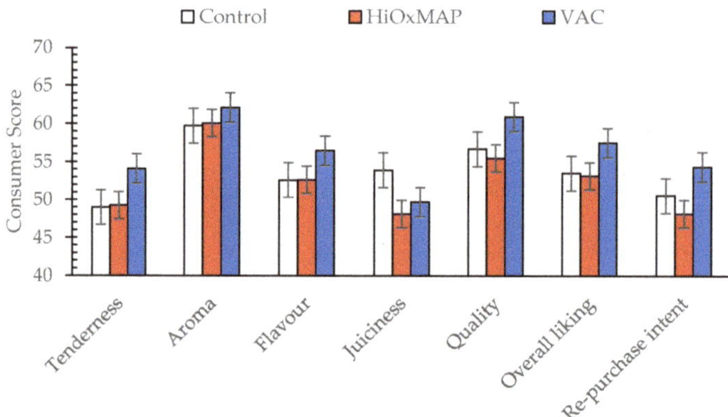

Figure 1. Effects of packaging (control, no packaging; HiOxMAP, Modified Atmosphere Packaging in 80% O_2, 20% CO_2 for 7 days; VAC, vacuum packaged for 7 days) on consumer scores for sensory attributes of porcine longissimus. Data are presented as least squares means ± s.e.d. Tenderness, $p < 0.05$; Aroma, $p > 0.05$; Flavour, $p < 0.10$; Juiciness, $p < 0.05$; Quality grade, $p < 0.05$; Overall liking, $p < 0.05$; Re-purchase intent, $p < 0.05$.

Packaging and ageing treatments did not result in a change in sensory scores for aroma of pork in the current study. This is in agreement with a study by Lund, Lametsch, Hviid, Jensen and Skibsted [10] who examined the effect of vacuum and HiOxMAP on pork aroma and flavour attributes using a trained sensory panel (not consumer panel). However, vacuum packaging increased tenderness scores. Attributes of VAC samples for overall like, quality grade and purchase intention corresponded with tenderness scores. Channon, et al. [40] concluded that tenderness of pork aged 7 days in vacuum bags improved by 4.7 units relative to 2 day aged pork samples and the overall liking score also improved by 3.0 units. The results presented in this study were also consistent with those from Jonsäll, Johansson and Lundström [7], where pork aged 8 days was more tender than 4 day aged pork. In terms of packaging, HiOxMAP negatively impacted tenderness in comparison with VAC, which agreed with findings in a previous study [10]. In addition, HiOxMAP treatment of pork resulted in a similar sensory score to that of the control samples. For the attribute of juiciness, both VAC and HiOxMAP samples had lower scores than the control samples, which may be explained by the higher purge loss.

3.2. WBSF and Compression

Both packaging and ageing influenced the tenderness of pork ($p < 0.01$), as presented in Table 2. WBSF was highest in control samples, intermediate for HiOxMAP samples at 3 and 7 days, and lowest for VAC samples at both 3 and 7 days. The WBSF value of control samples was 29.2 N, samples stored in VAC for 3 days had a WBSF of 20.1 N, and after 7 days ageing, the value was further reduced to 17.9 N. Pork aged in HiOxMAP either for 3 or 7 days had a similar WBSF of about 22.5 N, which supports the reduced sensory scores for tenderness for HiOxMAP samples.

Table 2. Effect of packaging (control, no packaging; HiOxMAP, high oxygen modified atmosphere packaging, 80% O_2, 20% CO_2; Vacuum, vacuum packaging) and ageing (0, 3, 7 days) on objective meat quality traits of pork loin. Data are presented as least squares means.

Treatment	Control	HiOxMAP		Vacuum		SED [1]	F-values			
Days Ageing	0 Days	3 Days	7 Days	3 Days	7 Days		Control	Pack	Age	Pack.Age
Cook loss (%)	19.28	19.54	18.97	16.73	18.45	0.590	<0.001	<0.001	0.337	0.017
Chewiness	10.00	10.12	10.29	9.74	9.15	0.452	0.111	0.021	0.505	0.238
Cohesiveness	0.435	0.412	0.436	0.427	0.421	0.0076	0.14	0.515	0.362	0.043
Hardness	34.18	34.53	34.82	31.59	30.65	0.874	0.001	<0.001	0.599	0.329
WBSF (N)	29.23	21.56	22.39	20.23	17.94	0.973	<0.001	<0.001	0.292	0.027

[1] SED for the interaction term.

There were effects of packaging type and ageing period on the hardness of pork loins ($p < 0.001$) as presented in Table 2. Control and HiOxMAP samples all had similar hardness whereas VAC samples at 3 and 7 days had lower hardness values, which supports the reduced sensory scores for tenderness for HiOxMAP samples. The effects of packaging and ageing on cohesiveness are shown in Table 2, and there was an interaction between packaging and ageing ($p < 0.05$) such that for HiOxMAP, cohesiveness increased between days 3 and 7 whereas there was no change for VAC samples. The influence of packaging and ageing on chewiness is also shown in Table 2 and there was an effect of packaging; VAC samples had lower chewiness relative to HiOxMAP ($p < 0.05$).

The physical forces exerted on meat during masticationg are shear force, compression (hardness, chewiness and cohesiveness) and tensile force. These mechanical forces are defined as objective assessments in estimating the tenderness of meat [41]. Data presented in this report indicate that packaging and ageing affected shear force and hardness, and to a lesser extent cohesive and chewiness. VAC samples showed a decrease in shear force values at days 3 and 7 day of ageing. Channon, Kerr and Walker [40] found that pork aged for 7 days in VAC resulted in a reduction in WBSF of pork in comparison with 2 day post-slaughter ageing. In addition, similar results were obtained for beef [15]. In'this study, pork aged in HiOxMAP presented a lower shear force than control samples, although not as low as VAC samples, likely because there was initially some tenderization in

HiOxMAP, then oxidation of calpain in the high oxygen environment prevented further tenderization and proteolysis [42]. This is similar to our previous results in lamb where sensory (WBSF not measured) tenderness increased in vacuum packed samples between 5 and 10 days of storage, but the tenderness of HiOxMAP samples decreased over the same period [42]. This also corresponded with results of a previous study [10] which indicated HiOxMAP inhibited the tenderization of pork. Sorheim, et al. [43] also indicated that HiOxMAP did not result in a decrease in the shear force value of beef stored up to 14 days, hence ageing was inhibited also in their study. Therefore, from an objective physical measurement perspective, ageing in vacuum packaging resulted in more tender meat in comparison with ageing in HiOxMAP.

3.3. Purge Loss

The influence of packaging and ageing on purge loss is presented in Figure 2. Purge loss was higher in VAC stored pork loin than HiOxMAP and was also higher for 7 days aged samples relative to 3 days aged samples ($p < 0.001$ for both). Also, there was a trend for an interaction between packaging and ageing ($p < 0.10$). Over the ageing period from 3 to 7 days ageing, purge loss of VAC pork increased from 6.8% to 10.5%, while purge loss of pork in HiOxMAP only slightly increased from 5.1% to 7.0%. Purge loss is associated with fluid loss from meat during storage [13], and is an indicator of the moisture content lost from the pork. This has been shown to mainly be associated with juiciness, tenderness and flavour of meat. Higher purge loss was observed in pork aged in VAC than in pork stored in HiOxMAP, which was consistent with a study by [13] who showed that the amount of purge loss of beef stored in vacuum was higher compared to beef stored in HiOxMAP. The increase in purge loss with an increased ageing time also agreed with other results [13]. The high weight loss in vacuum packed samples could be explained by fluid loss under the influence of vacuum pressure [13]. However, these results were contradictory with those of a pork study in which pork aged in HiOxMAP was found to have a greater purge loss than pork stored in vacuum condition, and the purge loss increased significantly with the increased ageing time from 3–7 days [8,10]. In 2009, Nam et al. [8] reported that pork with a low purge loss during storage has been shown to be more tender and have a better flavour. Interestingly, in the current study, HiOxMAP samples, which appeared to have a lower purge loss received a lower score for tenderness and flavour. In terms of juiciness, VAC stored pork with higher purge loss had a higher score than HiOxMAP, while control samples with no purge loss showed highest score on attribute of juiciness. These results in combination indicate that purge loss in this study could not be used as an independent variable to explain the effect of packaging and ageing on pork eating quality.

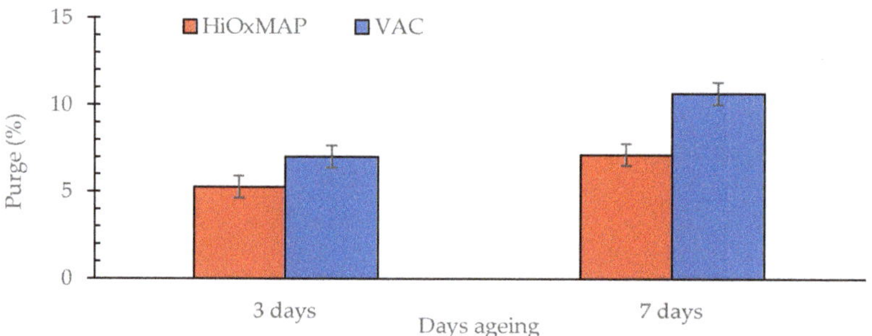

Figure 2. Effect of packaging (HiOxMAP, high oxygen modified atmosphere packaging, 80% O_2, 20% CO_2; VAC, vacuum packaging) and ageing (3 or 7 days) on purge loss. Data are presented as least squares means ± s.e.d for the interaction. Packaging, $p < 0.001$; Ageing, $p < 0.001$; Packaging.Ageing, $p < 0.10$).

3.4. Cooking Loss

Differences were observed in cooking loss among packaging × ageing treatments as displayed in Table 2 ($p < 0.05$). Cooking loss for control and pork loin steaks aged in HiOxMAP for 3 days were the highest, HiOxMAP and VAC samples stored for 7 days were intermediate and 3 day VAC samples had the lowest cook loss. Cooking loss is defined as the loss of fluid, along with soluble substances, from meat during the cooking process, and it has been established that it could affect tenderness and juiciness [44]. In this study, a lower cooking loss was observed for pork stored in vacuum packaging in comparison with HiOxMAP and control samples. The reduction in cooking loss of vacuum packed samples was correlated with both a lower shear force and a higher sensory score for tenderness of vacuum packed pork. It has been previously noted that panelists do not prefer pork which had high cooking loss and it was regarded as tough pork [8]. Lagerstedt et al. [13] found that beef aged for 7 days in HiOxMAP had a higher cooking loss than beef aged in vacuum bags. Although the control samples in the current study had a higher cooking loss, a higher sensory score was given for the juiciness attribute of the control samples, compared with pork stored in vacuum and HiOxMAP. A possible explanation for the control samples attracting a higher sensory score for juiciness is the lower purge.

3.5. Colour

The effect of time and packaging on L*a*b* values and oxy/met ratio is given in Figure 3. The L* values of HiOxMAP, VACb (vacuum packed pork colour after opening pack and blooming for 30 min) and VACnb (vacuum packed pork colour after opening pack and immediate colour measurement, no blooming) were not different after 3 days of storage ($p > 0.05$). After prolonged storage, HiOxMAP packaging produced the palest colour samples and this was significantly different after 7 and 14 days of storage ($p < 0.05$). An increase in lightness is attributed to relative contents of chemical forms of myoglobin and increased light scattering due to protein denaturation [23]. The increased lightness in HiOxMAP meat during storage indicates that HiOxMAP meat loses contrast over time, producing less pink colour in the meat [45].

The a* values of pork loins were highest at 3 days of storage when packed in HiOxMAP ($p < 0.001$), but gradually fell as storage time progressed. The a* values of HiOxMAP and VACb after 14 days of storage were not significantly different ($p > 0.05$). However, HiOxMAP loins still produced higher redness compared to VACnb after 14 days of storage ($p < 0.001$). Redness level is an important factor in producing the desired pink colour in pork [46,47].

The b* values of pork loins followed a similar trend to a* values, where HiOxMAP loins produced the highest b* values after 3 days of storage ($p < 0.0001$), but gradually decreased as storage time progresses. The b* values of VACb and HiOxMAP at day 14 were not different ($p > 0.05$) but VACnb and HiOxMAP were different ($p < 0.001$). O'Sullivan, Byrne, Martens, Gidskehaug, Andersen and Martens [46] and Chizzolini, et al. [48] indicated that b* values were important in determining the onset of brown pigmentation, and that unsatisfactory appearance in meat were associated with a more pronounced yellow tint which largely depends on the relative balance of a* and b*.

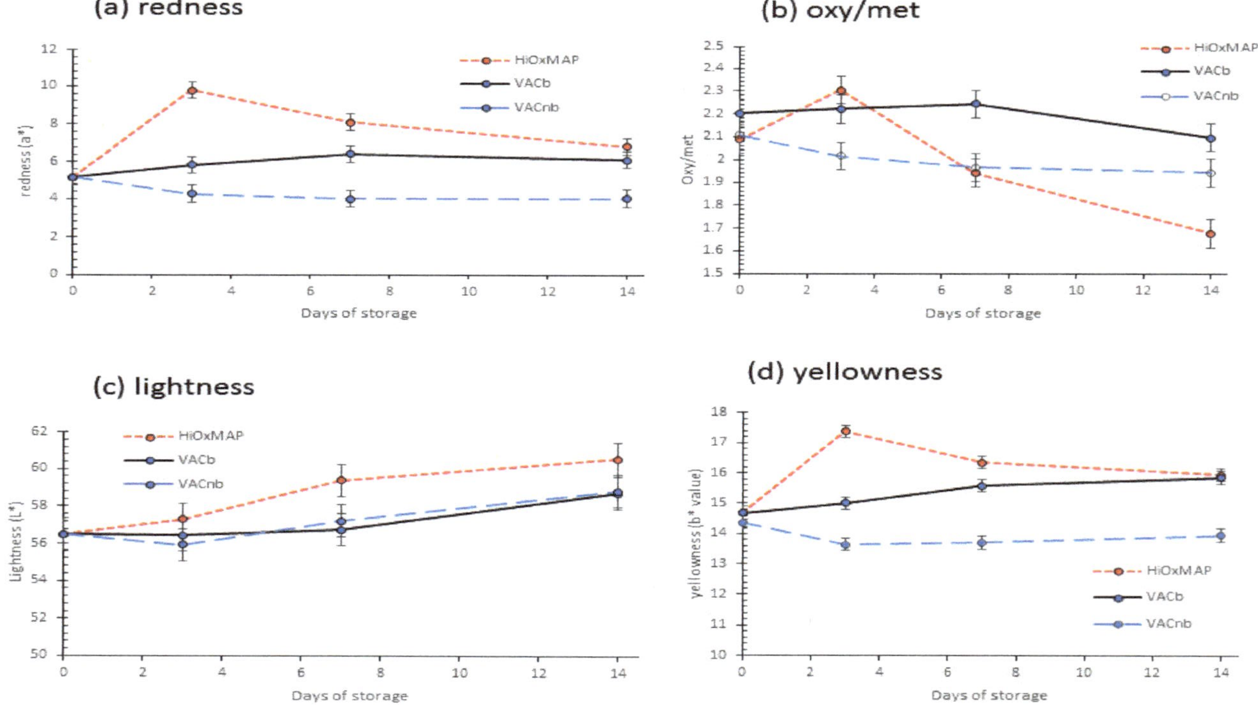

Figure 3. Effect of packaging (HiOxMAP, high oxygen modified atmosphere packaging vs VACnb, vacuum packing, colour measured after no bloom vs VACb, vacuum packing, colour measured after 30 min bloom) and day of storage (0, 3, 7, 14 days) on (**a**) the surface redness (a*), Packaging, Ageing, Packaging.Ageing, $p < 0.001$ for all, (**b**) oxy/met (R630/R580) values, Packaging, Ageing, Packaging.Ageing, $p < 0.001$ for all, (**c**) lightness (L*), Packaging, $p < 0.001$; Ageing, $p < 0.001$; Packaging.Ageing, $p > 0.05$ and (**d**) yellowness (b*), Packaging, Ageing, Packaging.Ageing, $p < 0.001$ for all. Each point is a least squares mean ± s.e.d. for the interaction.

For the oxy/met ratio, HiOxMAP produced the highest value after 3 days of storage ($p < 0.05$) but declined after this time period. At 7 days storage, the oxy/met ratio of HiOxMAP meat was lower than VACb, and was lower than both VACb and VACnb after 14 days of storage ($p < 0.001$). The oxy/met ratio is measured from reflectance and is a measurement which closely relates to what the human eye and brain can see [23]. This value gives an indication of changes in the predominant pigment of myoglobin that is present on the surface of meat—particularly oxymyoglobin (responsible for the red colour) and metmyoglobin (responsible for the brown colour). A reduction in oxy/met ratio is correlated to browning of meat, as prolonged exposure of air as well as other oxidation mechanisms promotes oxidation of oxymyoglobin to metmyoglobin [17]. The Australian lamb industry specifies that if the oxy/met ratio is < 3, the meat would be unacceptable to consumers [49]. According to our results, pork has lower oxy/met ratios than lamb and thus the cutoff for acceptability is yet to be established for pork. But it can certainly be concluded that the pork packaged under HiOxMAP had inferior colour relative to vacuum packed meat after 7–14 days of storage.

Brewer and McKeith [45] suggested that categorising meat colour on the basis of L*, a* and b* values to varying degrees of pink colour is possible, and many studies have suggested that meat quality can be attributed to colour measurements [46,48,50,51]. The combination of L*a*b* measurements to categorise meat colour seems to have strong interactions with one another – with each colour attribute making a contribution to the overall appearance quality of pork. However, pork meat which is generally considered to be very light pink and pale in appearance is associated with high L*, low a* and high b* values, whilst pork meat which is dark pink is associated with low L*, high a* and high b* values [45].

The results have indicated that HiOxMAP packaging caused an increased rate of decline in the oxy/met ratio on the surface of pork loins, indicating that metmyoglobin formation is accelerated in a high oxygen environment which produced the undesirable pale, brown colour. Blooming of VAC loins seemed to have a significant effect in increasing oxy/met ratio compared to no blooming ($p < 0.001$), suggesting the beneficial effects of VAC and 30 min blooming on minimising metmyoglobin formation and subsequently general appearance of pork loins in long term storage.

The high L* and low a* values that were measured on HiOxMAP loins after 14 days of storage indicate that the colour of the steak is very light pink in colour, and quite pale compared to its colour measurements after 3 days of storage. Colour measurement values are even tending towards colours that can be seen in PSE (pale, soft, exudative) pork —which is described as L* > 57 [48]. This gives further insight that VACb or VACnb are the best methods of maintaining colour for long term storage of meat. However, as this is a method of storage, which requires the removal of meat from the packaging, it will require an extra step of intervention by removing the meat and placing it at a low temperature in a different atmosphere; this requires extra labour and time, and is not desirable for industry compared to HiOxMAP packaging, which are packed in the format purchased by the consumers. Thus, it is recommended that vacuum skin packing should be considered, which would be equivalent to VACnb.

3.6. TBARS

Figure 4 displays the effects of VAC and HiOxMAP on TBARS values after simulayed retail storage. There was an interaction between packaging and ageing as well as main effects of packaging and ageing ($p < 0.001$ for all). HiOxMAP had a significant impact on lipid oxidation development, with TBARS values being higher at 3, 7 and 14 days of storage ($p < 0.001$) relative to VAC. VAC samples did not show an overall increase in TBARS values—with day 0 samples and day 14 samples being similar ($p > 0.05$) and were consistent with previous studies [9,52,53]. This indicates that vacuum stored pork loins had minimal lipid oxidation, i.e., almost zero development of oxidation products during 14 days of storage. On the other hand, increases in lipid oxidation in HiOxMAP samples occurred with an increase of MDA mg/kg from 0.12 to 0.37 mg/kg—almost a three-fold increase during the 14-day period storage. Lipid oxidation is therefore strongly affected by HiOxMAP and high oxygen environments.

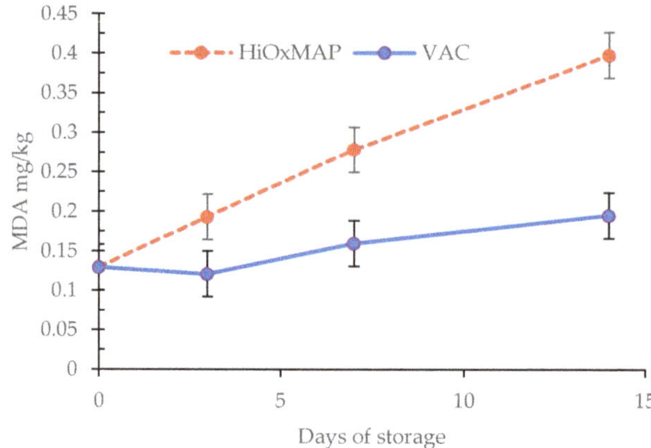

Figure 4. Effect of packaging (HiOXMAP, high oxygen modified atmosphere packaging vs VAC, vacuum packing) and day of storage (0, 3, 7, 14 days) on TBARS (MDA, Malonaldehyde, mg/kg of meat) values. Each point is a least squares mean ± s.e.d for the interaction. Packaging, $p < 0.001$; Ageing, $p < 0.001$; Packaging.Ageing, $p < 0.001$).

Lipid oxidation is closely correlated to pigment oxidation due to production of free radicals and ROS [54], and the results are consistent with these findings. Lipid oxidation promotes myoglobin oxidation [55] and decreases of surface redness as well as a decrease in oxy/met ratio indicates that as lipid oxidation progresses in HiOxMAP product, metmyoglobin production also increased.

Rancidity is described as a flavour, which develops due to various factors that create undesirable and unacceptable characteristics in meat. The connection of lipid oxidation and rancidity has not been fully established. Campo, et al. [56] attempted to create a link of TBARS values with sensory qualities of rancidity in beef, and found that the point of TBARS value of 2 is a limiting point for where rancid flavour overpowers beef flavour. The results largely varied, however, due to various limiting factors such as personal thresholds and experience of panellists. Furthermore, various other VOCs (volatile odour compounds), which are produced by other mechanisms other than lipid oxidation are also involved in creating the general rancid flavour in pork [57]. According to Greene and Cumuze [58], the TBARS value at which rancid odour is first observed by panelists is 0.5 to 1.0 mg MDA/kg but malonaldehyde content contributes only to a small part of the total odour complex.

The values of TBARS found in this study have indicated that malonaldehyde production in HiOxMAP and VAC meat did not reach unacceptable levels of lipid oxidation—with the largest TBARS value being 0.37 MDA mg/kg. However, the elevated rate of lipid oxidation found in HiOxMAP product provides further explanation of the deterioration of colour properties, as well as indicating that shelf life qualities were negatively impacted compared to VAC packaging.

4. Summary and Recommendations

Pork loin samples subjected to HiOxMAP exhibited inferior consumer acceptability for tenderness, flavour, overall liking, quality and re-purchase intention relative to samples vacuum packed for 7 days. Samples vacuum packed for 7 days also had lower shear force and hardness values compared to samples undergoing HiOxMAP for 7 days. In addition, HiOxMAP maintained a high a* value in samples during 3 days of storage and higher L* value at the end of storage. Blooming of vacuum packaged meat after long term storage resulted in greater colour stability, as shown by the same a* values, lower L* values and less browning (higher oxy/met) compared to HiOxMAP packaging. HiOxMAP packaging resulted in higher rates of lipid oxidation compared to VAC, which indicated a reduction in shelf life quality.

This research, based on a small number of samples, has shown that pork loins have lower 'colour and oxidation' shelf life when packed in HiOxMAP compared to vacuum packed samples. The results have also shown that pork loins stored in HiOxMAP have elevated lipid oxidation, reduced contrast in colour and increased browning. The findings suggest that pork loins are suited for packaging in HiOxMAP for less than 7 days, possibly only 3 days, and VAC packaging should be considered for storage for any time period longer than 3–7 days. Hence, the use of vacuum packing of pork for retail shelves should be considered as the preferred option, over HiOxMAP.

Using similar packaging and consumer sensory methodology, it has been shown that beef and sheep meat packed in HiOxMAP have reduced consumer sensory scores of 10–12 for beef [59] and 10–15 for sheep meat [42], over 9–10 days of retail display. The packaging of pork in HiOxMAP in our study resulted in a reduction of 5–6 sensory scores over 7 days, which although lower than seen in beef and pork, was over a shorter period. Hence, it is evident that in order for the pork industry to accurately and effectively manage stock and supply within the supply chain to ensure that high quality pork is delivered to the consumers, the effect of HiOxMAP on eating quality is an important consideration. In order to extend the results to the wider industry, more research is required on a greater range of muscles and under a wider range of conditions.

Author Contributions: Conceptualization, methodology, investigation, R.D.W., Y.P., K.A., A.P., H.C.; formal analysis, R.D.W., Y.P., K.A., A.P.; resources, R.D.W., Y.P., K.A., A.P., H.C., M.H.; data curation, Y.P., K..A., R.D.W., A.P.; writing—original draft preparation, K.A., Y.P.; writing—review and editing, R.D.W., Y.P., K.A., A.P., H.C., M.H.; visualization, Y.P., K.A., R.D.W.; supervision, R.D.W., A.P., M.H.; Validation, R.D.W., A.P., H.C., M.H.; project administration, R.D.W.; funding acquisition, R.D.W., H.C.

Funding: This research was funded by Pork Co-operative research centre student honours scholarships GL 000095-G19 and G20 also by the University of Melbourne subject AGRI90032, Major Research project.

Acknowledgments: The support of our sponsor the Pork Cooperative Research Centre for supplying the necessary research materials and invaluable insights within the Australian Pork industry is appreciated. We would also like to thank the staff, Michelle Rhee and Katherine Mistica at the University of Melbourne's Faculty of Veterinary and Agricultural Sciences for providing support and laboratory services to conduct our experiments in a safe and timely manner. We would also like to appreciate the academic support and resources provided by Rozita Vaskoska, Sigfredo Fuentes, Damir Torrico. We would like to offer a big thanks to friends Meiyin Lin, Luyuan Zhou, Shruti Sastri, Claudia Glez Viejo and Jordan Le who always gave us support, encouragement and help.

References

1. Simultaneous Improvement of Meat Quality and Growth and Carcass Traits in Pigs. Available online: http://wcgalp.org/system/files/proceedings/2002/simultaneous-improvement-meat-quality-and-growth-and-carcass-traits-pigs.pdf (accessed on 16 July 2001).

2. Warner, R.D.; Greenwood, P.L.; Pethick, D.W.; Ferguson, D.M. Genetic and environmental effects on meat quality. *Meat Sci.* **2010**, *86*, 171–183. [CrossRef] [PubMed]

3. Producing Consistent Quality Meat from the Modern Pig. Available online: https://www.taylorfrancis.com/books/e/9781351114493/chapters/10.1201/9781351114493-9 (accessed on 15 June 2018).

4. Channon, H.A.; Hamilton, A.J.; D'Souza, D.N.; Dunshea, F.R. Estimating the impact of various pathway parameters on tenderness, flavour and juiciness of pork using Monte Carlo simulation methods. *Meat Sci.* **2016**, *116*, 58–66. [CrossRef] [PubMed]

5. Channon, H.A.; D'Souza, D.N.; Dunshea, F.R. Developing a cuts-based system to improve consumer acceptability of pork: Impact of gender, ageing period, endpoint temperature and cooking method. *Meat Sci.* **2016**, *121*, 216–227. [CrossRef] [PubMed]

6. Delivering Consistent Quality Australian Pork to Consumers—A Systems Approach. Available online: https://www.researchgate.net/publication/239732001_Delivering_consistent_quality_australian_pork_to_consumers_-_a_systems_approach (accessed on 1 November 2011).

7. Jonsäll, A.; Johansson, L.; Lundström, K. Effects of red clover silage and RN genotype on sensory quality of prolonged frozen stored pork (M. *Longissimus dorsi*). *Food Qual. Prefer.* **2000**, *11*, 371–376. [CrossRef]

8. Nam, Y.J.; Choi, Y.M.; Lee, S.H.; Choe, J.H.; Jeong, D.W.; Kim, Y.Y.; Kim, B.C. Sensory evaluations of porcine longissimus dorsi muscle: Relationships with postmortem meat quality traits and muscle fiber characteristics. *Meat Sci.* **2009**, *83*, 731–736. [CrossRef]

9. Cayuela, J.M.; Gil, M.D.; Bañón, S.; Garrido, M.D. Effect of vacuum and modified atmosphere packaging on the quality of pork loin. *Eur. Food Res. Technol.* **2004**, *219*, 316–320. [CrossRef]

10. Lund, M.N.; Lametsch, R.; Hviid, M.S.; Jensen, O.N.; Skibsted, L.H. High-oxygen packaging atmosphere influences protein oxidation and tenderness of porcine longissimus dorsi during chill storage. *Meat Sci.* **2007**, *77*, 295–303. [CrossRef]

11. Lavieri, N.; Williams, S.K. Effects of packaging systems and fat concentrations on microbiology, sensory and physical properties of ground beef stored at 4 ± 1 °C for 25 days. *Meat Sci.* **2014**, *97*, 534–541. [CrossRef]

12. Dransfield, E.; Jones, R.C.D. Tenderising in m longissimus dorsi of beef veal rabbit lamb and pork. *Meat Sci.* **1980**, *5*, 139–147. [CrossRef]

13. Lagerstedt, Å.; Ahnström, M.L.; Lundström, K. Vacuum skin pack of beef—A consumer friendly alternative. *Meat Sci.* **2011**, *88*, 391–396. [CrossRef] [PubMed]

14. McMillin, K.W. Where is MAP Going? A review and future potential of modified atmosphere packaging for meat. *Meat Sci.* **2008**, *80*, 43–65. [CrossRef] [PubMed]

15. Geesink, G.; Robertson, J.; Ball, A. The effect of retail packaging method on objective and consumer assessment of beef quality traits. *Meat Sci.* **2015**, *104*, 85–89. [CrossRef] [PubMed]

16. Mancini, R.A.; Hunt, M.C. Current research in meat color. *Meat Sci.* **2005**, *71*, 100–121. [CrossRef] [PubMed]

17. Hunt, M.C.; King, A.J. *AMSA Meat Colour Measurement Guidelines*; Barbut, S., Claus, J., Cornforth, D.P., Hanson, D., Lindahl, G., Mancini, R.A., Milkowski, A., Mohan, A., Pohlman, F., Raines, C.R., et al., Eds.; American Meat Science Association: Champaignb, IL, USA, 2012; pp. 1–135.

18. Warner, R.D.; Kearney, G.; Hopkins, D.L.; Jacob, R. Retail colour stability of lamb meat is influenced by packaging, breed type, muscle and iron concentration. *Meat Sci.* **2017**, *129*, 29–37. [CrossRef] [PubMed]

19. Hood, D.E.; Riordan, E.B. Discolouration in pre-packaged beef: Measurement by reflectance spectrophotometry and shopper discrimination. *J. Food Sci. Technol.* **1973**, *8*, 333–343. [CrossRef]

20. Jakobsen, M.; Bertelsen, G. Colour stability and lipid oxidation of fresh beef. Development of a response surface model for predicting the effects of temperature, storage time, and modified atmosphere composition. *Meat Sci.* **2000**, *54*, 49–57. [CrossRef]

21. Nerin, C.; Tovar, L.; Djenane, D.; Camo, J.; Salafranca, J.; Beltran, J.A.; Roncales, P. Stabilization of beef meat by a new active packaging containing natural antioxidants. *J. Agric. Food Chem.* **2006**, *54*, 7840–7846. [CrossRef]

22. Trevisan, M.; Browne, R.; Ram, M.; Muti, P.; Freudenheim, J.; Carosella, A.M.; Armstrong, D. Correlates of Markers of Oxidative Status in the General Population. *Am. J. Epidemiol.* **2001**, *154*, 348–356. [CrossRef]

23. Warner, R. *Measurement Of Meat Quality\Measurements of Water-holding Capacity and Color: Objective and Subjective in Encyclopedia of Meat Sciences*, 2nd ed.; Devine, C., Dikeman, M., Eds.; Academic Press: Oxford, UK, 2014; pp. 164–171. [CrossRef]

24. Warner, R.D.; Ponnampalam, E.N.; Kearney, G.A.; Hopkins, D.L.; Jacob, R.H. Genotype and age at slaughter influence the retail shelf-life of the loin and knuckle from sheep carcasses. *Aust. J. Exp. Agric.* **2007**, *47*, 1190–1200. [CrossRef]

25. Mortimer, S.I.; Jacob, R.H.; Kearney, G.; Hopkins, D.L.; Warner, R.D. Genetic variation in colour stability traits of lamb cuts under two packaging systems. *Meat Sci.* **2019**, *157*, 107870. [CrossRef]

26. Jacob, R.H.; D'Antuono, M.F.; Gilmour, A.R.; Warner, R.D. Phenotypic characterization of colour stability of lamb meat. *Meat Sci.* **2014**, *96*, 1040–1048. [CrossRef] [PubMed]

27. Warner, R.D.; Ferguson, D.M.; McDonagh, M.B.; Channon, H.A.; Cottrell, J.J.; Dunshea, F.R. Acute exercise stress and electrical stimulation influence the consumer perception of sheep meat eating quality and objective quality traits. *Aust. J. Exp. Agric.* **2005**, *45*, 553–560. [CrossRef]

28. Channon, H.A.; Taverner, M.R.; D'Souza, D.N.; Warner, R.D. Aitchbone hanging and ageing period are additive factors influencing pork eating quality. *Meat Sci.* **2014**, *96*, 581–590. [CrossRef] [PubMed]

29. Bouton, P.E.; Harris, P.V.; Shorthose, W.R. Effect of ultimate pH upon the water-holding capacity and tenderness of mutton. *J. Food Sci.* **1971**, *37*, 358–360. [CrossRef]

30. Bouton, P.E.; Harris, P.V. A comparison of some objective methods used to assess meat tenderness. *J. Food Sci.* **1972**, *37*, 218–221. [CrossRef]

31. Witte, V.C.; Krause, G.F.; Bailey, M.E. A new extraction method for determining 2-thiobarbituric acid values of pork and beef during storage. *J. Food Sci.* **1970**, *35*, 582–585. [CrossRef]

32. Warner, R.D.; Ferguson, D.M.; Cottrell, J.J.; Knee, B.W. Acute stress induced by the preslaughter use of electric prodders causes tougher beef meat. *Aust. J. Exp. Agric.* **2007**, *47*, 782–788. [CrossRef]

33. Ferguson, D.M.; Warner, R.D.; Walker, P.J.; Knee, B. Effect of cattle marketing method on beef quality and palatability. *Aust. J. Exp. Agric.* **2007**, *47*, 774–781. [CrossRef]

34. Watson, R.; Gee, A.; Polkinghorne, R.; Porter, M. Consumer assessment of eating quality—Development of protocols for Meat Standards Australia (MSA) testing. *Aust. J. Exp. Agric.* **2008**, *48*, 1360–1367. [CrossRef]

35. Colditz, I.G.; Ferguson, D.M.; Greenwood, P.L.; Doogan, V.J.; Petherick, J.C.; Kilgour, R.J. Regrouping unfamiliar animals in the weeks prior to slaughter has few effects on physiology and meat quality in Bos taurus feedlot steers. *Aust. J. Exp. Agric.* **2007**, *47*, 763–769. [CrossRef]

36. Ha, M.; McGilchrist, P.; Polkinghorne, R.; Huynh, L.; Galletly, J.; Kobayashi, K.; Nishimura, T.; Bonney, S.; Kelman, K.R.; Warner, R.D. Effects of different ageing methods on colour, yield, oxidation and sensory qualities of Australian beef loins consumed in Australia and Japan. *Food Res. Int.* **2019**, *125*, 108528. [CrossRef] [PubMed]

37. Thompson, J.M.; Polkinghorne, R.J.; Gee, A.M.; Motiang, D.; Strydom, P.E.; Mashau, M.; Ng'ambi, J.; DeKock, R.; Burrow, H.M. *Beef Palatability in the Republic of South Africa: Implications for Niche-Marketing Strategies*; ACIAR Technical Reports No. 72; Australian Centre for International Agricultural Research: Canberra, Australia, 2010.

38. Thompson, J.M.; Polkinghorne, R.; Hwang, I.H.; Gee, A.M.; Cho, S.H.; Park, B.Y.; Lee, J.M. Beef quality grades as determined by Korean and Australian consumers. *Aust. J. Exp. Agric.* **2008**, *48*, 1380–1386. [CrossRef]

39. Huffman, K.L.; Miller, M.F.; Hoover, L.C.; Wu, C.K.; Brittin, H.C.; Ramsey, C.B. Effect of beef tenderness on consumer satisfaction with steaks consumed in the home and restaurant. *J. Anim. Sci.* **1996**, *74*, 91–97. [CrossRef] [PubMed]

40. Channon, H.A.; Kerr, M.G.; Walker, P.J. Effect of Duroc content, sex and ageing period on meat and eating quality attributes of pork loin. *Meat Sci.* **2004**, *66*, 881–888. [CrossRef] [PubMed]

41. Tornberg, E. Effects of heat on meat proteins—Implications on structure and quality of meat products. *Meat Sci.* **2005**, *70*, 493–508. [CrossRef]

42. Frank, D.C.; Geesink, G.; Alvarenga, T.I.R.C.; Polkinghorne, R.; Stark, J.; Lee, M.; Warner, R. Impact of high oxygen and vacuum retail ready packaging formats on lamb loin and topside eating quality. *Meat Sci.* **2017**, *123*, 126–133. [CrossRef]

43. Sorheim, O.; Wahlgren, N.M.; Nilsen, B.N.; Lea, P. Effects of high oxygen packaging on tenderness and quality characteristics of beef longissimus muscles. In Proceedings of the 51st International Congress of Meat Science and Technology, Helsinki, Finland, 8–13 August 2004.

44. Aaslyng, M.D.; Bejerholm, C.; Ertbjerg, P.; Bertram, H.C.; Andersen, H.J. Cooking loss and juiciness of pork in relation to raw meat quality and cooking procedure. *Food Qual. Prefer.* **2003**, *14*, 277–288. [CrossRef]

45. Brewer, M.S.; McKeith, F.K. Consumer-rated Quality Characteristics as Related to Purchase Intent of Fresh Pork. *J. Food Sci.* **1999**, *64*, 171–174. [CrossRef]

46. O'Sullivan, M.G.; Byrne, D.V.; Martens, H.; Gidskehaug, L.H.; Andersen, H.J.; Martens, M. Evaluation of pork colour: Prediction of visual sensory quality of meat from instrumental and computer vision methods of colour analysis. *Meat Sci.* **2003**, *65*, 909–918. [CrossRef]

47. Van Oeckel, M.J.; Warnants, N.; Boucqué, C.V. Measurement and prediction of pork colour. *Meat Sci.* **1999**, *52*, 347–354. [CrossRef]

48. Chizzolini, R.; Novelli, E.; Badiani, A.; Rosa, P.; Delbono, G. Objective measurements of pork quality: Evaluation of various techniques. *Meat Sci.* **1993**, *34*, 49–77. [CrossRef]

49. Warner, R.D.; Jacob, R.H.; Hocking Edwards, J.E.H.; McDonagh, M.; Pearce, K.; Geesink, G.; Kearney, G.; Allingham, P.; Hopkins, D.L.; Pethick, D.W. Quality of lamb meat from the Information Nucleus Flock. *Anim. Prod. Sci.* **2010**, *50*, 1123–1134. [CrossRef]

50. Karamucki, T.; Gardzielewska, J.; Rybarczyk, A.; Jakubowska, M.; Natalczyk-Szymkowska, W. Usefulness of Selected Methods of Colour Change Measurement for Pork Quality Assessment. *Czech J. Food Sci.* **2011**, *29*, 212–218. [CrossRef]

51. Chmiel, M.; Słowiński, M.; Dasiewicz, K. Lightness of the color measured by computer image analysis as a factor for assessing the quality of pork meat. *Meat Sci.* **2011**, *88*, 566–570. [CrossRef]

52. Cannon, J.E.; Morgan, J.B.; Schmidt, G.R.; Tatum, J.D.; Sofos, J.N.; Smith, G.C.; Delmore, R.J.; Williams, S.N. Growth and fresh meat quality characteristics of pigs supplemented with vitamin E. *J. Anim. Sci.* **1996**, *74*, 98–105. [CrossRef]

53. Nam, K.C.; Du, M.; Jo, C.; Ahn, D.U. Cholesterol oxidation products in irradiated raw meat with different packaging and storage time Journal Paper No. J-18976 of the Iowa Agriculture and Home Economics Experiment Station, Ames, IA 50011. Project No. 3322, supported by S-292 Regional Project and Iowa Egg Council. *Meat Sci.* **2001**, *58*, 431–435. [CrossRef]

54. Faustman, C.; Sun, Q.; Mancini, R.; Suman, S.P. Myoglobin and lipid oxidation interactions: Mechanistic bases and control. *Meat Sci.* **2010**, *86*, 86–94. [CrossRef]

55. Lin, T.-Z.; Hultin, H.O. Oxidation of myoglobin in vitro mediated by lipid oxidation in microsomal fractions of muscle. *J. Food Sci.* **1977**, *42*, 136–140. [CrossRef]

56. Campo, M.M.; Nute, G.R.; Hughes, S.I.; Enser, M.; Wood, J.D.; Richardson, R.I. Flavour perception of oxidation in beef. *Meat Sci.* **2006**, *72*, 303–311. [CrossRef]

57. Casaburi, A.; Piombino, P.; Nychas, G.-J.; Villani, F.; Ercolini, D. Bacterial populations and the volatilome associated to meat spoilage. *Food Microbiol.* **2015**, *45*, 83–102. [CrossRef]

58. Greene, B.E.; Cumuze, T.H. Relationships between TBA Numbers and Inexperienced panelists assessments of oxidized flavor in cooked beef. *J. Food Sci.* **1981**, *47*, 52–58. [CrossRef]

59. Polkinghorne, R.J.; Philpott, J.; Perovic, J.; Lau, J.; Davies, L.; Mudannayake, W.; Watson, R.; Tarr, G.; Thompson, J.M. The effect of packaging on consumer eating quality of beef. *Meat Sci.* **2018**, *142*, 59–64. [CrossRef] [PubMed]

Article

Consumers' Perceptions and Sensory Properties of Beef Patty Analogues

Jordan Taylor [1], Isam A. Mohamed Ahmed [2], Fahad Y. Al-Juhaimi [2] and Alaa El-Din A. Bekhit [1,*]

1   Department of Food Science, University of Otago, P.O. Box 56 Dunedin, New Zealand;
    jordan.taylornz@gmail.com
2   Department of Food Science and Nutrition, College of Food and Agricultural Sciences, King Saud University,
    Riyadh 11362, Saudi Arabia; iali@KSU.EDU.SA (I.A.M.A.); faljuhaimi@ksu.edu.sa (F.Y.A.-J.)
*   Correspondence: aladin.bekhit@otago.ac.nz; Tel.: +64-3-479-4994

Received: 4 December 2019; Accepted: 4 January 2020; Published: 7 January 2020

Abstract: The present study was carried out to gain consumer insights on the use of tempeh (a fermented soy bean product) to improve the healthiness of beef patties and to determine the acceptable level of tempeh (10%, 20%, or 30%) in the patty. The study consisted of conducting two focus groups ($n = 15$), a pilot sensory evaluation, and a full consumer sensory study. The focus groups were asked about their consumption of beef patties, attitudes towards processed meat, attitudes towards negative aspects of red meat consumption, and attitudes towards tempeh consumption, as well as sensory perceptions of the cooked patties and their visual acceptance of raw patties. Focus group discussions suggested that there was a market for the product if consumers were informed of tempeh health benefits. Participants seemed more willing to choose how to balance their diet with an antioxidant source than buy a beef patty with added antioxidants. The focus group participants rated the visual attributes of raw patties from all treatments and it was found that the 20% tempeh and 30% tempeh patties were ranked lower ($p < 0.05$) than the others. Overall, the sensory experiments showed that the inclusion of 10% tempeh was the most acceptable level of addition. There were no significant ($p > 0.05$) differences between the control and 10% tempeh patties for overall acceptability or acceptance of flavor. However, 10% tempeh patties were found to be more tender and juicier than the control ($p < 0.05$). A proper knowledge and awareness of consumers about the benefits of tempeh could allow the development of beef containing tempeh products.

Keywords: acceptability; beef; consumer perceptions; patties; sensory; tempeh

---

1. Introduction

Red meat has been consumed by humans for thousands of years and has played an important part in human evolution. It is a nutrient-rich food which is high in protein; minerals, such as iron, zinc, and selenium; and many vitamins [1]. However, in recent years, there has been a negative consumer reaction to red meat [2]. This has partially been due to its saturated fat content, but also due to the causal link between red meat consumption and the incidence of colorectal cancer. Red meat was found to be correlated with the risk of having colon cancer, whereas fish intake was not associated with the risk and a slight negative association was observed with poultry consumption [3]. The difference between meat types and risk of colon cancer could be due to the high levels of haem in red meat. The catalytic activity of haem iron in promoting oxidative processes may be linked to colon cancer formation. Sesink et al. [4] found that fat alone did not affect the levels of cations found in feces; however haem increased fecal cation concentrations in low, medium, and high fat diets, which demonstrates that haem in the presence of fat impairs the absorption of cations and causes epithelial damage. There was a significant interaction ($p < 0.001$) between haem and fat, which affected the fecal cation concentration [4], supporting a hypothesis for a haem-induced lipid oxidation mechanism as a potential contributor to colon cancer.

Much research has been conducted to investigate the addition of non-meat additives or extenders to improve nutritional properties [5–13], shelf life [14,15], sensory properties [5,16], and physical parameters [5,16] and make use of by-products of other food industries [5] to improve the quality of meat patties.

Partial substitution of meat by plant products is regarded as an emerging strategy to reduce meat consumption [17,18] and improve the healthiness of meat products. This strategy is successful and accepted by consumers because it is not aiming at eliminating meat from the diet, but targets the implementation of simple adjustment of the products without compromising important attributes desirable in meat [19]. The sensory properties of meat products substituted with plant products are important for the acceptability of products. In particular, taste and texture are highly important characteristics for acceptance [20,21]. The format of the meal [21] and repeat exposure [22] are important for the acceptability of meat substitutes and meat analogues (meat products where a portion of the meat is substituted by another food ingredient, also known as hybrid meat products). Evaluations of meat products substituted with plants [21] or insects [23] have indicated the potential of achieving the same level of acceptability by consumers, but gender differences may exist.

Consumers have become increasingly concerned about fat consumption and have often associated red meat with a high fat content. There are several classes of fat and each contributes differently to the risk of cardiovascular disease [24]. Cardiovascular disease is associated with atherosclerotic plaques which build up on the inside of coronary arteries that provide blood to the heart muscle (myocardium) [25]. These plaques are mainly composed of cholesterol and low density lipoprotein (LDL) particles, which are the main cholesterol carriers. Low density lipoproteins are oxidized and consumed by macrophages [25], which can consequently lead to increased oxidative stress and diseases. The objective of this study was to gain insights on consumers' perception of the addition of tempeh to beef patties with the aim of producing a healthier beef product, by conducting focus groups, as well as pilot and full consumer sensory trials.

2. Materials and Methods

All experiments were approved by the University of Otago Human Ethics committee (09/156). The experiment design included the following:

(1) Two focus groups (*n* = 15) of subjects that were screened (Section 2.3.1) before joining the focus groups. The focus groups explored the panelists' attitudes towards the consumption of beef patties, their knowledge of processed meat products, a sensory evaluation of beef patties containing tempeh, and their perception of the new beef analogue after providing information on the health benefits of including tempeh in the product;
(2) pilot sensory analysis study (*n* = 14) to determine the optimum tempeh % (10%, 20%, or 30%) to include in a full-scale consumer sensory study;
(3) Consumer sensory study (*n* = 118) to determine consumers' perception and acceptability of beef patties containing 10% tempeh compared with a control (100% beef) and a comparably reduced beef patty that contained 10% bread crumb.

2.1. Sample Preparation

Soy beans (*Glycine max*) were purchased locally from the Taste nature organic store (Dunedin, New Zealand) and were soaked for 2 days at 4 °C. The beans were cooked for 10 min in a pressure cooker at 100 °C, drained, dried in towels, and treated with white vinegar at a concentration of 2 mL/100 g of beans. A starter culture (*Rhizopus oligosporus*) was then added at a concentration of 1 g culture/kg beans. The mixture was packed in perforated ziplock bags (170 × 180 mm) and incubated in a snaplock container (Klip it, 255 × 120 × 55 mm, 1.75 L, Sistema Plastics, New Zealand) with 1 M potassium nitrate solution to create a humid atmosphere (92% relative humidity) at 31 °C in an incubator (Labserve,

Ontherm Scientific Ltd., Hutt City, New Zealand) for 24 h. The produced tempeh was vacuum packed and frozen at −80 °C for further use.

Fresh beef semitendinosus muscles (ST; eye of round, $n = 6$ with total weight of 22 kg) of a normal pH (range 5.55–5.64) were obtained from a local supplier (Alliance Wholesale meats, Dunedin). The meat was separated into lean and fat and then diced. Diced meat and 10% fat were added to a Kenwood blender with a mincing attachment (Alp 5 blade and mincing plate, 4.5 mm diameter die). The patties were prepared (about 4–5 kg for each treatment) as described in Table 1. Five experimental groups (Table 1) were as follows: non treated control sample (control); samples with 10% of the weight replaced with 10% bread crumb (Bread crumb 10%); and samples with part of the weight replaced with tempeh at the level of 10% (Tempeh 10%), 20% (Tempeh 20%), or 30% (Tempeh 30%). Patties were made by shaping 120 g of the mixture with a patty former. Fresh samples were used for color stability trials and other analyses, and the patties were vacuum packed and stored at −80 °C.

Table 1. Composition of the five patty treatments.

Treatment	Lean Meat (%)	Fat (%)	Tempeh (%)	Bread Crumb (%)	Salt (%)
Control	89	10	-	-	1
Bread crumb 10%	79	10	-	10	1
Tempeh 10%	79	10	10	-	1
Tempeh 20%	69	10	20	-	1
Tempeh 30%	59	10	30	-	1

2.2. Focus Group

Consumer perceptions, including attitudes and sensory perception, are important characteristics of a patty. Focus groups were used for exploratory research to obtain consumers' insights on the use of plant materials to improve the healthiness of beef patties, determine the level of tempeh inclusion, and serving conditions (preference for at home use vs. take out). This information was used later in designing the consumer sensory analysis and to investigate consumer attitudes towards a novel product such as beef patties containing tempeh.

A focus group is an interview based on a set of predetermined open ended questions which aims to generate discussion amongst participants to gain insights into consumer behavior [26]. It is based on a small number of issues, with the aim of understanding how the behavior of individuals is influenced by their beliefs, attitudes, and feelings [26–28].

2.2.1. Participant Recruitment

Flyers were used to recruit participants for two focus groups. They were placed around the University of Otago and Otago Polytechnic campuses, at supermarkets, a public library, and fish and chip shops. Flyers were placed over a period of two weeks and fifteen participants were chosen in total to take part in two focus groups after a brief screening over the phone. Respondents were screened based on three questions, in order to exclude those who would not be eligible for the focus group. The questions were as follows:

1. Are you willing to participate in a recorded discussion on this topic? The recorded data will be handled ethically according to the university policy on private data;
2. Do you have any ethical or religious objections to eating beef?
3. Are you allergic to gluten and/or soy?

Participants who answered yes to question one and no to questions two and three were invited to participate in the focus groups.

The first focus groups attempted to cover the research aspects from diverse age and professional groups, whereas the second focus group sought the opinions of young university students as it was clear that they represent a large fraction of consumers. The characteristics of the focus group participants are shown in Table 2.

Table 2. Summary of the characteristics of focus group participants.

Focus Group	Number of Participants	Number of Participants	Number of Participants	Number of Participants	Number of Participants	Number of Participants	Gender Male/Female
	Age ≤18	Age 19–25	Age 25–30	Age 30–40	Age 40–50	Age ≥50	
1	0	1	3	1	1	1	4/3
2	2	6	0	0	0	0	4/4
Total	2	7	3	1	1	1	8/7

2.2.2. Organization of the Focus Group

The focus groups were held on two separate days. The durations of the two focus groups were ninety minutes and eighty minutes for the first and second focus groups, respectively. The focus groups were moderated by the authors. The focus group sessions were held in a sensory lab to encourage interaction and allow for the best audio recording environment. A tape recorder with an external microphone was used to record the answers of participants. Participants read an information sheet and signed a consent form before the sessions. The participants engaged in an ice-breaker discussion with each other and with the moderator for five minutes before the focus group officially started. The focus group session was divided into five parts and guided by the focus group protocol. At the conclusion of the focus group, participants put their name in a basket for a random draw for a prize of a $50 grocery voucher.

2.2.3. Sample Preparation

The five patty treatments were prepared as described in Section 2.2. Patties were then placed in a refrigerator at 4 °C on a tray lined with wax paper; wax paper was put on the surface to prevent drying until they were later cooked on the same day. One patty of each formulation was also placed on polystyrene trays wrapped with glad wrap and stored at 4 °C in a refrigerator until they were later shown to participants for an evaluation of raw patties. The patties were cooked in canola oil on a Kambrook Banquet electric fry pan for two minutes on each side. They were then put into a fan forced oven for ten minutes at 180 °C, which was sufficient to produce an internal temperature >75 °C. The patties were removed, cut into quarters, and wrapped in aluminium foil. They were put into labelled trays and held in an oven at 80 °C until serving within 4–5 min from preparation.

2.2.4. Focus Group Protocol

The focus group protocol consisted of five parts that dealt with the consumption of beef patties, attitudes towards processed meat, attitudes towards negative aspects of red meat consumption, and attitudes towards tempeh consumption, as well as sensory perceptions of the cooked patties and visual acceptance of the raw patties.

Part 1: Attitudes Towards the Consumption of Beef Patties

The first set of questions was about the participants' normal consumption habits with regards to takeaways, especially beef patties "commonly known as burgers". Participants were asked about when and where they normally consume beef patties/burgers and what factors influenced their choice of takeaways and/or burgers.

Part 2: Consumer Perception and Knowledge of Processed Meats

At the start of this section, the participants were shown an information sheet on processed meat and the addition of non-meat ingredients to meat products. This section was important for evaluating the consumers' knowledge of meat extenders and their attitude towards products containing non-meat components as patties may contain up to 30% tempeh. Questions were aimed at exploring whether participants realized how many non-meat ingredients are included in processed meat, understanding why producers do it, and if it seems deceptive or not.

Part 3: Preliminary Sensory Analysis

During this part of the session, the participants analysed the patties for sensory attributes. Quarter sections of patties, temperature tested with a thermometer, were brought from the warming oven in a separate kitchen and were simultaneously served as 3-digit coded samples to participants. The attributes assessed were the intensity of beef odor (slight–strong), intensity of other (non-beef) odor, tenderness (tough–tender), juiciness (very dry–very juicy), chewiness (very chewy–very soft), beef flavor intensity (very slight–very strong), intensity of other (non-beef) flavor (very slight–very strong), acceptance of flavor (dislike extremely–like extremely), and overall acceptance (dislike extremely–like extremely). Attributes were rated on paper ballots with five-point word-anchored scales, with the exception of acceptance of flavor and overall acceptance, which were assessed on seven-point word-anchored scales. This section served as exploratory research for sensory analysis to choose an acceptable level for tempeh inclusion in a patty.

Part 4: Effect of Information of Health Benefits on the Consumer Perception of Novel Beef Patties

This section began by providing participants with an information sheet on published research suggesting the negative aspects of red meat consumption [29] and potential health benefits of tempeh [30,31]. The objective was to see how health information impacts attitudes towards adding a vegetal antioxidant source to the beef patties. Questions during this section were based around previous knowledge of a link between red meat and cancer and if this link led to a change of diet. Participants were also asked if this would increase their likelihood of eating meat with an antioxidant source and if they had tried tempeh. Participants were asked whether they would make an attempt to consume antioxidants with red meat as a separate part of a meal or whether the inclusion of the antioxidant source (such as in the tempeh) in a patty would be a convenient option. They were also asked if they would be willing to purchase a patty containing tempeh.

Part 5: Evaluation of Raw Beef Patties

One patty of each formulation (freshly prepared) was displayed to participants in raw form on a polystyrene tray, as it would normally be presented for retail sale. The objective was to determine the attitudes towards the product in the form it would be sold at retail after the health information had been given. This was decided to be a better method for assessing the purchase intention than consumption of the cooked patties alone, as consumers base meat purchases on visual cues [32,33].

2.2.5. Focus Group Analysis

The focus group discussions were recorded with Audacity software (version 1.2.6) and were later transcribed. Participants and responses were coded and the transcripts were analysed by sorting participant quotes thematically according to the insights they provided into consumer attitudes, in order to write the discussion.

2.3. Sensory Analysis

2.3.1. Pilot Sensory Analysis Study

A smaller pilot sensory study (n = 14) was carried out before a full-scale sensory study in order to aid in the design of the larger experiment. Participants were students and staff members from the University of Otago.

2.3.2. Consumer Sensory Analysis Study Recruitment

The panelists for sensory analysis (n = 118) were recruited by different contact and advertisement methods. The sample size is considered adequate to avoid type I and type II errors that can arise in a consumer sensory test [34]. Panelists were recruited from a database kept by the Food Science Department, University of Otago and from fliers placed around the University campus, Otago

Polytechnic campus, and halls of residence; an advertisement at lectures; and by emails circulated by the administrators of university departments. Respondents arranged a time to come in and taste the beef patties and were asked a set of questions to screen out participants unable to participate based on personal beliefs, allergies, or a lack of familiarity with the product. The gender and age categories of panelists are shown in Table 3.

Table 3. Gender and age group composition of the consumer sensory experiment.

Gender	Age Category					
	18–24	**25–30**	**31–40**	**41–50**	**>50**	**Total**
Male	17	12	8	6	5	48
Female	26	8	16	10	10	70
Total	43	20	24	16	15	118

Study Design:

The questionnaire for participants was created in Compusense Five (version 4.8.8, Guelph, Ontario, Canada). The panelists were asked to declare their age and gender categories and for each sample question, they were asked for the attributes of overall acceptability, intensity of beef odor, tenderness, chewiness, juiciness, intensity of flavor, level of non-meat flavor, and acceptance of flavor. After answering these questions for all samples, the consumers were asked about the frequency of beef patty/hamburger and soy product consumption.

Sample Preparation:

Samples were prepared as described in Section 2.1 above for the focus groups. The results obtained from the focus groups and pilot sensory studies guided the treatments chosen for consumer sensory analysis, which included a control, and 10% bread crumb- and 10% tempeh-containing patties. The samples were prepared 1 week before the sensory analysis, frozen at −20 °C, and defrosted in a refrigerator overnight prior to the sensory analysis. The defrosted patties were cooked as described for the focus groups above. After cooking, they were cut into quarters, wrapped in tinfoil, and placed in casserole dishes inside an oven set to 100 °C.

Sensory Analysis:

The analysis was performed in sensory booths in the Sensory Science Research Centre at the Food Science Department of the University of Otago. Participants first read an information sheet and signed a consent form and were then served the samples. The samples were coded with randomized numbers according to the serving order devised by Compusense. The samples were served under ambient light in booths under positive pressure. Participants took one minute breaks and drank water between assessing samples to cleanse their palettes.

2.4. Statistical Analysis

The sensory data from the focus groups and pilot sensory study were analysed using a Kruskal–Wallis test (Minitab 16, Minitab, State College, PA, USA). A statistical analysis of sensory attributes obtained in the consumer sensory study was performed as a one way analysis of variance (ANOVA), with treatments as the independent variable, and significant differences were detected at a level of $p < 0.05$, identified by post hoc Tukey tests using Minitab software version 16 (Minitab, State College, PA, USA).

3. Results and Discussion

3.1. Focus Group

The focus groups were coded for ease of discussion (Supplementary Table S1). The main messages from the focus groups are discussed in the following themes.

3.1.1. Consumption of Takeaways

Most focus group participants consumed takeaways at least once a week and the younger university students generally consumed takeaways more often. The proximity of the takeaway outlet and the price seemed to be the most important factors for an increased consumption of takeaways. Young consumers were reported to be the most frequent takeaway consumers, despite their belief that takeaway food is unhealthy [35]. Younger adult participants have been reported to be more frequent consumers of takeaways than older adults, possibly due to a more positive perception of convenience foods [34,35]. Full time workers consumed takeaways twice as often as non-full time workers [35], although this was not the case in this focus group study.

Two of the second focus group consumers said that they normally buy beef patties/burgers at fast food outlets. Although, in the groups, there was a majority of fast food hamburger consumers, amongst some of the participants in both groups, there was a definite preference for homemade burgers. For example, P2G1 stated that *"they just taste nicer normally, homemade patties and stuff"*, and P9G2 expressed, *"But, I love homemade hamburgers the best"*. One of the reasons mentioned for this is that it was a *"pretty easy meal to prepare"* (P2G1) and *"quite filling as well"* (P1G1). In the second focus group the reason for making homemade burgers were that *"it tastes way better"* (P10G2). There was a noticeable lack of trust in fast food outlets for some of the young female consumers. One reason given for this lack of trust was *"because I know what's inside"* (P15G2), as the participant studied Human Nutrition papers and had a knowledge of some ingredients used in beef patties and their nutrient contents. Another did not trust that burgers were made from the ingredients that they were claimed to be made from: *"with mince you know it is mince … rather than I don't know like in chicken burgers, you know is it actually chicken"* (P10G2).

For one participant, it was previous work in the fast food industry which influenced their beliefs: *"You know, how long the meat sits there"* (P9G2). Female consumers preferred to eat burgers from the higher quality takeaway outlets and one stated that they would be *" … willing to pay a bit more for a really good burger"* (P4G1). For the desirable attributes of the hamburgers, younger females placed more emphasis on health, whilst the male consumers did not. One male participant said, *"I don't think that healthy comes into it when I eat hamburgers personally"* (P1G1). Conversely, two female consumers cited health as one of the desirable attributes for a burger. The meat content was also stated as being important: *"I definitely think that all meat kind of burgers not like probably 30% meat and the rest is other things"* (P15G2).

The consumption of fast food and takeaway food represents a real paradox as consumers mostly find this food to be unhealthy, but its consumption is increasing [36–38]. For example, in a previous study, 78% of Americans considered fast food as "not at all good or not too good", but more than half of Americans were reported as eating fast food at least once a week [36]. The main reasons reported for the frequent use of takeaways and fast food outlets are their convenience, low cost, consistent taste, and easy access, since many outlets are found in localities [36–38].

3.1.2. Adding Non-Meat Ingredients and Processed Meat

Overall, the participants were quite skeptical about processed meat and related this to the profiteering of producers: "They are not adding ingredients because they want to make a consumer happy (but) because they see that they can add value to the product" (P6G1). European consumers have also expressed the view that meat processors only work for their benefit, rather than that of the consumer [39]. Some consumers accept this as a way of getting lower priced meat products and are not so concerned. The younger male students in the second focus group are in this category. These products were recognised as a way of selling second-grade meat, which is in agreement with the perception of European consumers [39], although some consumers accept this as part of buying cheaper meat products. Overall, the preference was for non-processed meat forms and was similar to that of the consumers interviewed by de Barcellos et al. [39]. For the second focus group, the word processed had especially negative connotations: "I don't think I have ever bought the frozen patties because I just think they look so yuck … like it just looks so processed" (P9G2). Additionally, two of the consumers were influenced by having grown up with home-killed meat on a farm.

3.1.3. Sensory Testing of the Beef Patties

One of the panelists was very familiar with tempeh, whilst most were not and used a variety of words to describe the flavor that was foreign to them. It was described as *"vegetablely"* (P2G1), and smelling like *"warm pecans"* (P4G1) or having a *"beaniness"* (P4G1).

3.1.4. The Link Between Red Meat and Colon Cancer

A couple of the panelists had previously heard about a link between red meat and cancer. One was a Human Nutrition student, but was more familiar with the link of heterocyclic amines and processed meats to colorectal cancer. Overall, the panelists were quite skeptical about this link as there were many factors reported as being linked to cancer in the media. In the two separate groups, panelists said, *"everything causes cancer these days"* (P1G1) and *"but they link everything to cancer"* (P10G2). Participants generally did not care or were not willing to change their consumption habits due to this fact and one panelist said, *"it's more dangerous to dye your hair"* (P8G2). This is in contrast to a separate study, which found that beef consumers were quite health conscious [40], although this study had a more varied age structure. The previous study investigated European consumers and found that a number of consumers had concerns about beef carcinogenicity and its long-term health effects [40].

3.1.5. Consuming Patties with an Antioxidant Source or Balancing Yourself

Participants seemed more willing to balance their diet with an antioxidant source than buy a beef patty with added antioxidants. Some participants perceived this as unnatural and said, *"Normally I would rather, I think, buy something more natural"* (P14G2) and *"I don't want people chopping and changing my food"* (P11G2). Consumers react negatively towards a perceived 'interference' with food products, including the manipulation of beef, which is perceived as 'unnatural' and may explain these answers [39–42]. The ability to sell tempeh patties may be enhanced by the inclusion of a health claim on the packaging as these claims are able to positively influence the consumer perception of a health benefit [43].

However, frequent takeaway consumers are significantly less likely to try to achieve dietary requirements of fruits and vegetables [44], which may make it more difficult to sell tempeh patties through a takeaway outlet. This was stated by the participants: *"to go to eat to McDonalds to have a healthy hamburger, it seems a bit paradoxal"* (P6G1). Participants expressed that a potential consumer would need to be informed of the health benefits in order to be willing to purchase the product: *"If you outlined the ingredients its all like, you need to have that in it maybe, otherwise it would be like why change?"* (P8G2).

3.1.6. Consuming Tempeh

There were contrasting attitudes towards consuming tempeh. Three of the participants had tried it, but those who had not were generally not accepting of the description of tempeh. The description was unappealing for younger consumers not familiar with the product and elicited responses such as *"That doesn't sound good"* (P14G2) and *"that doesn't sound appealing"* (P9G2). For an idea of what tempeh was, two participants asked if it was similar to tofu. Food neophobia has a negative effect on the acceptance of functional food products such as beef patties containing tempeh [45]. Food neophobia was not explored in this focus group; however, these consumers may be more neophobic than the public in general. Food neophobia in this group could be due to a lack of exposure to novel and foreign foods.

3.1.7. Evaluation of the Beef Patties

The first focus group did not find any differences among all the samples (Figure 1A). For the second focus group, there were significant differences in the tenderness, where the 10% bread crumb was found to be chewy and the least tender, and the 30% tempeh was perceived to be soft and the most tender (Figure 1B). There was no difference ($p > 0.05$) in the intensity of beef flavor among all treatment groups (Figure 1A,B). The participants were asked to choose and rate visual attributes of raw patties from all treatments. The 20% tempeh and 30% tempeh patties were consistently rated lower ($p < 0.05$) than the others (Figure 2). The 10% tempeh patties were not different ($p > 0.05$) from the 10% bread crumb patties and were not significantly different from the control in terms of color or appearance.

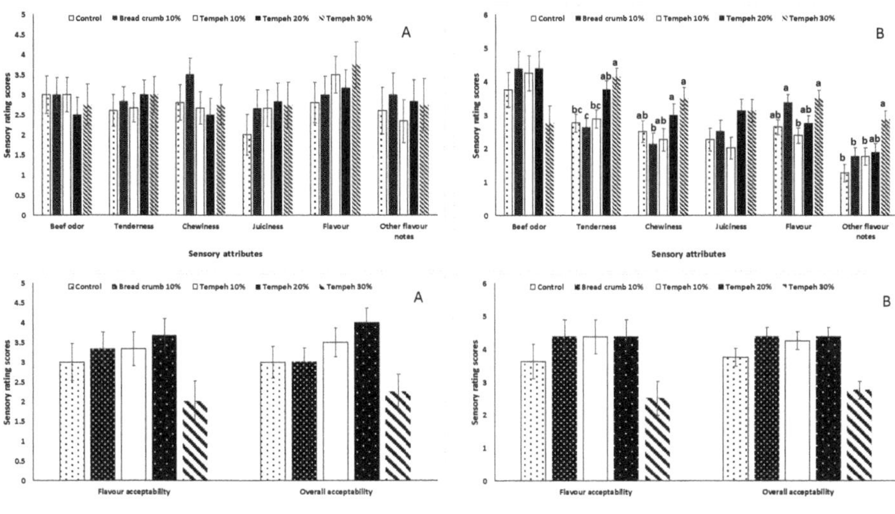

Figure 1. Sensory attribute scores of five beef patties substituted by tempeh at different levels. The patties were evaluated during focus groups (A = focus group 1 and B = focus group 2). [a–c] Within each sensory attribute, bars that do not share the same superscripts are significantly different ($p < 0.05$).

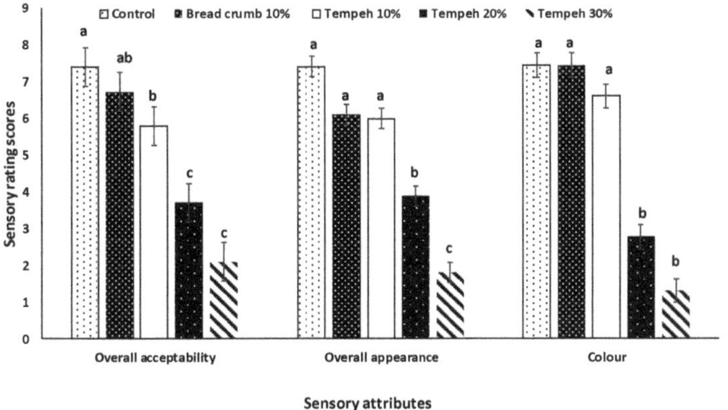

Figure 2. Sensory attribute scores of five raw beef patties substituted by tempeh at different levels. The patties were evaluated by focus groups ($n = 15$). a–c Within each sensory attribute, bars that do not share the same superscripts are significantly different ($p < 0.05$).

There is a relatively large decrease in ratings of visual attributes when increasing tempeh incorporation from 10% to 20% (Figure 2). Ratings of appearance are important as this attribute influences consumer purchase decisions [29,30]. Data from the second focus group suggests that tempeh 10% is the tempeh-containing patty most likely to be purchased. A pilot sensory study was necessary following the focus groups, in order to provide a larger sample size to select an appropriate patty.

3.2. Pilot Sensory Study

From the pilot sensory study, it was observed that the 10% tempeh patties had rating scores closer to the control than 20% and 30% tempeh, regardless of significance (Table 4). Most importantly, they were closest in overall acceptability and there were decreased ratings with additional tempeh incorporation (Table 4). This evidence, combined with the focus group quantitative data, which suggested that 10% tempeh patties were the most similar to the control, led to the decision to take the 10% tempeh patty to the full consumer sensory trial. The 10% tempeh patty was more likely to be accepted and less likely to have significant differences in sensory attributes in the larger sample size of the full-size sensory trial than patties containing higher levels of tempeh incorporation.

Table 4. Mean scores for sensory attributes for five beef patties substituted by tempeh at different levels in the pilot sensory study. The patties were evaluated by a small number of panelists ($n = 14$). [a–c] Within each row, means that do not share the same letters are significantly different ($p < 0.05$).

	Control	Bread Crumb 10%	Tempeh 10%	Tempeh 20%	Tempeh 30%	*p* Value
Overall acceptability	4.21	4.21	3.93	3.5	2.93	0.155
Intensity of non-beef odors	1.71 [a]	2.57 [a,b]	2.43 [a,b]	2.57 [a,b]	3.00 [b]	0.045
Intensity of beef odor	4.66 [a]	4.10 [a,b]	3.94 [a,b]	3.02 [b]	2.11 [c]	0.000
Tenderness	2.79 [a]	4.29 [c]	2.86 [a,b]	4.00 [c]	3.79 [b,c]	0.000
Chewiness	2.57 [a]	3.86 [b]	2.36 [a]	3.57 [b]	3.57 [b]	0.000
Juiciness	2.57	3.21	2.36	3.14	2.50	0.088
Flavor intensity	3.57	3.29	3.43	3.5	3.3 [6]	0.802
Non-beef flavor	1.86 [a]	2.50 [a,b]	2.64 [a,b]	2.64 [a,b]	3.43 [b]	0.007
Acceptance of flavor	4.29	4.36	3.29	3.71	3.00	0.110

3.3. Consumer Sensory Study

Differences were perceived by participants in terms of flavor and texture sensory attributes, but not for overall hedonic attributes. Mean sensory scores of participants for the overall acceptability and acceptance of flavor did not significantly ($p > 0.05$) differ between the three beef patty treatments (Table 5). The overall acceptability values were lower than in some previous studies with other extenders for tempeh-containing patties [11,46,47]. The substitution of 10% tempeh was more acceptable than the substitution of 3% tomato peel, 2% hazelnut pellicle, or 9% flaxseed flour [5,8,10]; however, it was similar to the substitution of 10% okara powder, an olive oil/corn oil/fish oil blend, 0.5% carrageenan, 6% olive cake, or 7.5% okara, which did not affect the overall acceptability of beef patties [8,11,46,48,49].

The acceptance of tempeh patty flavor was higher than that for patties extended with 4% hazelnut pellicle or 37.5% wet okara, but lower than that for patties extended with carrageen, textured soy protein, tri sodium phosphate, or 10% plum puree [10–12,46,47]. Partial substitution of the fat with an olive oil/corn oil/fish oil blend or 3% flaxseed flour similarly did not significantly affect the acceptability of flavor [8,48]. Decreases in flavor acceptability have been observed with the substitution of 3% hazelnut pellicle, 1.5% texturized soy protein, 30% wet okara, and 6% flaxseed flour, and increases have been recorded with the addition of 10% plum puree [8,12,46,47]

Similar to tempeh-containing patties, there were no significant differences in the perception of intensity of beef odor with 15% date fiber [13].

The control was higher ($p < 0.05$) in terms of beef odor than both 10% bread crumb and 10% tempeh (Table 5). Despite the lower beef odor, the overall flavor intensity of 10% bread crumb and 10% tempeh did not differ from the control (Table 5).

Table 5. Mean scores for sensory attributes of five beef patties substituted by tempeh at different levels in the pilot sensory study. The patties were evaluated by a small number of panelists ($n = 14$). [a–c] Within each row, means that do not share the same letters are significantly different ($p < 0.05$).

	Control	Bread Crumb 10%	Tempeh 10%	*p* Value
Overall acceptability	5.42	5.44	5.38	0.97
Beef odor	4.19 [a]	3.53 [b]	3.78 [b]	0.00
Tenderness	3.23 [c]	4.64 [a]	4.14 [b]	0.00
Chewiness	2.65 [c]	4.05 [a]	3.43 [b]	0.00
Juiciness	3.66 [b]	3.95 [ab]	4.10 [a]	0.04
Flavor intensity	4.31	4.32	4.25	0.81
Non-beef flavor	2.43 [b]	3.41 [a]	3.11 [a]	0.00
Acceptance of flavor	5.62	5.60	5.42	0.64

The 10% bread crumb treatment was rated the most tender, followed by 10% tempeh and then the control, and all were significantly ($p < 0.05$) different (Table 5). This is in agreement with literature where increased tenderness occurred with the addition of 15% date fiber and 10% carbohydrate-lipid composites [13,50]. The same trend was observed for chewiness. Juiciness was rated the highest for 10% tempeh, while the control was rated significantly ($p < 0.05$) lower and 10% bread crumb did not differ significantly ($p > 0.05$) from either treatment (Table 5). Increases in juiciness with a 10% substitution of carbohydrate-lipid composites or 10% tomato paste have also been reported [50,51]. However, the substitution of up to 30% sorghum flour did not produce any significant difference in juiciness [52]. The use of soy bean products in beef products has been extensively investigated to improve the healthiness of beef products, improve the production economics, or modify the sensory attributes of the products. The use of textured soy protein (TSP) or soy protein concentrate (SPC) at a substitution level of 20% or 30% in beef patties was investigated using a consumer panel and a family consumer panel [53]. The 20% TSP-containing beef patties were rated similar to whole beef patties and the scores for both of these treatments were higher than those found with 30%TSP, and 20% and 30% SPC treatments [53]. Similarly, substitution beef patties with 20% or 30% TSP did not affect consumers' acceptability, despite the formation of a beany flavor and taste caused by the TSP addition [54]. Additionally, the use of TSP at 25% did not affect the flavor of beef patties [55]. Contrary to these findings, the use of 15% or 30% hydrated TSP in beef patties was found to reduce the beef flavor and overall acceptability of the patties [56], but improved the tenderness of the patties. The same trend was reported for 20% hydrated soy bean [57]. Overall, the various results reported above are likely to be related to differences in the soy bean product, the addition level, and possibly the background of the consumer panel. Unlike many of the studies mentioned above, which used trained sensory assessors [8,10–13,46,47,50–52], this study used a larger untrained consumer sensory trial, which is important for testing the market potential.

4. Conclusions

Information from the focus group suggested that consumers are not very concerned with a link between red meat and colon cancer, although several participants had heard of this link. They were skeptical about the media reporting of cancer risks. This gives the impression that there is little potential market for a novel product, such as the tempeh patty; however, the participants were mainly young people, who probably did not have much consideration of a healthy diet. Quantitative data did not show great differences between the patties tested. For appearance, however, 10% tempeh patties were rated closer to the control and bread crumb patties, which suggests that they are more likely to be purchased than other tempeh-containing patties. The 10% tempeh patties had better eating properties. For example, these patties were more tender, juicier, and had more flavor, but they were lower in the intensity of beef odor.

Overall, the 10% tempeh patty was the tempeh-containing patty with the most positive attributes. It was not significantly different than a control patty for overall acceptance and acceptability of flavor and is comparable to a control for visual attributes and more acceptable visually than patties containing more tempeh. The present study did not investigate the impact of tempeh substitution on the volatiles and flavor of the cooked beef patties and future work will address this issue.

Supplementary Materials: The following are available online at http://www.mdpi.com/2304-8158/9/1/63/s1, Table S1: Focus group themes for the consumption of takeaways, knowledge of the addition of additives to processed meat products, and relationship between colon cancer and the consumption of processed meat products.

Author Contributions: Conceptualization, J.T., I.A.M.A., F.Y.A.-J., and A.E.-D.A.B.; methodology, J.T., I.A.M.A., and A.E.-D.A.B.; formal analysis, J.T. and A.E.-D.A.B.; investigation, J.T., and A.E.-D.A.B.; resources, F.Y.A.-J., and A.E.-D.A.B.; data curation, J.T., and I.A.M.A.; writing—original draft preparation, J.T., and A.E.-D.A.B.; writing—review and editing, all authors; supervision, A.E.-D.A.B.; funding acquisition, I.A.M.A. All authors have read and agreed to the published version of the manuscript.

Funding: The authors extend their appreciation to the International Scientific Partnership Program ISPP at King Saud University for funding this research work through ISPP# 0073.

Conflicts of Interest: The authors declare no conflict of interest.

References

1. Farouk, M.M.; Yoo MJ, Y.; Hamid, N.S.; Staincliffe, M.; Davies, B.; Knowles, S.O. Novel meat-enriched foods for older consumers. *Food Res. Int.* **2018**, *104*, 134–142. [CrossRef] [PubMed]
2. Kusch, S.; Fiebelkorn, F. Environmental impact judgments of meat, vegetarian, and insect burgers: Unifying the negative footprint illusion and quantity insensitivity. *Food Qual. Prefer.* **2019**, *78*, 103731. [CrossRef]
3. Giovannucci, E.; Rimm, E.B.; Stampfer, M.J.; Colditz, G.A.; Ascherio, A.; Willett, W.C. Intake of fat, meat and fiber in relation to risk of colon cancer in men. *Cancer Res.* **1994**, *54*, 2390–2397.
4. Sesink, A.L.A.; Termont, D.S.M.L.; Kleibucker, J.H.; Van der Meer, R. Red meat and colon cancer: Dietary haem but not fat, has cytotoxic and hyperproliferative effects on rat colonic epithelium. *Carcinogenesis* **2000**, *21*, 1909–1915. [CrossRef]
5. Garcia, M.L.; Calvo, M.M.; Selgas, M.D. Beef hamburgers enriched in lycopene using dry tomato peel as an ingredient. *Meat Sci.* **2009**, *83*, 45–49. [CrossRef]
6. Danowska-Oziewicz, M. Nutritional quality of low-fat pork patties manufactured with the use of soy protein isolate. *Int. J. Food Sci. Technol.* **2010**, *45*, 193–199. [CrossRef]
7. Mansour, E.H.; Khalil, A.H. Characteristics of low fat beef burgers as influenced by various types of wheat fibres. *J. Sci. Food Agric.* **1999**, *79*, 493–498. [CrossRef]
8. Bilek, A.E.; Turhan, S. Enhancement of the nutritional status of beef patties by adding flaxseed flour. *Meat Sci.* **2009**, *82*, 472–477. [CrossRef]
9. Dzudie, T.; Scher, J.; Hardy, J. Common bean flour as an extender in beef sausages. *J. Food Eng.* **2002**, *52*, 143–147. [CrossRef]
10. Turhan, S.; Sagir, I.; Ustun, S. Utilisation of hazelnut pellicle in low-fat beef burgers. *Meat Sci.* **2005**, *71*, 312–316. [CrossRef] [PubMed]
11. Turhan, S.; Temiz, H.; Sagir, I. Utilisation of wet okara in low-fat beef patties. *J. Muscle Foods* **2006**, *18*, 226–235. [CrossRef]
12. Turhan, S.; Temiz, H.; Sagir, I. Characteristics of beef patties using okara powder. *J. Muscle Foods* **2009**, *20*, 89–100. [CrossRef]
13. Hashim, I.B.; Khalil, A.H. Quality characteristics of beef patties extended with date fibre. In Proceedings of the 54th International Conference of Meat Science and Technology, Cape Town, South Africa, 10–15 August 2008.
14. Banon, S.; Diaz, P.; Rodriguez, M.; Delores Garrido, M.; Price, A. Ascorbate, green tea and grape seed extracts increase the shelf life of low sulfite beef patties. *Meat Sci.* **2007**, *77*, 626–633. [CrossRef] [PubMed]
15. Ismail, H.A.; Lee, E.J.; Ko, K.Y.; Paik, H.D.; Ahn, D.U. Effect of antioxidant application methods on the color, lipid oxidation and volatiles of irradiated ground beef. *J. Food Sci.* **2009**, *74*, 25–31. [CrossRef] [PubMed]
16. Das, A.K.; Anjaneyulu, A.S.R.; Kondaiah, N. Development of reduced beany flavor full-fat soy paste for comminuted meat products. *J. Food Sci.* **2006**, *71*, 395–400. [CrossRef]

17. Spencer, M.; Cienfuegos, C.; Guinard, J.X. The Flexitarian Flip™ in university dining venues: Student and adult consumer acceptance of mixed dishes in which animal protein has been partially replaced with plant protein. *Food Qual. Prefer.* **2018**, *68*, 50–63. [CrossRef]
18. Spencer, M.; Guinard, J.X. The flexitarian Flip™: Testing the modalities of flavour as sensory strategies to accomplish the shift from meat-centered to vegetable-forward mixed dishes. *J. Food Sci.* **2018**, *83*, 175–187. [CrossRef]
19. Lang, M. Consumer acceptance of blending plant-based ingredients into traditional meat-based foods: Evidence from the meat-mushroom blend. *Food Qual. Prefer.* **2020**, *79*, 103758. [CrossRef]
20. Elzerman, J.E.; Hoek, A.C.; van Boekel, M.A.J.S.; Luning, P.A. Consumer acceptance and appropriateness of meat substitutes in a meal context. *Food Qual. Prefer.* **2011**, *22*, 233–240. [CrossRef]
21. Neville, M.; Tarrega, A.; Hewson, L.; Foster, T. Consumer-orientated development of hybrid beef burger and sausage analogues. *Food Sci. Nutr.* **2017**, *5*, 852–864. [CrossRef]
22. Hoek, A.; Elzerman, J.E.; Hageman, R.; Kok, F.J.; Luning, P.A.; de Graaf, C. Are meat substitutes liked better over time? A repeated in home use test with meat substitutes or meat in meals. *Food Qual. Prefer.* **2013**, *28*, 253–263. [CrossRef]
23. Megido, R.C.; Gierts, C.; Blecker, C.; Brostaux, Y.; Haubruge, É.; Alabi, T.; Francis, F. Consumer acceptance of insect-based alternative meat products in Western countries. *Food Qual. Prefer.* **2016**, *52*, 237–243. [CrossRef]
24. Kris-Etherton, P.; Feming, J.; Harris, W.S. The debate about n-6 polyunsaturated fatty acid recommendations for cardiovascular health. *J. Am. Diet. Assoc.* **2010**, *110*, 201–204. [CrossRef]
25. Mann, J.; Chisolm, A. Chapter 20: Cardiovascular diseases. In *Essentials of Human Nutrition*, 3rd ed.; Mann, J., Truswell, A.S., Eds.; Oxford University Press: New York, NY, USA, 2007; pp. 282–285, 288–289.
26. Tremblay, M.C.; Hevner, A.R.; Berndt, D.J. The use of focus groups. In *Design Science Research, Design Research in Information Systems*; Integrated Series in Information Systems; Springer: New York, NY, USA, 2010; pp. 121–143.
27. Rabiee, F. Focus-group interview and data analysis. *Proc. Nutr. Soc.* **2004**, *63*, 655–660. [CrossRef] [PubMed]
28. Blackham, T.M.; Stevenson, L.; Abayomi, J.C.; Davie, I.G. Consumers' knowledge and attitudes to takeaway food in Merseyside. *Proc. Nutr. Soc.* **2016**, *75*, E91. [CrossRef]
29. Richi, E.B.; Baumer, B.; Conrad, B.; Darioli, R.; Schmid, A.; Keller, U. Review Health Risks Associated with Meat Consumption: A Review of Epidemiological Studies. *Int. J. Vitam. Nutr. Res.* **2015**, *85*, 70–78. [CrossRef]
30. Berghofer, E.; Grzeskowiak, B.; Mundigler, N.; Sentall, W.B.; Walcak, J. Antioxidative properties of fababean-, soybean- and oat tempeh. *Int. J. Food Sci. Nutr.* **1998**, *49*, 45–54. [CrossRef]
31. Nassar, A.G.; Mubarak, A.E.; El-Beltagy, A.E. Nutritional potential and functional properties of tempe produced from mixture of different legumes. *Int. J. Food Sci. Technol.* **2008**, *43*, 1754–1758. [CrossRef]
32. Troy, D.J.; Kerry, J.P. Consumer Perception and the role of science in the meat industry. *Meat Sci.* **2010**, *86*, 214–216. [CrossRef]
33. Sanders, S.K.; Morgan, J.B.; Wulf, D.M.; Tatum, J.D.; Williams, S.N.; Smith, G.C. Vitamin E supplementation of cattle and shelf life of beef for the Japanese market. *J. Anim. Sci.* **1997**, *75*, 2634–2640. [CrossRef]
34. Hough, G.; Wakeling, I.; Mucci, A.; Chambers, I.V.E.; Mendez Gallardo, I.; Rangel Alves, L. Number of consumers necessary for sensory acceptability tests. *J. Food Qual. Prefer.* **2006**, *17*, 522–526. [CrossRef]
35. Hunter, W.; Worsley, T. Understanding the older food consumer: Present day behaviours and future expectations. *Appetite* **2009**, *52*, 147–154. [CrossRef] [PubMed]
36. Ming, T.T.; Bin Ismail, H.; Rasiah, D. Hierarchical chain of consumer-based brand equity: Review from the fast food industry. *Int. Bus. Econ. Res. J.* **2011**, *10*, 67–80.
37. Tamuliene, V. Consumer attitude to fast food: The case study of Lithuania. *Res. Rural Dev.* **2015**, *2*, 255–261.
38. Min, J.; Jahns, L.; Xue, H.; Kandiah, J.; Wang, Y. Americans' perceptions about fast food and how they associate with its consumption and obesity risk. *Adv. Nutr.* **2018**, *9*, 590–601. [CrossRef]
39. de Barcellos, M.D.; Kugler, J.O.; Grunert, K.G.; Van Wezemael, L.; Perez-Cueto, F.J.A.; Ueland, O.; Verbeke, W. European consumers' acceptance of beef processing technologies: A focus group study. *Innov. Food Sci. Emerg. Technol.* **2010**, *11*, 721–732. [CrossRef]
40. Van Wezemael, L.; Verbeke, W.; de Barcellos, M.D.; Scholderer, J.; Perez-Cueto, F. Consumer perceptions of beef healthiness: Results from a qualitative study in four European countries. *BMC Public Health* **2010**, *10*, 835–844. [CrossRef]

41. Vidigal MC, T.R.; Minim, V.P.; Simiqueli, A.A.; Souza PH, P.; Balbino, D.F.; Minim, L.A. Food technology neophobia and consumer attitudes toward foods produced by new and conventional technologies: A case study in Brazil. *LWT-Food Sci. Technol.* **2015**, *60*, 832–840. [CrossRef]

42. Zhong, Y.; Wu, L.; Chen, X.; Huang, Z.; Hu, W. Effects of food-additive-information on consumers' willingness to accept food with additives. *Int. J. Environ. Res. Public Health* **2018**, *15*, 2394. [CrossRef]

43. Bech-Larsen, T.; Grunert, K.G. The perceived healthiness of functional foods: A conjoint study of Danish, Finnish and American consumers' perceptions of functional foods. *Appetite* **2003**, *40*, 9–14. [CrossRef]

44. Smith, K.J.; McNaughton, S.A.; Gall, S.L.; Blizzard, L.; Dwyer, T.; Venn, A.J. Takeaway food consumption and its associations with diet quality and abdominal obesity: A cross-sectional study of young adults. *Int. J. Behav. Nutr. Phys. Act.* **2009**, *6*, 29. [CrossRef] [PubMed]

45. Labrecque, J.; Doyon, M.; Bellavance, F.; Kolodinsky, J. Acceptance of Functional Foods: A comparison of French, American and Canadian consumers. *Can. J. Agric. Econ.* **2006**, *54*, 647–661. [CrossRef]

46. Angor, M.M.; Al-Abdullah, B.M. Attributes of low-fat beef burgers aimed at enhancing product quality. *J. Muscle Foods* **2010**, *21*, 317–326. [CrossRef]

47. Yildiz-Turp, G.; Serdaroglu, M. Effects of using plum puree on some properties of low fat beef patties. *Meat Sci.* **2010**, *86*, 896–900. [CrossRef]

48. Martinez, B.; Miranda, J.M.; Vasquez, B.I.; Fente, C.A.; Franco, C.M.; Rodriguez, J.L.; Cepeda, A. Development of a hamburger patty with healthier lipid formulation and study of its nutritional, sensory and stability properties. *Food Bioprocess Technol.* **2012**, *5*, 200–208. [CrossRef]

49. Hawashin, D.M.; Al-Juhaimi, F.; Mohamed Ahmed, I.A.; Ghafoor, K.; Babiker, E.E. Physicochemical, microbiological, and sensory evaluation of beef patties incorporated with destoned olive cake powder. *Meat Sci.* **2016**, *122*, 32–39. [CrossRef]

50. Garzon, G.A.; McKeith, F.K.; Gooding, J.P.; Felker, F.C.; Palmquist, D.E.; Brewer, M.S. Characteristics of low-fat beef patties formulated with carbohydrate-lipid composites. *J. Food Sci.* **2003**, *68*, 2050–2056. [CrossRef]

51. Candogan, K. The effect of tomato paste on some quality characteristics of beef patties during refrigerated storage. *Eur. Food Res. Technol.* **2002**, *215*, 305–309. [CrossRef]

52. Huang, J.-C.; Zayas, J.F.; Bowers, J.A. Functional properties of sorghum powder as an extender in beef patties. *J. Food Qual.* **1999**, *22*, 51–61. [CrossRef]

53. Twigg, G.G.; Kotula, A.W.; Young, E.P. Consumer acceptance of beef patties containing soy protein. *J. Anim. Sci.* **1977**, *44*, 218–223. [CrossRef]

54. Babji, A.S.; Abdullah, A.; Fatimah, Y. Taste panel evaluation and acceptance of soy-beef burger. *Pertanika* **1986**, *9*, 225–233.

55. Liu, M.N.; Human, D.I.; Egbert, W.R.; Mccaskey, T.A.; Liu, C.W. Soy protein and oil effects on chemical, physical and microbial stability of lean ground-beef patties. *J. Food Sci.* **1991**, *56*, 906–912. [CrossRef]

56. Deliza, R.; Saldivar, S.O.S.; Germani, R.; Benassi, V.T.; Cabral, L.C. The effects of colored textured soybean protein (tsp) on sensory and physical attributes of ground beef patties. *J. Sens. Stud.* **2002**, *17*, 121–132. [CrossRef]

57. Kassem, G.M.A.; Emara, M.M.T. Quality and acceptability of value-added beef burger. *World J. Dairy Food Sci.* **2010**, *5*, 14–20.

Article

A Mixed Method Approach for the Investigation of Consumer Responses to Sheepmeat and Beef

Melindee Hastie *, Hollis Ashman, Damir Torrico †, Minh Ha and Robyn Warner

School of Agriculture and Food, Faculty of Veterinary and Agricultural Sciences, The University of Melbourne, Melbourne, VIC 3010, Australia; hollis.ashman@unimelb.edu.au (H.A.); Damir.Torrico@lincoln.ac.nz (D.T.); minh.ha@unimelb.edu.au (M.H.); robyn.warner@unimelb.edu.au (R.W.)
* Correspondence: hastiem@student.unimelb.edu.au; Tel.: +61-474214123
† Present address: Faculty of Agriculture and Life Sciences, Lincoln University, Lincoln 7647, New Zealand.

Received: 18 December 2019; Accepted: 23 January 2020; Published: 24 January 2020

Abstract: Coupling qualitative and quantitative consumer research methodologies enables the development of more holistic and comprehensive perspectives of consumer responses. In this study, consumer responses to beef and sheepmeat were investigated using a mixed method approach combining perceptual mapping (qualitative), and sensory (quantitative) methodologies. Qualitative insights indicated Australian and Asian consumers differ in perception of familiarity and 'premiumness' of meat products. Specific findings included: Australians consume grilled or roasted meat as a centre of the plate 'hero' ingredient, while Asians prefer stovetop cooking methods where meat is one ingredient in a complex dish. Labelling meat as 'Australian' was important for Australian consumers but not for Asian consumers. Quantitative data demonstrated that older consumers (31–70 years) scored sheepmeat higher than younger consumers (18–30 years) for healthiness ($p = 0.004$), juiciness ($p = 0.029$), odour liking ($p = 0.005$) and tenderness ($p = 0.042$). Older consumers also had a lower willingness to pay than younger consumers for "premium" quality meat; 30–40 vs. 40–50 AUD (Australian dollar) per kg respectively for sheepmeat, and 40–50 vs. 50–60 AUD per kg respectively for beef. In conclusion, the approach used effectively integrated consumer attitudes, usage information and sensory assessments with socio-demographic factors to generate insights for the refinement of market strategies and product offerings.

Keywords: consumers; attitude; sensory; beef; sheepmeat; premium; holistic product development

---

1. Introduction

Australia, one of the countries with the highest meat consumption rate per capita in the world, has experienced a steady decline in red meat consumption over the past 30 years, with beef consumption dropping from 39 to 26 kg/person/year and sheepmeat from 25 to 8.5 kg/person/year [1]. The decline has been linked to consumers reducing their red meat intake in response to health warnings and the increasing affordability of alternative protein sources [1].

Australia now exports more red meat than the local market consumes [2,3], and the nearby developing markets of Asia are increasingly important for the Australian red meat industry. Australia is not alone in seeking to maximise its Asian export opportunity, and competes with other exporting nations such as Brazil, India, USA and New Zealand for a share of this market [3,4]. In the commoditised global red meat market, Australia is less competitive on a price basis; thus, "premiumisation" of Australian beef, sheepmeat and pork products is currently employed as a key strategy for market differentiation and enables access to the growing numbers of affluent and discerning consumers in both Asia and Australia, facilitating realisation of higher financial returns [2–5].

However, there is evidence of cultural differences in consumer responses to red meat. Simply "lifting and launching" products, and marketing formats that work in Australia and applying them in the Asian market, may not be maximising the export opportunity [6–9]. Consumers, ultimately, are the purchase decision makers. Understanding their attitudes to, and perceptions of, red meat along with the attributes linked to consumer choice and "premiumness" can inform the development of more targeted marketing strategies and optimised product development, thereby maximising the opportunity for Australian red meat products in a highly competitive environment [10–14].

Consumer responses to red meat are influenced by a complex range of dynamic, interrelated factors. Font-I-Furnols and Guerrero [15], in referring to consumers' responses to pork, categorised these factors as: (1) product's sensory qualities; (2) marketing factors (information received by the consumer in the form of advertising, labels, etc.); and (3) psychological factors (the consumer's motivations, expectations and perceptions). However, most existing research tends to be isolated and fragmented, residing either within the consumer science realm (focused on the attitudes or purchase decision drivers), or the sensory science realm (focused on the consumption experience) [16–18]. Combining methodologies from these two realms allows the researcher to develop a more holistic and detailed view of the consumer's response and determine if modifications of the current product offering are required to increase its appeal to the target consumer, as previously done for pork by Bredahl et al. [14], and Dransfield et al. [19].

The overall objective of this study was to explore the effectiveness of a "mixed method" approach to investigate both local and export market consumer responses and attitudes towards beef and sheepmeat attributes, and to understand how they might differ for consumer groups according to their sociodemographic features. Revealing these differences will inform the development of differentiated marketing and product design for Australian red meat in accordance with target consumers' preferences.

2. Materials and Methods

This study combines:

(a) a qualitative analysis of consumers' attitudes and perceptions through facilitated discussions and "perceptual mapping exercises" as described by Beckley et al. [20], Nestrud and Lawless [21] and Risvik et al. [22], with

(b) a quantitative sensory assessment of eating quality based on MSA (Meat Standards Australia) sensory protocols as described by Watson et al. [23] and, subsequently, a priori consumer segmentation analysis based on demographic factors, a method previously described by Dolnicar et al. [24].

All subjects gave their informed consent for inclusion before they participated in the study. The study was conducted in accordance with the Declaration of Helsinki, and the protocol was approved by the Human Research Ethics Committee of The University of Melbourne (HREC 1545786.1).

2.1. Overview of Qualitative Approach

Variables examined in the qualitative study included consumer cultural heritage (Asian or Australian) and meat species (beef or sheepmeat). Perceptual mapping exercises were completed by both consumer groups for both meat species separately. Each session focused on a defined set of stimuli samples according to an experimental design that encompassed a range of meat attributes, e.g., cut, colour, fat content, freshness (see Section 2.1.3). Additionally, at the end of the perceptual mapping sessions, participants for the sheepmeat sessions completed a sheepmeat descriptor mapping exercise (Section 2.1.5), and beef session participants completed a beef concept mapping exercise (see Section 2.1.6).

2.1.1. Panels

Seven panels were run in total at the University of Melbourne. A total of 67 panelists were recruited from the university population. The group's age ranged from 18–58 years and included 31 male participants and 36 female participants. Most participants were postgraduate students (84%) with ages ranging from 22–58 years. Panelists were asked to self-identify as Australian or Asian before being assigned to either a sheepmeat or beef panel. Asian ethnicity was defined as having been born in

Asia and having been a resident in Australia for less than two years. Separate panels were run for Asian and Australian consumers, and these panels focused on either sheepmeat or beef. For sheepmeat, a total of 26 Australians participated in three sessions, and 18 Asians participated in two sessions. For beef, one session was run with six Australian participants and one session was run with 17 Asian participants. Each session was run for approximately 1.5 h and was led by a trained facilitator with at least one note taker/observer; sessions were also video recorded for later review.

2.1.2. Discussion Guide and Facilitation

A discussion guide was used throughout the beef and sheepmeat panels (Supplementary A), providing the facilitator with timings and topics to be covered during the session. In brief, the sheepmeat discussion guide included: 5 min for the introduction of researchers and participants; 15 min for seeking information on familiarity with sheepmeat, frequency of consumption, eating occasions and understanding of the terms hogget, mutton, and lamb (In Australia, lamb is defined as an ovine animal under 12 months of age with no permanent incisors in wear; hogget is older than a lamb, has lost its milk teeth, can have up to two permanent teeth in wear and is usually between 12 months and 2 years of age; mutton is older again and will have more than two permanent incisors [25,26]); 45 min for a stimuli mapping exercise (see Section 2.1.4); 20 min target for sheepmeat descriptor mapping (see Section 2.1.5).

A similar discussion guide structure was used for beef except that the final 20 min sheepmeat descriptor mapping exercise (see Section 2.1.5) was replaced with a beef concept mapping exercise (see Section 2.1.6). All facilitators were experienced with the perceptual mapping methodology, were aware not to use leading questions and ensured everyone within the group had the opportunity to provide their opinions on mapping decisions. They were also free to explore new topics if they arose during the discussion.

2.1.3. Stimulus Selection

For each group, the participants were presented with the same set of either sheepmeat or beef images in a designated order. Images were sourced from the University of Melbourne's collection or adapted from cited sources and selected to reflect the stimuli design of experiment to test the range of the sensory or product space. In brief, the images demonstrated a variety of retail and primal cuts from sheep and beef (including some familiar market cuts from Asia and some from Australia). All images were of uncooked meat on a plain white background printed in colour on a high gloss photo paper measuring 130 mm × 80 mm. The images were unlabelled, but participants were informed that all images were either beef, or sheepmeat according to the session. Images incorporated a range of meat colours (ranging from light pink to dark red, assessed visually to reflect the range of Australian retail meat) and ranged in freshness (freshly cut meat to slightly spoiled meat), with varying amounts of fat (both subcutaneous and IMF (intramuscular fat)) and varying amounts of preparation (e.g., untrimmed joint, stringed roasting joint, chops, deboned, French trimmed). The image/stimuli descriptions, design of experiment, personation order and source are detailed in Supplementary B.

2.1.4. Perceptual Mapping

During the perceptual mapping exercise, consumers were asked to sort and place the set of stimulus samples onto a predetermined two-dimensional map according to their similarities and differences. To enable comparison between groups, the x and y dimensions of the map were labelled by the facilitator before group mapping commenced. The x-axis ranged from every day to premium and the y-axis from unfamiliar to familiar. The selection of these axes was based on the fact that "premiumisation" is a key strategy for the marketing of Australian products [2–4]. Familiarity was chosen as it is understood that consumer acceptance can be enhanced by product familiarity [27–29] and, conversely, unfamiliarity can be a barrier to consumer acceptance [30]. While the final map is informative, the discussion undertaken within the group during the exercise also generates insights into how the consumers assess product attributes, and it is from this discussion that most insights were distilled. Sheepmeat stimuli were used initially with replicate sessions to validate insight development, and beef sessions were subsequently used to compare/test those insights. Mapping was completed on a tabletop with the axis marked out using masking tape. For all sessions, the same first sample was placed by the facilitator at the centre of the map. For the sheepmeat sessions, the image selected was a leg of lamb and in the case of beef, porterhouse steak from the striploin. Both cuts were chosen to represent the middle of the range in terms of "premiumness" and are common formats in the Australian market. All subsequent samples were presented by the facilitator to be placed on the map by the group (one at a time and in predetermined order as described in Supplementary B).

2.1.5. Sheepmeat Descriptor Mapping

At the end of the sheepmeat mapping session, all groups were presented with a random set of descriptors laid out on a tabletop, each printed individually on a plain white card of similar dimensions to the mapping stimulus used previously. The descriptors were derived from production and eating quality factors and were selected to cover a range of premiumness and familiarity for the two consumer groups and are described in detail below.

Production Descriptors

"Lamb", "Spring lamb", "Natural", "Dry-aged", "Australian", "Lean", "Organic", "Healthy" and "Fresh" were selected from labels already in common use on the Australian and Chinese market and were expected to be familiar terms for both groups.

"Hogget" and "Mutton" are not in common use in Australia due to the fact that lamb is the main commercial source of sheepmeat; they are, however, important product categories for export [25,31], and therefore were included.

"Green" was selected as a familiar label for Asian participants as there are a number of established green label certification systems operating in Asia and there is evidence of Asian consumer preference for these labels [32–34].

"Mutton" is also a term often used in the Indian subcontinent to describe goat meat, and "Goat" was included to ensure consumers were differentiating goat from sheep mutton [35].

Other production terms included "Small portion" and "Scientific".

Eating Quality Descriptors:

Eating quality descriptors from the literature included consumer-friendly terms describing sheepmeat; "Juicy", "Succulent", "Tasty", "Fatty", "Plain" "Sweet", "Tender", "Bitter", "Crispy" and "Firm" [36–38]. In addition, "Neutral", Hot" and "Cool" were included to represent traditional Asian food classification systems. [39–41].

In order to identify the most important descriptors for each of the consumer groups, participants were asked to select the top three descriptors they would like to see on their meat as a group. They were then asked to place these descriptors onto a large target image mounted on the wall in order of importance (the centre of the target being the most important product).

2.1.6. Beef Concept Mapping

Concept phrases were developed to test a range of provenance factors with the aim of identifying the provenance attributes considered "premium" by the two consumer groups. The selected provenance (extrinsic) factors were developed from the descriptor mapping results for sheepmeat, and provenance stories already being used in the market for "premium" meat products. Concept phrases were printed individually onto photo paper cut to the same dimensions as the perceptual mapping exercise images. They were presented to the group one at a time in a prescribed order (randomised on the basis of statement complexity), and the group was asked to place these concepts onto the same axes used for the perceptual mapping exercise (Section 2.1.4). Concepts, presentation order and their associated provenance attributes are detailed in Table 1.

Table 1. Beef concepts, presentation order and their associated provenance attributes.

Order of Presentation	Concept	Provenance Factors
1	Premium pasture-fed beef from Blackmore's Wagyu, Cape Grim or Minderoo	Established Premium Australian beef Brands
2	Fresh Australian Beef	Country of origin
3	Traditional Australian breeds like Brahman or Angus	An incongruent/unfamiliar breed claim, Angus and Braham breeds do not originate from Australia
4	Certified Organic Australian Beef	Generic Australian organic beef statement
5	Raised on a small family farm grass-fed using biodiverse pastures, hormone free and sustainable farming practices	Welfare and sustainability claims
6	Aged using traditional craftmanship practices like dry aging for 35 days to tenderise and create a distinctive melt in your mouth flavour	Artisan/craftmanship and eating quality claims
7	Unique breeds like older Longhorn that have a chance to develop more flavour, with a delicate beefy flavour and a slightly acid finish without having a very high fat content	Unique breed, sustainability, health and eating quality claims
8	Lean, heart-healthy beef, raised to have monosaturated fats to lower your blood pressure and cholesterol, but still have lots of flavour	Health and eating quality claim
9	Highest quality premium meat, recommended by celebrities, and chefs as their favourite	Generic claim of premium based on celebrity endorsement without any provenance information

2.1.7. Analysis of Results

At the end of each session, transcripts were typed up by the assigned note-takers and reviewed by the research team. To reduce bias in the analyses of session outputs, the facilitators, observers and note-takers met after each session, reviewed the final set of notes, and compared the maps from each of the sessions. Observations were discussed by the group, comparisons were made between sessions, and insights were generated and validated within the research team. Insights were then collated into themes and compared across the two cultural groups.

2.2. Quantitative Sensory Methodology

2.2.1. Samples

Subsequent to the panels (at a later date and using different participants), a sensory experiment was conducted where consumers assessed the eating quality of either beef or sheepmeat. Two samples were selected for each group to provide a range of marbling (intramuscular fat) and ageing method characteristics. The samples were aged by either the conventional method of wet ageing (vacuum packed then stored in a chiller to age at 0.5–2.0 °C) or the "premium" method of dry ageing—where unpackaged meat is hung in low temperature (2–4 °C), in a humidity controlled environment (RH approx. 85%) for a defined ageing period [42]. Beef samples used were: beef striploin (longissimus lumborum wet ($n = 4$) or dry-aged ($n = 4$) for 35 days. The wet-aged beef was MSA marble score 500 and the dry-aged beef was marble score 300 [43]. Sheepmeat samples included wet-aged boneless 'lamb' loins (longissimus thoracis et lumborum $n = 6$, wet-aged for 14 days) and bone in 'mutton' loins (longissimus thoracis et lumborum, $n = 6$, dry-aged for 35 days), from Dorper sheep. All samples except for the dry-aged mutton were sourced from a specialty butcher in Melbourne, Victoria, Australia. The dry-aged mutton was sourced from a specialist dry-ageing facility located in the Adelaide Hills district of South Australia, Australia. In addition, a 'link' (assumed mid-range eating quality sample served at the commencement of every tasting session in order to familiarise the consumer with the product and protocol, results not used) was purchased from a supermarket in Melbourne, Victoria, Australia. Beef link samples were wet-aged MSA graded beef porterhouse steaks ($n = 10$) fabricated from longissimus lumborum, and for sheepmeat, the link sample was wet-aged lamb mid-loin chops ($n = 30$) fabricated from longissimus thoracis et lumborum. All samples were received at the University of Melbourne meat research centre under refrigerated conditions the day before the sensory session. Upon receipt, all samples were prepared as boneless portions, ready for grilling at the sensory session. The sheepmeat link samples and dry-aged sheepmeat samples were deboned, and the subcutaneous fat was trimmed to a similar thickness for all samples. The deboned beef and sheepmeat samples were then cut into steaks of the same thickness (2.5 cm for beef and 1.5 cm for sheepmeat). The prepared sheepmeat and beef samples and link samples were then vacuum packed and kept refrigerated at 2 °C until required for testing.

2.2.2. Consumers

Consumers ($n = 75$) were solicited as they walked past sensory booths. Upon consenting to participate in a tasting, they self-selected for either beef or sheepmeat. This self-selection strategy facilitates the participation of consumers familiar with the product being tested. Before commencing the sensory sessions, participants were provided with verbal and written instructions covering demographic survey completion, palate cleansing procedures, completion of the sensory assessment form and tasting procedures. In addition, the first tasting (a link sample, data not used) provided a training opportunity for those participants not experienced with the sensory methodology. Experienced staff checked paperwork throughout the sessions and were available to answer any participant queries. All participants completed a demographic survey based on the survey format of Hwang et al. [44] before sensory testing commenced.

2.2.3. Sensory Testing

Up to four consumers participated in each sensory session, and each consumer tasted three samples. All sessions were started with a link sample and the sample presentation (wet-aged or dry-aged) was alternated between second and third order between sessions. Cooking method was based on the MSA protocols described previously by Watson et al. [23], Thompson et al. [45] and Polkinghorne et al. [46]. In brief, samples were grilled before serving, on a preheated silex clamshell grill with the top plate set to 185 °C and the bottom plate set to 195 °C. Samples were placed on the preheated grill, the top plate closed 30 s later, and grilled until they reached a medium level of doneness (approx. 3 min to reach an internal temperature of 60 °C). Once cooked, samples were rested for 2 min (while covered with foil), cut into equal size portions (a visually estimated 20 + 5 g, based on practice sessions carried out in the laboratory), then each portion was individually placed on a plate with foil placed over the meat and immediately served. All participants were provided with water crackers and 10% apple juice/water to cleanse their palate, before tasting commenced and between samples. Upon receiving the sample, consumers were asked to remove the foil and rate odour liking (Od) and then to taste the sample. While tasting, consumers recorded their response to tenderness (T), juiciness (J), flavour (F) and overall liking (OL) using the questionnaire format described in Watson et al. [23]. Consumers indicated their liking for each of the eating quality attributes by marking on 100 mm lines (score range = 0–100) anchored with the words 'very tough/dry 'to 'very tender/juicy' and 'dislike' to 'like extremely' for flavour/odour/overall liking. In addition to the eating quality assessments, consumers were also asked to rate each sample for health and premiumness by marking a similar scale anchored with words from 'not healthy/premium' to 'very healthy/premium'. These marks were subsequently converted to an eating quality score out of 100 by measuring the distance of the mark from the point of origin of the scale in mm. After completing these assessments, they rated the quality grade of each sample by checking a box labelled "unsatisfactory", "good everyday quality", "better than everyday quality "or "premium quality".

At the end of the sensory session, consumers were also asked to indicate how much they would be willing to pay (WTP) for each quality grade by selecting a price category from 0–10, 10–20, 20–30, 30–40, 40–50, 50–60, 60–70 or 70–80 AUD (Australian dollar) per kg. In addition, immediately after selecting a price category, consumers indicated likelihood to purchase the product (LTP) at the selected price point by marking a 100 mm scale anchored with words 'not at all' to 'very likely' (score range = 0–100).

The MSA eating quality scores of SEQ (sheepmeat eating quality score) and MQ4 (meat quality combined score for beef) were determined as described below according to Pethick [47] and MLA [48], respectively.

$$SEQ = 0.3\ (T) + 0.1\ (J) + 0.3\ (F) + 0.3\ (OL),$$

$$MQ4 = 0.3\ (T) + 0.1\ (J) + 0.3\ (F) + 0.3\ (OL),$$

where T = tenderness score out of 100, J = juiciness score out of 100, F = flavour liking score out of 100 and OL = overall liking score out of 100.

2.2.4. Statistical Analysis

Statistical analyses of sensory test data were performed using REML in GenStat for Windows (16th Edition, VSN International, Hemel Hempstead, UK). Sheepmeat and beef data were analysed separately. The factors in the initial model were ageing method (wet or dry), and the random model included consumer ID. The effect of demographic data on sensory scores was investigated for each demographic factor. After visual inspection of the consumer age vs. sensory data using box plots (data not shown), two consumer clusters (differing according to their sensory scores) were identified. The data were then segmented into these two age categories being 18–30 years old (29% of data) and 31–70 years old (71% of data) to reflect the clusters and enable a comparison between younger and older consumers. The final parsimonious model for each species only included consumer age (18–30 years vs. 31–70 years) and the random model included consumer ID.

3. Results

3.1. Qualitative Results

3.1.1. Perceptual Mapping of Sheepmeat and Beef

On reviewing the session transcripts and maps for sheepmeat, it was revealed that aside from the factors included in the experimental design, i.e., familiarity, premiumness, cut, colour, fat content, trimming, etc., there were a number of extrinsic factors, such as perceived tenderness, convenience, labelling, value for money and eating occasion, that consumers utilised when assessing the "premiumness" of a sheepmeat product. These factors, along with the design of experiment factors were captured as themes and are detailed in Supplementary C, Table SC1, along with notes on how each of the groups assessed the factors. Figure SC1 contains exemplars of the Australian and Asian consumer maps for sheepmeat. In general, the themes developed from the sheepmeat perceptual mapping exercise were also found in the beef perceptual mapping exercise, therefore the results were tabulated in a similar manner to enable comparison between the sessions (Table SC2 of Supplementary C). Figure SC2 provides exemplary maps from the beef perceptual mapping sessions for Australian and Asian consumer groups.

The perceptual mapping results were combined with the sheepmeat target exercise results and the beef concept results to generate the insights described in Section 3.3.

3.1.2. Sheepmeat Descriptor Mapping Results

Table 2 summarises the results of the sheepmeat descriptor mapping. For the two Australian groups, "Australian" was the most important descriptor for the preferred sheepmeat product, demonstrating the importance of a product's origin for Australian consumers. Group 1 then selected "Spring lamb" followed by "Tender/fresh" (they were not able to reach consensus on the third most important factor), while group 2 selected "Organic" followed by "Tasty". Both Asian groups selected "Fresh" in their top three descriptors (most important for group 1 and second most important for group 2). They also selected "Tender", "Tasty" and "Organic" and in common with the Australian groups, "Aged" was only selected by group 2 and when questioned whether a product could be aged and fresh, they were clear it could be; later conversations with these consumers revealed the term fresh was used as opposed to frozen.

Table 2. Top three product descriptors selected by the panel in order of preference, (selected from 28) descriptors for sheepmeat.

Order of Preference *	Australian Group 1	Australian Group 2	Asian Group 1	Asian Group 2
1	Australian	Australian	Fresh	Organic
2	Spring Lamb	Organic	Tender	Fresh
3	Tender/fresh	Tasty	Tasty	Aged

* 1 = most important, 2 = second most important, 3 = third most important.

3.1.3. Concept Phrase Testing Results for Beef

The concept mapping results for beef are detailed in Supplementary C Table SC3. While both groups mapped "Fresh Australian Beef" into the familiar everyday quadrant, there were marked contrasts in how Asian and Australians mapped the remaining concepts. For the Asian group, all other concepts were unfamiliar but mapped as premium. For Australians, the deliberately unfamiliar concept "Traditional Australian breeds like Brahman or Angus" was mapped into the unfamiliar everyday quadrant, but the remaining concepts were mapped in the familiar premium quadrant. It should also be noted that the mapping by Australian consumers was a slower process than for the Asian consumers, due to the amount of discussion in the Australian group regarding the validity of each claim.

3.2. Insights Generated from Qualitative Assessments of Sheepmeat and Beef

As hypothesised, it was found that Australians and Asians assess visual meat quality attributes and concepts differently; for instance, Australians rely very heavily on labelling/cut identification to assess a product's eating quality, and they showed a preference for locally produced product and a degree of skepticism concerning the premiumness of unfamiliar cuts, most often mapping unfamiliar cuts to the "everyday" side of the map. The Asian group, on the other hand, were comfortable choosing meat for home use based on visual quality cues, mapping meat without labels relatively easily and plotting unfamiliar cuts as premium based only on visual quality cues.

The Australian requirement for labels could be attributed in part to home usage practices; during this study, it was found Australians tend to use meat as the centrepiece of a meal with a range of cooking methods utilised at home; e.g., high-temperature, short duration cooking methods such as grilling or BBQ, medium temperatures, medium time methods such as oven roasting and low temperature, long time cooking such as braising and slow cooking. This range of cooking methods generates different eating experiences with varying cuts of meat and getting it wrong could lead to an unsatisfactory eating experience. For instance, oven roasting beef round steak produces a very tough and probably inedible dish, so this cut is better suited to low temperature long time cooking methods. The fear of getting it wrong, combined with the common Australian retail practice of providing cooking instructions, has created a label dependency for this group. For the Asian group, purchase of meat was traditionally at the "wet market" and based on visual quality cues (not labels), and at home preparation typically involved incorporation of small pieces of meat into a complex dish, usually via a stovetop cooking method (e.g., stir fry or slow-cooked curry) with tenderness determined by format (such as thin slicing) and cooking time.

3.2.1. The Channel for Premium

The differences in cooking styles for the two groups raised questions around premiumisation strategies. When meat is a centre of the plate "hero" ingredient, it is relatively easy to understand how a high-quality luxury eating experience might be delivered by a special piece of meat (e.g., dry-aged or wagyu). However, where meat is part of a complicated dish, how does the "premium" quality of the meat stand out? Inquiries into home vs. restaurant use highlighted that the channels for "premium" meat might differ for the two groups. For instance, Australians (at least those who are experienced home cooks) are comfortable purchasing premium meat from high end butchers and preparing it at home, especially for special occasions, whereas many Asians (not so familiar with Australian cooking styles) and those Australians who are not experienced cooks would prefer these premium eating experiences happen in the restaurant, (where professionals have taken on the responsibility of preparing these costly cuts of meat and created a whole environment for enjoying the eating experience). It was also noted that the butcher plays a special role in the selection of premium meat for Australian consumers, many describing discussions with the butcher where they seek advice on meat selection and the best way of preparing the meat at home.

3.2.2. Labelling Requirements for Home Use and Retail

Australians' strong preference for products labelled as "Australian" and their need for labelling to enable cut identification, both suggest labels identifying the cut and recommended cooking methods are a minimum requirement for Australian retail, and preference would be given by consumers to those products that are labelled "Australian".

For Asian consumers, labelling was useful, but its absence was not a barrier to potential purchase. While it was clear that being labelled as Australian was important for Australian consumers, it was equally clear this was not as important to the Asian consumer. For this group, an "Australian" label would not drive the perception of premiumness. The unfamiliar beef concepts covering provenance factors of health, brand, uniqueness craftmanship and eating quality were all well received by this group and would enhance the premiumness of the meat. These results suggest there is considerable scope to leverage stories of provenance for the export of red meat from Australia to Asia. Australian consumers displayed a more skeptical attitude to the beef concepts, mapping familiar concepts to premium and unfamiliar concepts to everyday. From the group discussion, it was clear that claims needed to be validated by Australian consumers before acceptance.

3.2.3. Response to Cut, Colour, Bone and Fat Content

The two groups placed different levels of importance on the way the meat was cut; Asians preferred to cut meat to the desired format at home while Australians preferred convenient cuts, such as mince, strips and cubes, for cooking on weeknights and special cuts for weekends and entertaining. The acceptability of bone content also differed between the groups, with Australians in general considering bone as waste, while Asians who selected meat cuts with bone were often planning to use the bone in soup or stock. There also appeared to be differences in what was considered acceptable fat content for the two groups, with Australians expressing a preference for leaner meat and actively avoiding visible fat for health reasons. The Asian group differentiated subcutaneous fat and intramuscular fat (IMF), and too much subcutaneous fat was undesirable but high IMF content was related to a premium eating experience.

The Asian consumers appeared more sensitive to the colour of the stimuli images than the Australian consumers, and they used it as an indicator of meat quality. Dark brick red meat was described as organic and fresh, while grey or pale tones indicated the meat was not fresh. From the sheepmeat descriptor mapping results (Table 2), freshness was identified as an important attribute for Asians, and the term fresh was used in contrast to frozen. Interestingly, the Asian consumers reported that meat could be aged and still considered "fresh". The undesirability of frozen product poses a considerable barrier for export of meat from Australia to Asia, as short shelf life combined with long distances has made freezing a necessity.

It was also observed that both groups, when assessing sheepmeat, tended to extrapolate expectations of tenderness and flavour from previous experience with beef. For example, dry-aged sheepmeat is a novel product in the Australian market, and while neither group had experience with this product, both assessed the dry-aged sheepmeat image as premium.

3.3. Quantitative Result

3.3.1. Demographics

For the entire consumer group ($n = 75$), the gender split was males 45%, females 55%; 0.8% of the group fell into the 18 to 19 year old category, 10% into the 20–25 year old category, 12% into the 26–30 year old category, 12% in the 31-39 year old category, 46% into the 40–60 year old category, and 11% in the 61–70 year old category. For the purposes of statistical analysis, the age categories were condensed into two categories of 18–30 and 31–70 years old which represented 29% and 71% of the consumers, respectively. Most consumers described their cultural heritage as Australian (70%), followed by British descent (12%), Asian descent (8%), European descent (4%), and Other (5%). The demographic distribution was similar between the beef and sheep tasting groups, and detailed demographic data can be found in Supplementary D.

3.3.2. Eating Quality, Healthiness and Premiumness

When the data were analysed separately for each species for the effect of ageing method, the sheepmeat sensory data had no significant differences between the two samples (wet- and dry-aged; $p > 0.05$) for any of the eating quality parameters (tenderness, overall liking, flavour, juiciness, odour liking, SEQ), or for healthiness and premiumness (data not reported). For beef, the wet-aged samples were more tender than the dry-aged samples ($p = 0.022$, data not presented) with no other sensory traits being different between wet- and dry-aged beef ($p > 0.05$).

As no significant effects were found due to ageing method for sheepmeat and only for tenderness of beef, the data were further interrogated a priori for indications of consumer segmentation. Due to location challenges, an Asian cohort of appropriate size was not available for the sensory sessions, so analysis of cultural heritage effects was not possible. Other demographic effects were investigated and are described below.

For sheepmeat, it was found that females scored tenderness higher and tended to rate premiumness higher than males ($p = 0.028$ and $p = 0.058$ respectively, data not presented). Consumer age also affected consumer scores or sheepmeat for tenderness, juiciness, odour liking, SEQ, and healthiness, and tended to affect scores for overall liking, flavour, and premiumness, with older consumers (31–70 years) scoring these qualities higher than younger consumers (18–30 years), (Table 3). No significant age category effects were found for beef ($p > 0.05$), (Table 3).

Table 3. Effect of consumer age category (18–30 or 31–70 years) on consumer scores for eating quality (tenderness, overall liking, flavour, juiciness, odour liking and MQ4/SEQ), healthiness, and premiumness according to meat species (beef, sheep).

Attribute	Sheepmeat				Beef			
	18–30 years	31–70 years	*p*-Value	SED	18–30 years	31–70 years	*p*-Value	SED
Tenderness	66.7	77.9	0.042	5.28	77.5	70.7	0.202	5.28
Overall liking	69.2	78.0	0.053	4.40	75.8	71.8	0.334	4.17
Flavour	66.9	77.1	0.057	5.22	78.0	69.5	0.112	5.25
Juiciness	70.5	79.5	0.029	3.97	78.3	70.8	0.154	5.13
Odour liking	55.8	74.6	0.005	6.27	72.6	68.7	0.522	6.06
MQ4/SEQ *	67.9	77.9	0.026	4.28	77.2	70.8	0.156	4.43
Healthiness	60.1	75.9	0.004	5.10	75.1	705	0.339	4.69
Premiumness	68.8	75.8	0.081	3.89	75.8	69.6	0.227	5.00

Consumer scores ranged from 0–100; the higher the value, the more the attribute was appreciated. SED is the standard error of differences. * MQ4 = Meat quality combined score for beef. SEQ = Sheepmeat eating quality score. See Section 2.2.3 for calculation of both.

3.3.3. Quality Gradings, WTP (Willingness to Pay) and LTP (Likelihood to Purchase)

During the sensory sessions consumers were asked to select a quality grade ("unsatisfactory", "good everyday quality", "better than everyday quality" or "premium quality") for each sample tasted. At the end of the sensory session, willingness and likelihood to purchase for each of the quality grades were indicated by all participants (these results are summarised in Table 4). Wet-aged beef was most likely to be rated as "better than everyday quality", while the dry-aged beef was most likely to be rated as "good everyday quality", and a similar pattern was observed for the sheepmeat samples. Consumers were willing to pay, on average, up to 50–60 AUD per kg for premium quality beef and 30–40 AUD per kg for premium quality sheepmeat, with prices decreasing with quality grade. Likelihood to purchase data indicated that consumers were less likely to purchase unsatisfactory meat even if it was priced considerably lower than higher quality meat, and this was especially evident for sheepmeat where mean consumer likelihood estimates for unsatisfactory meat were only 16% compared to 32% for beef.

The WTP and LTP consumer data for each of the quality grades were also segmented by age using the categories 18–30 and 31–70 years, and differences were found for the age groups (Figure 1). Older consumers were more likely to rate sheepmeat as a higher quality grade than younger consumers, with older consumers most frequently selecting "better than everyday quality", and "premium quality", while younger consumers most frequently selected "good everyday quality". This effect was not found in the beef data. Older consumers (31–70 years) also indicated they would pay less than the younger consumers (18–30 years), for "premium quality" meat (30–40 AUD per kg vs. 40–50 AUD per kg, respectively, for sheepmeat and 40–50 AUD per kg vs. 50–60 AUD per kg, respectively, for beef) (Figure 1). The older consumers also differed from the younger consumers in WTP for "unsatisfactory" graded sheepmeat, with older consumers willing to pay ~10 AUD per kg less than younger consumers (0–10 AUD per kg vs. 10–20 AUD per kg, respectively). The age groups did not differ for WTP for "good everyday quality" and "better than everyday quality" grades for both sheepmeat and beef. Likelihood to pay results also indicated some differences between the two age groups, with younger consumers less likely to purchase "unsatisfactory" grade sheepmeat than older consumers (12% likelihood vs. 20% likelihood).

Table 4. Frequency of selecting a quality grade ("unsatisfactory", "good everyday quality", "better than everyday quality", or "premium quality") within each meat species (sheep or beef) according to ageing method (wet or dry) and associated willingness to pay data, indicated by median price category (0–10, 10–20, 20–20, 30–40, 40–50, 50–60 or 60–70, Australian dolllars (AUD) per kg) and the average likelihood of purchasing.

Meat Species	Quality Grade	Relative Frequency of Quality Grade Selection (%)		Median Price Category (AUD Per kg)	Average Likelihood of Purchasing (%)
		Dry-aged	Wet-aged		
Sheep	Unsatisfactory	0	3	0–10	16
	Good everyday quality	47	22	20–30	53
	Better than everyday quality	33	56	30–40	53
	Premium quality	19	19	30–40	58
Beef	Unsatisfactory	3	8	10–20	32
	Good everyday quality	42	21	20–30	58
	Better than everyday quality	25	44	30–40	66
	Premium quality	31	28	50–60	67

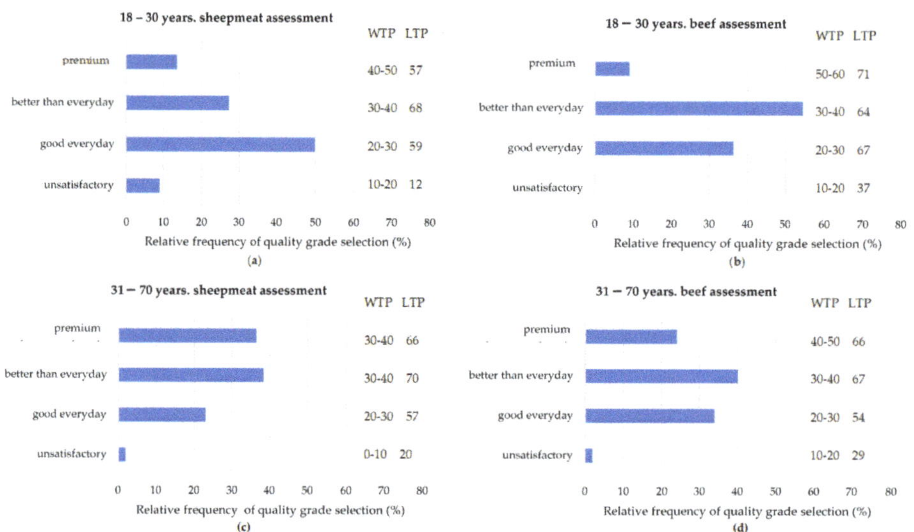

Figure 1. Relative frequency of quality grade selection (%, "premium", better than everyday", "good everyday", "unsatisfactory"), willingness to pay (WTP; in AUD per kg.) and likelihood to purchase (LTP; %) for each quality grade split on age category and species. (**a**) 18–30 years, sheepmeat assessment; (**b**) 18–30 years, beef assessment; (**c**) 31–70 years, sheepmeat assessment; (**d**) 31–70 years, beef assessment.

4. Discussion

Consumers are not a single homogenous group, and there is a substantial body of work demonstrating segmentation within consumer study populations. These segments exhibit differences in attitudinal and sensory responses to red meat [6,8,49–57]. Prescott and Bell [6] proposed that consumer preference is the result of food experience and this, in turn, is determined by cultural heritage. Indeed, Font-I-Furnols et al. [56] found that while Spanish consumers generally prefer the meat from lighter lambs, a small minority of Spanish consumers demonstrated a clear preference for meat from heavier lambs. German and British consumers were found to prefer meat from heavier lambs [53].

In this study, qualitative research revealed underlying differences in Asian and Australian consumer responses to red meat and these differences between cultures were similar for the two meat species investigated. For instance, Australian consumers' preference for "Australian" labels was not reflected by their Asian counterparts, nor did Asians exhibit the Australians need to know the cut identity before they could assess eating quality.

The two groups also responded differently to unfamiliar cuts and concepts. Australians tended to move the unfamiliar into the everyday space, and in general were more skeptical of the unfamiliar, whereas the Asian group tended to map unfamiliar cuts and concepts into the premium space. This was especially apparent with the beef concepts; most were unfamiliar for the Asian group, but this presented no barrier to mapping them as premium. These results are in contradiction to previous views that unfamiliarity can be a barrier to consumer acceptance [28,30], thus highlighting the importance of investigations on preferences of specific groups of consumers and products. Ashman [58], based on the work of Von Hippel [59], reported that Australians are more conservative than Chinese consumers when considering new food products and estimated the 'lead user' population of Australia was only 12% compared to the Chinese 'lead user' population of 47%, which may explain some of the differences observed in this study. Furthermore, visual meat quality cues were found to be a potential barrier for Asian consumers. Pale colours were associated with frozen products, which had a lower acceptability. The Australian consumers in this study did not demonstrate the same sensitivity to colour.

While there were overlaps in what constitutes "premiumness" for the different cultures, there was evidence that the premium experience may be accessed via different channels with Asians going to the restaurant to experience premium meat as a professionally prepared hero ingredient. Although Australians also go to the restaurant, they are more likely to purchase expensive premium products for special occasions from the butcher, to cook at home. From a marketing perspective, the data suggest that both butchers and foodservice professionals play an important role in delivering a "premium" meat experience/product for the Australian market, while foodservice facilities specialising in meat (as the hero ingredient) should be used to deliver the premium experience for Asian consumers.

Concept testing results indicated that when marketing products for Asian export markets, information indicating craftsmanship, rarity and personal quality (e.g., "small family farm produced") provides a compelling story for these consumers and significantly enhances the perception of premium.

Quantitative data indicated no significant differences in consumer eating quality, healthiness and premiumness scores for sheepmeat samples, and in the case of beef, only tenderness was different with the wet-aged sample (marble score = 500), preferred over the dry-aged sample (marble score = 300), likely because marbling increases meat tenderness and hence consumer tenderness scores [60].

The a priori segmentation analysis of the quantitative sheepmeat data indicated female consumers gave higher scores for sheepmeat tenderness and tended to score premiumness higher than male consumers. Older consumers (31–70 years) also scored sheepmeat higher than the younger consumers (18–30 years) for tenderness, juiciness, odour liking and health, while no effects were found for beef. There is evidence of differences in the drivers of liking according to meat species, with tenderness identified as the most important driver for beef, and flavour the most important driver for sheepmeat [61]. Sheepmeat has an intense flavour, and the aroma/flavour of sheepmeat can be a barrier to consumer acceptance, especially when unfamiliar, and thus consumers are not habituated [7,9]. It is proposed the sheepmeat assessments found in this study reflect the polarising nature of the consumers response to its aroma/flavour. Possibly, younger consumers were not as accustomed as the older consumers to sheepmeat. Watkins et al. [9] reported that consumers need to be habituated to eating sheepmeat in order to like it. This effect was also reflected in the quality grades assigned to sheepmeat by the two different age groups; older consumers (31–70 years) were more likely to rate sheepmeat as "better than everyday quality" and "premium quality", while younger consumers (18–30 years) were more likely to rate it one grade lower, as "good everyday quality".

Additionally, the a priori segmentation of WTP data indicates that the older consumer group was not prepared to pay as much as the younger consumer group for "premium quality" meat, 30–40 AUD per kg vs. 40–50 AUD per kg for sheepmeat, and 40–50 AUD per kg vs. 50–60 AUD per kg for beef.

Similar gender and age effects to those found in this study have been reported previously. Kubberod et al. [55] demonstrated that young females preferred meat with a low intensity flavour and aroma compared to their male counterparts. Lyford et al. [8] reported that younger consumers (age range 25–35 years) in Australia, USA, Japan and Ireland were willing to pay more than older consumers for quality meat as did, Strydom et al. [62], in a study of South African consumers.

A final observation on the consumer sampling utilised for both the qualitative and quantitative panels was that it led to underrepresentation of some demographic groups. For example, as discussed above, the Asian cohort was not large enough to enable a cross-cultural comparison in the quantitative sensory sessions. Targeted consumer recruitment according to demographic profiles would enhance the mixed method approach described.

5. Conclusions

Coupling qualitative and quantitative methods provided an effective approach to investigate consumer preference for different meat products. This combined approach was able to differentiate consumer preferences from cultural and demographic perspectives while providing valuable insights into the underlying reasons for the differences. These insights can be used to inform product positioning strategies and to increase the competitiveness of Australian meats in both domestic and overseas markets.

Qualitative research demonstrated that factors influencing acceptability and rating of meat products differed between Australian and Asian consumers. These factors include colour, labelling information (including product quality, origin and craftmanship), domestic utility, bone and fat content, familiarity and access channel. Quantitative sensory analysis demonstrated that consumers were not able to differentiate the wet- and dry-aged sheepmeat samples for eating quality, premiumness or healthiness while the wet- and dry-aged beef samples were only differentiated for tenderness. A priori segmentation analysis of the demographic suggests gender and age may affect acceptability of sheepmeat among Australian consumers, and further study is warranted.

The mixed method approach described could be further improved by the use of stratified consumer populations that reflect target market demographics, especially as Asia encompasses a number of large markets with diverse religious backgrounds and cuisines. This study has highlighted the diversity of Australia's local and export market consumers and the need for further targeted investigations on consumer response to Australian meat products.

Supplementary Materials: The following are available online at http://www.mdpi.com/2304-8158/9/2/126/s1, Supplementary A. Discussion guides, including Table SA1. Sheepmeat familiarity, Table SA2. Beef familiarity. Supplementary B. Stimulus design and source, including Table SB1. Stimulus descriptions and attributes design of experiment for sheepmeat, Table SB2. Sheepmeat descriptions and source, Table SB3. Stimulus descriptions and attribute design of experiment for beef, Table SB4. Beef stimulus descriptions and source. Supplementary C. Sheepmeat and beef perceptual mapping results, including Figure SC1. Exemplar maps from sheepmeat mapping exercise, Figure SC2. Exemplar map from group mapping beef exercise, Table SC1. Themes and group responses for sheepmeat perceptual mapping, Table SC2. Themes and group responses for beef perceptual mapping, Table SC3. Australian vs. Asian mapping of beef concepts in the quadrants of unfamiliar everyday, familiar everyday, unfamiliar premium, and familiar premium. Supplementary D. Demographic summary for sensory testing, including Table SD1. Number of participants per demographic category according to the meat species tested (sheepmeat or beef).

Author Contributions: Conceptualisation, M.H. (Melindee Hastie), H.A., M.H. (Minh Ha); Methodology, M.H. (Melindee Hastie), H.A., M.H. (Minh Ha); Formal analysis, M.H. (Melindee Hastie), R.W., M.H. (Minh Ha), D.T.; Resources, M.H. (Melindee Hastie), M.H. (Minh Ha), H.A., R.W.; Data curation, M.H. (Melindee Hastie); Writing—original draft preparation, M.H. (Melindee Hastie); Writing—review and editing, M.H. (Melindee Hastie), H.A., M.H. (Minh Ha), D.T., R.W., Visualisation, M.H. (Melindee Hastie); supervision, R.W., H.A., D.T.; Project administration, R.W.; Funding acquisition, R.W. All authors have read and agreed to the published version of the manuscript.

Funding: This research was funded by Faculty of Veterinary and Agricultural Science, University of Melbourne, Australia, Meat and Livestock Australia and the Western Australian Department of Primary Industries and Regional Development as part of the project reference; P.PSH0863, Dry ageing Australian sheep meat for potential markets, technical guidelines and financial benefits.

Acknowledgments: The authors would like to acknowledge the assistance of colleagues including Allison McNamara, Colette Day and Evan Bittner for facilitating the QMA sessions and Colette Day and Behannis Mena for assistance with the sensory sessions.

References

1. Rees, C.; Mullumby, J. Trends in Australian meat consumption. *Agric. Commod.* **2017**, *7*, 82–85.
2. MLA. Market Snapshots - Beef. Available online: https://www.mla.com.au/globalassets/mla-corporate/prices--markets/documents/os-markets/red-meat-market-snapshots/2018-mla-ms_australia_beef.pdf (accessed on 22 October 2019).
3. MLA. Global Market Snapshot - Sheepmeat. Available online: https://www.mla.com.au/globalassets/mla-corporate/prices--markets/documents/os-markets/export-statistics/oct-2018-snapshots/all-sheepmeat-markets-snapshots-oct2018.pdf (accessed on 22 October 2019).
4. MLA. Mla's Spotlight on Southern Asia. Available online: https://www.mla.com.au/news-and-events/industry-news/mlas-southern-asia-focus/ (accessed on 22 October 2019).
5. Bittner, E.P.; Ashman, H.; Hastie, M.; van Barneveld, R.J.; Hearn, A.H.; Thomson, N.; Dunshea, F.R. Innovation in an expanding market: Australian pork is not a commodity. *Anim. Prod. Sci.* **2017**, *57*, 2339–2344. [CrossRef]
6. Prescott, J.; Bell, G. Cross-cultural determinants of food acceptability: Recent research on sensory perceptions and preferences. *Trends Food Sci. Technol.* **1995**, *6*, 201–205. [CrossRef]
7. Prescott, J.; Young, O.; O'Neill, L. The impact of variations in flavour compounds on meat acceptability: A comparison of Japanese and New Zealand consumers. *Food Qual. Prefer.* **2001**, *12*, 257–264. [CrossRef]
8. Lyford, C.; Thompson, J.; Polkinghorne, R.; Miller, M.; Nishimura, T.; Neath, K.; Allen, P.; Belasco, E. Is willingness to pay (wtp) for beef quality grades affected by consumer demographics and meat consumption preferences? *Australas. Agribus. Rev.* **2010**, *18*, 1.
9. Watkins, P.J.; Frank, D.; Singh, T.K.; Young, O.A.; Warner, R.D. Sheepmeat flavor and the effect of different feeding systems: A review. *J. Agric. Food Chem.* **2013**, *61*, 3561–3579. [CrossRef] [PubMed]
10. Van Trijp, H.C.M.; Punter, P.H.; Mickartz, F.; Kruithof, L. The quest for the ideal product: Comparing different methods and approaches. *Food Qual. Prefer.* **2007**, *18*, 729–740. [CrossRef]
11. Grunert, K.G. What's in a steak? A cross-cultural study on the quality perception of beef. *Food Qual. Prefer.* **1997**, *8*, 157–174. [CrossRef]
12. Steenkamp, J.-B.E.M.; van Trijp, H.C.M. Quality guidance: A consumer-based approach to food quality improvement using partial least squares. *Eur. Rev. Agric. Econom.* **1996**, *23*, 195–215. [CrossRef]
13. Costa, A.I.A.; Dekker, M.; Jongen, W.M.F. An overview of means-end theory: Potential application in consumer-oriented food product design. *Trends Food Sci. Technol.* **2004**, *15*, 403–415. [CrossRef]
14. Bredahl, L.; Grunert, K.G.; Fertin, C. Relating consumer perceptions of pork quality to physical product characteristics. *Food Qual. Prefer.* **1998**, *9*, 273–281. [CrossRef]
15. Font-I-Furnols, M.; Guerrero, L. Consumer preference, behavior and perception about meat and meat products: An overview. *Meat Sci.* **2014**, *98*, 361–371. [CrossRef]
16. Grunert, K.G. The common ground between sensory and consumer science. *Curr. Opin. Food Sci.* **2015**, *3*, 19–22. [CrossRef]
17. Hoppert, K.; Mai, R.; Zahn, S.; Hoffmann, S.; Rohm, H. Integrating sensory evaluation in adaptive conjoint analysis to elaborate the conflicting influence of intrinsic and extrinsic attributes on food choice. *Appetite* **2012**, *59*, 949–955. [CrossRef]
18. Schifferstein, H.N. Employing consumer research for creating new and engaging food experiences in a changing world. *Curr. Opin. Food Sci.* **2015**, *3*, 27–32. [CrossRef]
19. Dransfield, E.; Ngapo, T.M.; Nielsen, N.A.; Bredahl, L.; Sjoden, P.O.; Magnusson, M.; Campo, M.M.; Nute, G.R. Consumer choice and suggested price for pork as influenced by its appearance, taste and information concerning country of origin and organic pig production. *Meat Sci.* **2005**, *69*, 61–70. [CrossRef]
20. Beckley, J.H.; Paredes, M.D.; Lopetcharat, K. *Product Innovation Toolbox: A Field Guide to Consumer Understanding and Research*; Wiley-Blackwell: Hoboken, NJ, USA, 2012.

21. Nestrud, M.A.; Lawless, H.T. Perceptual mapping of citrus juices using projective mapping and profiling data from culinary professionals and consumers. *Food Qual. Prefer.* **2008**, *19*, 431–438. [CrossRef]

22. Risvik, E.; McEwan, J.A.; Colwill, J.S.; Rogers, R.; Lyon, D.H. Projective mapping: A tool for sensory analysis and consumer research. *Food Qual. Prefer.* **1994**, *5*, 263–269. [CrossRef]

23. Watson, R.; Gee, A.; Polkinghorne, R.; Porter, M. Consumer assessment of eating quality - development of protocols for meat standards australia (MSA) testing. *Aust. J. Exp. Agric.* **2008**, *48*, 1360–1367. [CrossRef]

24. Dolnicar, S.; Grün, B.; Leisch, F. Market segmentation analysis. In *Market Segmentation Analysis; Understanding It, Doing It, Making It Useful*; Springer: Singapore, 2018; pp. 11–22.

25. Meat Notice 2019-01—Amended Ovine Definition in the Export Control (Meat and Meat Products) Orders 2005. Available online: http://www.agriculture.gov.au/export/controlled-goods/meat/elmer-3/notices/2019/mn19-01 (accessed on 18 November 2019).

26. Ausmeat Sheepmeat Language. Available online: https://www.ausmeat.com.au/WebDocuments/SheepMeat_Language.pdf (accessed on 16 December 2019).

27. Borgogno, M.; Favotto, S.; Corazzin, M.; Cardello, A.V.; Piasentier, E. The role of product familiarity and consumer involvement on liking and perceptions of fresh meat. *Food Qual. Prefer.* **2015**, *44*, 139–147. [CrossRef]

28. Deliza, R.; MacFie, H.J.H. The generation of sensory expectation by external cues and its effect on sensory perception and hedonic ratings: A review. *J. Sens. Stud.* **1996**, *11*, 103–128. [CrossRef]

29. Torrico, D.D.; Fuentes, S.; Gonzalez Viejo, C.; Ashman, H.; Dunshea, F.R. Cross-cultural effects of food product familiarity on sensory acceptability and non-invasive physiological responses of consumers. *Food Res. Int.* **2019**, *115*, 439–450. [CrossRef] [PubMed]

30. Tuorila, H.; Hartmann, C. Consumer responses to novel and unfamiliar foods. *Curr. Opin. Food Sci.* **2020**, *33*, 1–8. [CrossRef]

31. East, I.J.; Foreman, I. The structure, dynamics and movement patterns of the Australian sheep industry. *Aust. Vet. J.* **2011**, *89*, 477–489. [CrossRef] [PubMed]

32. Xiaohua, Y.; Zhifeng, G.; Yinchu, Z. Willingness to pay for the "green food" in China. *Food Policy* **2014**, *45*, 80–87.

33. Paul, J. The greening of china's food - green food, organic food, and eco-labeling. In Proceedings of the Sustainable Consumption and Alternative Agri-Food Systems Conference, Liege University, Arlon, Begium, 27–30 May 2008.

34. Ecolabel Index. Available online: http://www.ecolabelindex.com/ecolabels/?st=region,asia#G (accessed on 18 November 2019).

35. Mutton. (n.d.) In Oxford English Dictionary Online (Electronic Resource). Available online: https://www.oed.com (accessed on 22 October 2019).

36. Jeremiah, L.E.; Tong, A.K.W.; Gibson, L.L. Influence of lamb chronological age, slaughter weight, and gender. Flavor and texture profiles. *Food Res. Int.* **1998**, *31*, 227–242. [CrossRef]

37. Frank, D.; Raeside, M.; Behrendt, R.; Krishnamurthy, R.; Piyasiri, U.; Rose, G.; Watkins, P.; Warner, R. An integrated sensory, consumer and olfactometry study evaluating the effects of rearing system and diet on flavour characteristics of Australian lamb. *Anim. Prod. Sci.* **2016**, *57*, 347–362. [CrossRef]

38. Munoz, A.M. Consumer perceptions of meat. Understanding these results through descriptive analysis. *Meat Sci.* **1998**, *49*, S287–S295. [CrossRef]

39. Tan, S.P.; Wheeler, E. Concepts relating to health and food held by Chinese women in London. *Ecol. Food Nutr.* **1983**, *13*, 37–49. [CrossRef]

40. Anderson, E.N. 'Heating' and 'cooling' foods in Hong Kong and Taiwan. *Social Anthr.f Food* **1980**, *19*, 237–268. [CrossRef]

41. Khare, R.S.; Rao, M.S.A. *Food, Society, and Culture: Aspects in South Asian Food Systems*; Carolina Academic Press: Durham, NC, USA, 1986.

42. Savell, J.W. Dry-Aging of Beef. Center for Research and Knowledge Management, National Cattlemen's Beef Association, Centennial, CO. 2008. Available online: https://www.beefresearch.org/CMDocs/BeefResearch/Dry%20Aging%20of%20Beef.pdf (accessed on 24 January 2020).

43. Ausmeat. *Handbook of Australian Meat*, 7th ed.; Ausmeat: South Brisbane, Australia, 2005.

44. Hwang, I.H.; Polkinghorne, R.; Lee, J.M.; Thompson, J.M. Demographic and design effects on beef sensory scores given by Korean and Australian consumers. *Aust. J. Exp. Agric.* **2008**, *4*, 1387–1395. [CrossRef]

45. Thompson, J.M.; Gee, A.; Hopkins, D.L.; Pethick, D.W.; Baud, S.R.; O'Halloran, W.J. Development of a sensory protocol for testing palatability of sheep meats. *Aus. J. Exp. Agric.* **2005**, *45*, 469–476. [CrossRef]

46. Polkinghorne, R.; Thompson, J.M.; Watson, R.; Gee, A.; Porter, M. Evolution of the meat standards Australia (MSA) beef grading system. *Aus. J. Exp. Agric.* **2008**, *48*, 1351–1359. [CrossRef]

47. Pethick, D. Personal communication. Murdoch University, Murdoch, WA 6150, Australia, 2017.

48. MLA. Meat Standards Australia Beef Information Kit. Available online: https://www.mla.com.au/globalassets/mla-corporate/marketing-beef-and-lamb/documents/meat-standards-australia/msa-beef-tt_full-info-kit-lr.pdf (accessed on 23 January 2020).

49. Verbeke, W.; Vackier, I. Profile and effects of consumer involvement in fresh meat. *Meat Sci.* **2004**, *67*, 159–168. [CrossRef] [PubMed]

50. Gracia, A.; de-Magistris, T. Preferences for lamb meat: A choice experiment for Spanish consumers. *Meat Sci.* **2013**, *95*, 396–402. [CrossRef] [PubMed]

51. Font I Furnols, M.; Tous, N.; Esteve-Garcia, E.; Gispert, M. Do all the consumers accept marbling in the same way? The relationship between eating and visual acceptability of pork with different intramuscular fat content. *Meat Sci.* **2012**, *91*, 448–453. [CrossRef]

52. Malek, L.; Umberger, W.J.; Rolfe, J. Segmentation of Australian meat consumers on the basis of attitudes regarding farm animal welfare and the environmental impact of meat production. *Anim. Prod. Sci.* **2018**, *58*, 424–434. [CrossRef]

53. Bernués, A.; Ripoll, G.; Panea, B. Consumer segmentation based on convenience orientation and attitudes towards quality attributes of lamb meat. *Food Qual. Prefer.* **2012**, *26*, 211–220. [CrossRef]

54. Bernues, A.; Olaizola, A.; Corcoran, K. Extrinsic attributes of red meat as indicators of quality in Europe: An application for market segmentation. *Food Qual. Prefer.* **2003**, *14*, 265–276. [CrossRef]

55. Kubberod, E.; Ueland, O.; Rodbotten, M.; Westad, F.; Risvik, E. Gender specific preferences and attitudes towards meat. *Food Qual. Prefer.* **2002**, *13*, 285–294. [CrossRef]

56. Font I Furnols, M.; San-Julian, R.; Guerrero, L.; Sanudo, C.; Campo, M.M.; Olleta, J.L.; Oliver, M.A.; Caneque, V.; Alvarez, I.; Diaz, M.T.; et al. Acceptability of lamb meat from different producing systems and ageing time to German, Spanish and British consumers. *Meat Sci.* **2006**, *72*, 545–554. [CrossRef] [PubMed]

57. Kayser, M.; Nitzko, S.; Spiller, A. Analysis of differences in meat consumption patterns. *Int. Food Agribus. Manag. Rev.* **2013**, *16*, 43–56.

58. Ashman, H. Industry transformation and making it work: A case study; unlocking the food value chain: Australian food industry transformation. In Proceedings of the ISPM Innovation Forum, Boston, MA, USA, 26 March 2018.

59. Von Hippel, E. Lead users: A source of novel product concepts. *Manag. Sci.* **1986**, *32*, 791–805. [CrossRef]

60. Thompson, J. Managing meat tenderness. *Meat Sci.* **2002**, *62*, 295–308. [CrossRef]

61. Pleasants, A.B.; Thompson, J.M.; Pethick, D.W. Model relating a function of tenderness, juiciness, flavour and overall liking to the eating quality of sheep meat. *Aus. J. Exp. Agric.* **2005**, *45*, 483–489. [CrossRef]

62. Strydom, P.; Burrow, H.; Polkinghorne, R.; Thompson, J. Do demographic and beef eating preferences impact on South African consumers' willingness to pay (wtp) for graded beef? *Meat Sci.* **2019**, *150*, 122–130. [CrossRef]

Article

Exploring Consumer Palatability of Australian Beef Fajita Meat Enhanced with Phosphate or Sodium Bicarbonate

Andrea Garmyn [1],*, Nicholas Hardcastle [1], Clay Bendele [1], Rod Polkinghorne [2] and Mark Miller [1]

[1] Department of Animal and Food Sciences, Texas Tech University, Lubbock, TX 79409, USA; nicholas.hardcastle@ttu.edu (N.H.); clay.bendele@ttu.edu (C.B.); mfmrraider@aol.com (M.M.)

[2] Birkenwood Pty. Ltd., 46 Church St, Hawthorn VIC 3122, Australia; rod.polkinghorne@gmail.com

* Correspondence: andrea.garmyn@ttu.edu; Tel.: +1-806-834-6559

Received: 10 December 2019; Accepted: 7 February 2020; Published: 11 February 2020

Abstract: The objective of this study was to determine the consumer eating quality of five Australian beef muscles (outside skirt/*diaphragm*, inside skirt/*transversus abdominis*, inside round cap/*gracilis*, bottom sirloin flap/*obliquus abdominis internus*, and flank steak/*rectus abdominis*) served as fajita strips. All the muscles were divided in half and enhanced (12%) with a brine solution containing either phosphate, a "clean label" ingredient sodium bicarbonate, or not enhanced. Muscle and enhancement independently influenced ($P < 0.01$) tenderness, juiciness, flavor, and overall liking. Overall, the bottom sirloin flap was liked the most ($P < 0.05$) when compared with all the other muscles, while the inside round cap was liked less but did not differ ($P > 0.05$) from the inside skirt or flank steak. Samples enhanced with sodium bicarbonate were the most ($P < 0.05$) tender and juicy; samples enhanced with phosphate were intermediate, and the control samples were the least tender and juicy, regardless of the muscle. Flavor and overall liking were similar ($P > 0.05$) between clean and phosphate-enhanced samples, and both were liked more than the control samples. Enhancement was necessary for acceptable eating quality of all the muscles evaluated in this study; however, the inside round cap was the least suitable. These results indicate that a "clean label" enhanced fajita product is possible without compromising cooking yield or consumer satisfaction.

Keywords: beef; consumer; eating quality; enhancement; fajita; muscle; phosphate; sensory testing; sodium bicarbonate

---

1. Introduction

The global beef industry strives to produce and deliver consistent, high-quality beef products that meet expectations for consumer eating quality. In Australia, beef eating quality is underpinned by the Meat Standards Australia (MSA) grading system [1], but this standard focuses on predicting the eating quality of fresh beef (i.e., no enhancement). Garmyn et al. [2] recently showed that enhancement significantly improved palatability in Australian *longissimus lumborum* and *gluteus medius* steaks with greater consumer ratings for tenderness, juiciness, flavor liking, overall liking, and satisfaction according to US consumers. Likewise, Lees et al. [3] found that infusion with kiwifruit extract improved consumer scores for eating quality of the *longissimus lumborum* and *biceps femoris*.

Value-adding or enhancement is still required of some lower quality cuts or cuts that would otherwise be considered unsatisfactory in the MSA grading system. Many researchers have explored the use of non-meat ingredients, such as phosphate and salt, to enhance beef quality and sensory characteristics [2,4–8]. These studies examined the enhancement of beef *longissimus dorsi*, muscles from the chuck (*complexus, serratus ventralis, splenius, subscapularis, supraspinatus,* and *triceps brachii*), *biceps femoris,* and *gluteus medius,* but the investigation of enhancement in beef muscles outside this list is somewhat limited.

Beef fajitas are commonly found on menus of Mexican restaurants in the US and utilize muscles, such as the inside or outside skirt steak or the flank steak, combined with marinade or seasoning. Huerta-Montauti et al. [9] investigated processing techniques of several muscles for thin meat alternatives, including the outside and inside skirt steaks for fajitas, but limited research has been done exploring the use of enhancement on flank steaks [10] or other muscles from the round or sirloin.

Given the consumer interest in clean labeling [11], finding an alternative functional ingredient that can be used in lieu of phosphate for meat enhancement also merits investigation. Studies have investigated the use of sodium bicarbonate ($NaHCO_{3-}$) as a phosphate replacement in pork [12] and chicken [13]. These studies have shown that $NaHCO_3$ can positively affect the physiochemical and eating quality attributes of meat and poultry compared with non-treated counterparts. In an attempt to classify phosphate alternatives' impact on beef eating quality, Hardcastle et al. [14] determined that Australian steaks from multiple muscles enhanced with $NaHCO_{3-}$ resulted in greater consumer palatability scores compared with those treated with sodium phosphate. Value-adding lower-quality muscles are of extreme interest, particularly if muscles destined for beef trim could be utilized in dishes such as fajitas or stir fry rather than ground beef. Therefore, our objective was to determine the consumer eating quality of five Australian beef muscles selected from carcasses representative of four MSA-predicted eating quality grades. Muscles were enhanced (12%) with either phosphate, sodium bicarbonate, or not enhanced. Ultimately, we investigated which of these muscles are suitable for fajita manufacture and examined the usefulness of the MSA sorting system for these muscles when using a preparation/cook method previously untested in the MSA grading scheme [15]. We hypothesized that differences in eating quality would exist between muscles and enhancement treatments, but that enhancement would mitigate any differences in eating quality due to the MSA grade.

2. Materials and Methods

2.1. Product Procurement

Cattle were harvested on one of four days within a two-week period at a commercial abattoir in Rockhampton, Australia. Tropical breed content and implant usage was monitored and reported for cattle according to their accompanying MSA vendor declaration when they were transferred to an MSA-licensed abattoir [16]. All cattle had substantial and equal *Bos indicus* influence, none had received hormone growth promotants, approximately 36% were grass-fed and 63% were grain-fed, and 60% were male and 40% of cattle were female. Carcasses were subjected to MSA grading, and data were recorded and utilized for sorting purposes. MSA marbling was scored from 100 to 1190 in increments of 10 based on the amount and distribution of marbling in the longissimus dorsi [17]. Ossification was scored from 100 to 590 in increments of 10 using the AUS-MEAT Carcase Maturity Chart [17]. In addition, hot carcass weight (HCW; kg), 12th rib fat thickness (mm), eye muscle area (cm^2), and hump height (mm) were collected and recorded for each carcass. In the MSA grading system, hump height serves as an indicator of tropical breed content. AUS-MEAT fat and meat color scores were also recorded [17]. Finally, *longissimus* muscle ultimate pH and temperature values were recorded using a hand-held, probe-type pH meter (model WP-80, TPS Pty Ltd, Springwood, Brisbane, Australia). Carcass data were used to select carcasses representative of four different predicted eating quality categories: 2 = unsatisfactory; 3* = "Selected"; 4* = "Classic"; and 5* = "Premium".

Five subprimals were collected from the right side of each carcass: outside skirt [Institutional Meat Purchase Specification (IMPS) 121C; *diaphragm*], inside skirt (IMPS 121D; *transversus abdominis*), inside round cap (IMPS 169B; *gracilis*), bottom sirloin flap (IMPS 185A; *obliquus abdominis internus*), and flank steak (IMPS 193; *rectus abdominis*). Subprimals were vacuum packaged individually, held in chilled storage at 0 to 1 °C until 14 d postmortem, and frozen at −20 °C. All subprimals were combined into a single consignment from Brisbane, Australia, to Texas Tech University, Lubbock, TX, USA, via cargo freight and road transport. Subprimals were held at frozen temperatures (−20 °C) during shipment and storage upon arrival at Texas Tech University, Lubbock, TX, USA.

2.2. Sample Processing

The muscles were thawed for 48 h at 2 to 4 °C before sample processing. Excess fat and sinew were removed from all the muscles prior to processing. All the muscles were cut in half parallel to the muscle fibers, with each half being alternatively assigned to enhancement treatment (control, clean, and phosphate). All the muscle halves were weighed to obtain the green weight. The control muscle halves were untreated. The clean muscle halves were enhanced with sodium chloride (NaCl; Morton Salt Inc., Chicago, IL, USA), food grade sodium bicarbonate ($NaHCO_3$; Church & Dwight Co. Inc., Ewing Township, NJ, USA), and water. The brine was prepared in 5 °C tap water with 4.16% NaCl and 4.00% $NaHCO_{3-}$. Brine was poured over the muscles in the vacuum tumbler for a target pickup of 112% (111.68% ± 3.82) of fresh muscle weight. All the muscles were placed in a vacuum tumbler (Koch Industries, Wichita, KS, USA) and a vacuum (508 mm Hg) was pulled. The muscles were batch tumbled at 10 RPM for 20 min. The samples were allowed to rest for 10 min, and the tumbled weight was obtained and recorded for each muscle half. The pickup percentage was calculated by taking the tumbled weight divided by the green weight, multiplied by 100.

The phosphate muscle halves were enhanced with NaCl (Morton Salt Inc., Chicago, IL, USA), sodium tripolyphosphate (STPP; Carfosel 408, Prayon Inc., Augusta, GA, USA), and water. The brine was prepared in 5 °C tap water with 4.16% NaCl and 4.00% STPP. Brine was poured over the muscles in the vacuum tumbler for a target pickup of 112% (111.59% ± 4.00) of fresh muscle weight. The muscles were tumbled as previously described for clean enhancement. The samples were allowed to rest for 10 min, and the tumbled weight was obtained and recorded for each muscle half. The pickup percentage was calculated by taking the tumbled weight divided by the green weight, multiplied by 100.

Regardless of enhancement treatment, a sample was obtained (weighing approximately 10 g) after treatment application and before vacuum packaging. This sample was vacuum packaged individually and frozen at −20 °C until analysis of final pH. All the remaining muscle samples were vacuum packaged individually and held at 2 to 4 °C. The samples were sorted into one of six testing days, and the muscles were used in consumer sensory testing within 9 days of processing.

2.3. Ultimate pH Determination

The samples were thawed for 24 h at 2 to 4 °C prior to analysis. Individual samples were mixed with distilled water for 1 min in a tabletop blender (Model 80335R, Hamilton Beach Brands, Glen Allen, VA, USA) to allow for homogenization. Homogenized samples were placed in a 150 mL beaker with a filter cone. The sample pH was measured with a bench-top probe-type pH meter (Model 14703; Denver Instrument Company, Bohemia, NY, USA), and the ultimate pH of each sample was determined as the average of two samples.

2.4. Consumer Sensory Evaluation

The Texas Tech University Institutional Review Board approved procedures for use of human subjects for consumer panel evaluation of meat sensory attributes (IRB#: 2017-598).

Each muscle sample was removed from packaging and weighed individually to obtain a raw weight. The muscles were cooked individually to 74 °C on a George Foreman clamshell grill (Model GRP99, Spectrum Brands. Inc., Middleton, WI, USA) with the lid closed and a plate temperature set to 218 °C. The muscle temperature was monitored using a digital, instant read Thermapen thermometer (Model Mk4, ThermoWorks, American Fork, UT, USA). When the muscles reached the required temperature, they were removed from the heat source. The peak temperature and cooked weight were recorded. The cooking loss percentage was calculated by subtracting the cooked weight from the raw weight, dividing by the raw weight, and multiplying by 100. In addition, total cooking time was recorded. The muscles were rested for at least 3 min prior to slicing. The muscles were sliced into 13 mm strips perpendicular to the muscle fibers, and the strips were cut in half lengthwise, resulting in strips that were approximately 5 cm long. The strips were transferred to pre-heated rectangular stainless-steel pans, which were maintained in insulated water bath warming units (Model W-3Vi; American Permanent Ware Company; Dallas, TX, USA) at ~60 °C throughout the test session. Each warming unit held nine pans.

Consumer panels were conducted in the Texas Tech University Animal and Food Sciences Building. Consumer panelists ($n = 360$) were recruited from Lubbock, Texas, and the surrounding local communities by scheduling community groups with populations of regular red meat eaters within the age range of 18–75. Each consumer was monetarily compensated and was only allowed to participate one time. Each session consisted of 60 people and lasted approximately 60 min.

Consumer testing was conducted according to MSA protocols [15], with previously described modifications for cooking method. Each consumer evaluated seven samples, including one warm-up sample, which was excluded from analysis, to orient consumers to the sample format. The warm-up samples were always served in the first position, followed by six test samples served in a predetermined, balanced order. The serving order of the six test products was controlled by a 6 × 6 Latin square design, ensuring that all the products were presented an equal number of times in each serving order position and before and after each other product. The test products were selected from the five muscles that were or were not enhanced. The predicted eating quality score (MSA grade) was also used for selection and allocation into the consumer testing design. Those products were equally represented and evenly distributed among the 60 consumers each evening. The testing occurred over the course of six evenings, with a similar and even product distribution in each testing session. Software-controlled routines ensured that the samples from each individual muscle were served in five different order positions and within different subsets of 12 consumers within each group of 60.

Each panelist was seated at a numbered booth and was provided with a ballot, plastic utensils, a toothpick, unsalted crackers, a napkin, an empty cup, a water cup, and a cup with diluted apple juice (10% apple juice and 90% water). Each ballot consisted of a demographic questionnaire, seven sample ballots, and a post-panel survey regarding beef purchasing habits. Before beginning each panel, consumers were given verbal instructions by Texas Tech personnel about the ballot and the process of testing samples. The panels were conducted in a large classroom that has standard fluorescent lighting (i.e., no red filters were used) with tables that were divided into individual sensory booths.

Each sample had 10 consumer observations (i.e., each muscle half yielding at least 20 strips served in duplicate to 10 predetermined consumers). Consumers scored palatability traits, including tenderness, juiciness, flavor liking, and overall liking, on 100-mm line scales verbally anchored at 0 (not tender, not juicy, dislike extremely) and 100 (very tender, very juicy, like extremely). Consumers were asked to rate the quality of each sample as unsatisfactory, good everyday quality, better than everyday quality, or premium quality. The 10 individual scores for each trait were averaged to generate mean sensory scores for each palatability trait and satisfaction prior to analysis. A composite score (MQ4) was calculated using the following equation: (tenderness × 0.3) + (juiciness × 0.1) + (flavor liking × 0.3) + (overall liking × 0.3) [15]. Weightings for tenderness have decreased and flavor liking increased from original weightings by [15] for a balanced contribution to the MQ4 value. The weightings give an indication of the relative importance of the four sensory attributes (tenderness, juiciness, flavor, and overall satisfaction) to the final meat quality score. According to the fixed weightings above, when a consumer eats beef, their satisfaction is influenced equally by tenderness, flavor liking, and overall liking, with a lesser contribution by juiciness.

All remaining pieces that were not consumed were vacuum packaged and chilled overnight at 2–4 °C. The samples were then snap frozen in liquid nitrogen. Frozen and cubed samples were homogenized in a food processor (Model Blixer 3 Series D, Robot Coupe, Ridgeland, MS, USA), blended into an ultrafine powder, and transferred into a labeled Whirl-Pak bag. The bags were stored in a freezer at −80 °C until subsequent analysis for percent moisture.

2.5. Percent Moisture

Moisture percentages were obtained for the cooked samples. The cooked sample moisture percentages were obtained in accordance with AOAC International protocol #950.46 [18]. Five grams (±0.05 g) of powdered sample were weighed into crucibles. The weight of each crucible and the crucible plus the sample weight were recorded. The samples were placed in a drying oven (Isotemp Oven, Thermo Fischer Scientific, Waltham, MA, USA) at 100 °C for 24 h. Upon completion of drying, the crucibles were removed from the oven and placed into desiccators for 30 min to cool. The crucibles were then weighed to determine the percentage of moisture in each sample. The percent moisture was obtained by taking the difference between the pre- and post-dried crucibles divided by the pre-dried crucibles and multiplied by 100. The total moisture was calculated by taking the sample weight and multiplying by the percent moisture for water holding capacity determination.

2.6. Statistical Analysis

The data were analyzed in SAS using PROC GLIMMIX (version 9.4, SAS Inst. Inc., Cary, NC, USA). For compositional analyses, the enhancement treatment, the muscle, and their interaction were included as fixed effects. For consumer sensory analyses, the enhancement treatment, the muscle, the MSA grade, and their interactions were included as fixed effects. The percent pickup was considered and tested as a potential covariate but was decided against due to its relationship both to the muscle (one of the treatment factors) and the enhancement treatment, as that would violate the statistical guidelines for covariate usage. Treatment least squares means were separated with the PDIFF option of SAS using a significance level of $P \leq 0.05$. Mean separation tests for all pairwise comparisons were performed using the PDIFF function, which requests that P-values for differences of all least squares means be produced. The PROC CORR of SAS was used to assess the relationship between compositional traits and consumer eating quality traits by generating Pearson correlation coefficients. The PROC FREQ of SAS was used to summarize consumer demographic information.

3. Results

3.1. Carcass Traits

All the carcasses were graded using MSA grading specifications. The carcass characterization can be found in Table 1.

Table 1. Mean, standard deviation, and range of carcass traits ($n = 22$).

Trait	Mean	Standard Deviation	Minimum	Maximum
Dentition	2.5	1.3	0	4
Carcass weight, kg	298.6	47.7	216	388.5
Hump height, mm	114.3	24.2	65	160
Eye muscle area, cm^2	73.3	11.6	49	90
Rib fat, mm	8.9	3.3	3	18
Ossification	158.6	13.2	140	180
Marbling	456.4	186.4	200	780
Meat color	2.7	0.7	1C	5
Fat color	1.1	1	0	3
pH	5.56	0.07	5.43	5.7
Temperature, °C	5	0.7	3	6.1

3.2. Processing Characteristics

As seen in Table 2, muscle and enhancement influenced ($P < 0.01$) post-processing pH, and muscle impacted ($P < 0.01$) all other processing characteristics, but the percent pickup was not different between the clean and phosphate enhancement treatments ($P = 0.52$). No interactions between muscle and enhancement were detected ($P \geq 0.10$) for processing characteristics.

Table 2. The main effects of muscle and enhancement on the processing characteristics of the Australian beef muscle samples ($n = 216$) [1].

Treatment	Post- Processing pH	Green Weight, kg	Tumbled Weight, kg	Percent Pickup, %
Muscle				
Outside skirt	5.78 [d]	0.29 [c]	0.33 [c]	15.9 [a]
Inside skirt	5.92 [bc]	0.47 [b]	0.53 [b]	12.8 [b]
Bottom sirloin flap	6.06 [a]	0.62 [a]	0.70 [a]	12.9 [b]
Flank steak	5.95 [ab]	0.33 [c]	0.35 [c]	7.2 [d]
Inside round cap	5.81 [cd]	0.51 [b]	0.56 [b]	9.6 [c]
SEM [2]	0.429	0.02	0.22	0.49
P-value	<0.01	<0.01	<0.01	<0.01
Enhancement				
Clean	6.43 [a]	0.45	0.5	11.8
Phosphate	5.84 [b]	0.44	0.5	11.5
Control	5.43 [c]	-	-	-
SEM	0.328	0.012	0.013	0.3
P-value	<0.01	0.96	0.96	0.52
P-value (muscle × enhancement)	0.16	0.65	0.73	0.1

[a–d] Within a column, least squares means without a common superscript differ ($P < 0.05$) due to muscle. [1] Muscle: outside skirt ($n = 42$), inside skirt ($n = 42$), bottom sirloin flap ($n = 44$), flank ($n = 44$), inside round cap ($n = 44$); Enhancement: $n = 72$/treatment. [2] Pooled (largest) SE of least squares means.

Tumbling muscles with an enhancement solution containing sodium bicarbonate resulted in the greatest post-processing pH, phosphate-enhanced muscles had an intermediate pH, and the control samples that were not enhanced had the lowest post-processing pH ($P < 0.05$). The green weight, tumbled weight, and percent pickup were not influenced ($P \geq 0.52$) by enhancement. Post-processing pH varied by muscle ($P < 0.01$). The bottom sirloin flap and flank steak had greater ($P < 0.05$) pH than the outside round cap or outside skirt, but flank steak and inside skirt did not differ ($P > 0.05$). For the green weight and the tumbled weight, the bottom sirloin flap samples were the heaviest ($P < 0.05$), followed by the inside round cap and inside skirt, which were intermediate; the flank steak and outside skirt had the lightest muscle weights. Although the percent pickup was similar between the enhancement treatments, the percent pickup was influenced by the muscle ($P < 0.01$). Because all the muscles were batch tumbled as opposed to tumbling by muscle, variation in the raw material size existed, which led to uneven solution uptake by the different muscles [19]. The outside skirt had the greatest ($P < 0.05$) percent pickup, followed by the bottom sirloin flap and inside skirt, which were similar ($P > 0.05$). The flank steak had the lowest percent pickup ($P < 0.05$) compared with all the other muscles.

3.3. Cooking Properties

Tables 3 and 4 illustrate the effects of muscle and enhancement on the cooking properties of the beef muscle samples. Cooking loss and cooking time were both influenced ($P \leq 0.02$) by muscle. The inside skirt had lower ($P < 0.05$) cooking loss than did the outside skirt and inside round cap but did not differ ($P > 0.05$) from the other muscles. As expected, the larger and thicker muscles took longer ($P < 0.05$) to cook than the smaller or thinner muscles. Cooking loss was impacted ($P < 0.01$) but cooking time was not influenced ($P = 0.71$) by the enhancement treatment. The muscles enhanced with phosphate had greater ($P < 0.05$) cooking loss than the clean enhancement or the control, which was somewhat unexpected given the elevated post-processing pH of the phosphate enhanced muscles compared with the pH of the control samples.

Table 4 shows the interactive effects of muscle and enhancement on the cooked moisture percentage ($P < 0.01$). For the outside skirt, inside skirt, and flank steak, the muscles tumbled with sodium bicarbonate had a greater ($P < 0.05$) cooked moisture percentage than the muscles tumbled with phosphate, which in turn had a greater ($P < 0.05$) moisture percentage than the non-enhanced control samples. For the bottom sirloin flap and inside round cap, the cooked moisture percentage did not differ ($P > 0.05$) between the samples enhanced with sodium bicarbonate or phosphate, but all the enhanced samples had a greater ($P < 0.05$) cooked moisture percentage compared with the non-enhanced control samples.

Table 3. The main effects of muscle and enhancement on the cooking properties (cooking loss and cooking time) of the Australian beef muscle samples ($n = 216$) [1].

Treatment	Raw Weight, g	Cooked Weight, g	Cooking Loss, %	Cooking Time, s
Muscle				
Outside skirt	279.8 [d]	196.8 [c]	29.2 [a]	379 [d]
Inside skirt	449.2 [c]	337.5 [b]	25.0 [b]	407 [d]
Bottom sirloin flap	634.5 [a]	462.5 [a]	26.7 [ab]	822 [b]
Flank steak	310.6 [d]	227.1 [c]	27.0 [ab]	556 [c]
Inside round cap	507.6 [b]	363.4 [b]	28.5 [a]	986 [a]
SEM [2]	17.84	12.92	0.99	39
P-value	<0.01	<0.01	0.02	<0.01
Enhancement				
Clean	459.0 [a]	341.5 [a]	25.4 [b]	621
Phosphate	449.9 [a]	318.6 [ab]	29.3 [a]	620
Control	400.1 [b]	292.3 [b]	27.1 [b]	649
SEM [2]	13.63	9.86	0.76	30
P-value	<0.01	<0.01	<0.01	0.71
P-value (muscle × enhancement)	0.58	0.36	0.32	0.89

[a–d] Within a column, least squares means without a common superscript differ ($P < 0.05$) due to muscle. [1] Muscle: outside skirt ($n = 42$), inside skirt ($n = 42$), bottom sirloin flap ($n = 44$), flank ($n = 44$), inside round cap ($n = 44$); Enhancement: $n = 72$/treatment. [2] Pooled (largest) SE of least squares means.

Table 4. The interactive effects ($P < 0.01$) of muscle and enhancement on the cooked moisture of the Australian beef muscle samples ($n = 216$) [1].

Muscle	Clean	Phosphate	Control
Outside skirt	62.5 [cdef]	57.8 [g]	50.6 [h]
Inside skirt	66.4 [a]	62.1 [def]	55.5 [g]
Bottom sirloin flap	64.8 [abc]	62.8 [cde]	57.1 [g]
Flank steak	64.2 [abcd]	61.8 [ef]	57.7 [g]
Inside round cap	65.3 [ab]	64.0 [bcde]	60.3 [f]
SEM [2]	0.97		

[a–h] Within all columns and rows, least squares means without a common superscript differ ($P < 0.05$). [1] Muscle: outside skirt ($n = 42$), inside skirt ($n = 42$), bottom sirloin flap ($n = 44$), flank ($n = 44$), inside round cap ($n = 44$); Enhancement: $n = 72$/treatment. [2] Pooled (largest) SE of least squares means.

3.4. Consumer Sensory

The demographic characteristics of participating consumers can be found in Table 5. Over half of the participants were aged 20–39 years old. Nearly half of the population in Lubbock, TX, USA, is less than 45 years old [20], so this percentage of consumers is representative of society. We also believe this percentage is suitable according to the product studied. Participants were evenly split between male and female. Most participants (97.2%) identified with Caucasian/white or Hispanic as their ethnic origin, with an even split between the two distinctions. For census purposes, persons who identify as Hispanic or Latino can identify as any race; however, in the latest census data available for Lubbock, TX, USA, 35% reported themselves as Hispanic or Latino, while 65% reported themselves as not Hispanic or Latino [20]. The most common household size consisted of 2–3 adults, representing over 80% of participants. Nearly half of the participants had no children living in their household. The level of education with the highest proportion of participants was for "some college/technical school" (32.4%), while high school and college graduates collectively accounted for another 50%. Additionally, the majority of consumers ate beef at least twice per week (82.8%). The most preferred degree of doneness was medium-rare, with medium and medium-well contributing another 49% collectively.

Table 5. The demographic characteristics of consumers (n = 360) who participated in the consumer sensory panels at Texas Tech University in Lubbock, TX, USA.

Characteristic	Response	% of Consumers
Age group	<20	5.8
	20–29	22.5
	30–39	31.9
	40–49	14.2
	50–59	13.9
	>60	11.7
Gender	Male	47.0
	Female	53.1
Ethnic origin	African American	1.7
	Asian	0.0
	Caucasian/white	49.0
	Hispanic	48.2
	Native American	0.6
	Other	0.6
Household size (adults)	1	9.5
	2	64.6
	3	18.1
	4	4.5
	5	2.8
	6+	0.6
Household size (children)	0	47.0
	1	11.1
	2	22.5
	3	13.9
	4	3.1
	5	2.5
Annual household income	<$20,000	9.8
	$20,000–$50,000	24.9
	$50,001–$75,000	23.0
	$75,001–$100,000	19.3
	>$100,000	
Level of education	Non-high school graduate	5.3
	High school graduate	22.9
	Some college/technical school	32.4
	College graduate	26.8
	Post-college graduate	12.6
Beef consumption	Daily	8.9
	4–5 times a week	33.2
	2–3 times a week	40.7
	Weekly	11.7
	Every other week	4.7
	Monthly	0.8
Preferred beef degree of doneness	Rare	4.8
	Medium-rare	35.4
	Medium	25.6
	Medium-well	23.0
	Well-done	11.2

The consumer sensory outcomes can be found in Table 6. No interactions were detected ($P > 0.05$) between muscle, enhancement, or MSA grade. Muscle and enhancement independently influenced ($P < 0.01$) tenderness, juiciness, flavor liking, overall liking, MQ4, and satisfaction. The MSA grade only impacted tenderness but had no effect on juiciness, flavor liking, overall liking, MQ4, or satisfaction. For tenderness and juiciness, the bottom sirloin flap was scored higher ($P < 0.05$) than any other muscle, followed by the outside skirt. The inside skirt and flank steak were scored similarly ($P > 0.05$) for tenderness and juiciness, and the inside round cap was the least ($P < 0.05$) tender and juicy compared with all the other muscles. Consumers liked the flavor of the bottom sirloin flap more ($P < 0.05$) than all other muscles, which did not differ ($P > 0.05$). Ultimately, the bottom sirloin flap was liked most ($P < 0.05$) overall compared with all other muscles, while the inside round cap was liked less, but did not differ ($P > 0.05$) from the inside skirt or flank steak. The bottom sirloin flap had the greatest composite MQ4 score, followed by the outside skirt, flank steak, and inside skirt, which were similar ($P > 0.05$) and had intermediate MQ4. Lastly, the inside round cap had the lowest ($P < 0.05$) composite MQ4 score compared with all other muscles. Satisfaction followed a similar trend to overall liking in terms of the differences between the muscles, but all muscles were classified as "good everyday quality" according to their satisfaction score.

Table 6. The main effects of muscle, enhancement, and Meat Standards Australia (MSA) grade on consumer scores ($n = 360$) for tenderness, juiciness, flavor liking, overall liking, composite score (MQ4), and satisfaction of the Australian beef muscles prepared for fajitas.

Treatment	Tenderness [1]	Juiciness [1]	Flavor Liking [1]	Overall Liking[1]	MQ4 [2]	Satisfaction [3]
Muscle [4]						
Outside skirt	64.0 [b]	60.0 [b]	54.7 [b]	56.5 [b]	58.6 [b]	3.22 [b]
Inside skirt	59.0 [c]	54.4 [c]	52.5 [b]	53.7 [bc]	55.1 [b]	3.13 [bc]
Bottom sirloin flap	73.7 [a]	66.7 [a]	67.5 [a]	68.6 [a]	69.6 [a]	3.65 [a]
Flank steak	57.3 [c]	51.2 [c]	54.8 [b]	55.0 [bc]	55.3 [b]	3.16 [bc]
Inside round cap	49.2 [d]	44.7 [d]	52.2 [b]	51.2 [c]	50.3 [c]	3.01 [c]
SEM [7]	1.8	1.8	1.7	1.7	1.5	0.058
P-value	<0.01	<0.01	<0.01	<0.01	<0.01	<0.01
Enhancement [5]						
Clean	70.8 [a]	64.1 [a]	61.9 [a]	64.0 [a]	65.5 [a]	3.48 [a]
Phosphate	64.5 [b]	59.4 [b]	62.9 [a]	63.2 [a]	63.1 [a]	3.43 [a]
Control	46.7 [c]	42.7 [c]	44.2 [b]	43.9 [b]	44.7 [b]	2.78 [b]
SEM [7]	1.4	1.5	1.3	1.3	1.3	0.46
P-value	<0.01	< 0.01	<0.01	<0.01	<0.01	<0.01
MSA Grade [6]						
5*—Premium	58.7 [b]	55.3	54.8	55.1	56.1	3.20
4*—Classic	64.3 [a]	58.1	57.8	58.8	60.1	3.21
3*—Selected	60.5 [ab]	54.7	56.1	56.6	57.5	3.24
2—Unsatisfactory	59.1 [b]	53.5	56.7	57.5	57.4	3.28
SEM[7]	2.0	2.1	1.9	1.9	1.8	0.065
P-value	0.04	0.29	0.46	0.29	0.17	0.67

[a–d] Within a column, least squares means without a common superscript differ ($P < 0.05$) due to muscle. [1] Scores: 0 mm = not tender, not juicy, dislike flavor extremely, dislike overall extremely; 100 mm = very tender, very juicy, like flavor extremely, like overall extremely. [2] MQ4 = tenderness*0.3 + juiciness*0.1 + flavor liking* 0.3 + overall liking*0.3. [3] Satisfaction score: 2 = unsatisfactory, 3 = good everyday quality, 4 = better than everyday quality, and 5 = premium quality. [4] Muscle: outside skirt ($n = 42$), inside skirt ($n = 42$), bottom sirloin flap ($n = 44$), flank ($n = 44$), and inside round cap ($n = 44$). [5] Enhancement: $n = 72$/treatment. [6] MSA grade: premium ($n = 68$), classic ($n = 50$), selected ($n = 68$), and unsatisfactory ($n = 30$). [7] Pooled (largest) SE of least squares means.

The samples enhanced with sodium bicarbonate were the most ($P < 0.05$) tender and juicy, the samples enhanced with phosphate were intermediate, and the control samples were the least tender and juicy, regardless of muscle or MSA grade. Despite differences in tenderness and juiciness between enhanced samples, flavor and overall liking were similar ($P > 0.05$) between the clean and phosphate-enhanced samples, and both were liked more than the control samples. MQ4 score and satisfaction of all the samples increased ($P < 0.05$) due to enhancement, resulting in a shift into the next quality category for satisfaction. MQ4 and satisfaction did not differ between the clean or phosphate enhancement treatments. So, rather than being considered "unsatisfactory", the enhanced samples were perceived as "good everyday quality".

Lastly, the MSA grade influenced ($P = 0.04$) tenderness scores only but not in a linear fashion. The classic (4*) samples were more tender than the premium (5*) and unsatisfactory (2) samples but did not differ ($P > 0.05$) from the selected (3*) samples. These results suggest that predicted eating quality scores may need to be adjusted when value-adding strategies, such as enhancement and alternative cooking methods, are employed.

3.5. Correlations

Table 7 illustrates the relationships between composition and cooking characteristics with eating quality traits. All the eating traits were positively related ($P < 0.01$) with post-processing pH, suggesting that greater pH resulted in higher eating quality. A similar trend was observed for the percent pickup ($P < 0.01$). The cooked moisture was also positively correlated ($P < 0.01$) to eating quality traits; this was likely due to the greater water holding capacity induced from greater pH from enhancement. Cooking time was positively linked ($P < 0.01$) to flavor liking, suggesting that the samples that took longer to cook had longer for flavor to develop, which was valued by consumers. Lastly, cooking loss was negatively associated ($P < 0.05$) with tenderness and juiciness.

Table 7. Pearson correlation coefficients of relationships between composition, cooking traits, and consumer eating quality scores.

Attribute	Post- Processing pH	Percent Pickup	Moisture	Cooking Time	Cooking Loss
Tenderness	0.53 **	0.65 **	0.34 **	0.00	−0.16 *
Juiciness	0.47 **	0.62 **	0.26 **	−0.04	−0.14 *
Flavor liking	0.43 **	0.58 **	0.32 **	0.24 **	0.04
Overall liking	0.46 **	0.60 **	0.33 **	0.16 *	−0.03

** Correlation coefficient differs from 0 ($P < 0.01$). * Correlation coefficient differs from 0 ($P < 0.05$).

To estimate the extent to which eating quality scores are linked to overall liking and satisfaction, correlation coefficients between palatability traits, MQ4, and satisfaction scores were determined (Table 8). Consumer overall liking was associated ($P < 0.01$) with consumer tenderness ($r = 0.88$) and juiciness ratings ($r = 0. 87$) but most highly related with flavor liking ($r = 0.96$). Individual palatability traits were strongly correlated to each other ($r \geq 0.80$), indicating that individual improvements of these traits could influence the perception of another trait. MQ4 was highly related ($P < 0.01$) to eating quality scores for tenderness, juiciness, flavor liking, and overall liking, as would be expected given that it is a composite score of those traits. Satisfaction was positively linked ($P < 0.01$) to all the eating quality traits, especially overall liking, and was highly correlated to MQ4 ($P < 0.01$).

Table 8. Pearson correlation coefficients of relationships between consumer eating quality scores.

Attribute	Tenderness	Juiciness	Flavor Liking	Overall Liking	MQ4
Juiciness	0.89 **				
Flavor liking	0.80 **	0.80 **			
Overall liking	0.88 **	0.87 **	0.96 **		
MQ4	0.94 **	0.91 **	0.95 **	0.98 **	
Satisfaction	0.83 **	0.82 **	0.91 **	0.93 **	0.92 **

** Correlation coefficient differs from 0 ($P < 0.01$).

4. Discussion

This study is the first to evaluate the influence of the enhancement of beef muscles for consumption as fajita meat with phosphate or an alternative functional ingredient. The results from this study show that enhancing various muscles either with phosphate or sodium bicarbonate improved the tenderness, juiciness, flavor liking, and ultimately overall liking compared with non-enhanced muscles. The clean enhancement solution garnered greater tenderness and juiciness scores compared with the phosphate enhancement, regardless of the muscle, but flavor liking did not differ between the two enhancement solutions, resulting in similar overall liking. Despite similar percent pickup, the clean enhancement samples had greater post-processing pH compared with the samples enhanced with phosphate. A greater pH for the clean samples was expected, as sodium bicarbonate possesses a greater buffering capacity in water than sodium phosphate salts [21]. This increase in pH led to lower cooking loss and greater cooked moisture percentage. Ultimately, this resulted in greater consumer juiciness without affecting flavor liking. We believe this likely contributed to the greater consumer tenderness scores of the clean vs. phosphate enhanced samples as well, given the high positive correlation coefficient (r = 0.89) between consumer tenderness and juiciness.

Consumers also detected palatability differences between the muscles in the current study. Huerta-Montauti et al. [9] evaluated several muscles from the beef carcass to find muscles suitable as beef fajita options. Common muscles between the two studies included the outside skirt, inside skirt, and bottom sirloin flap. In the current study, the bottom sirloin flap received greater palatability scores across all traits and was liked the most compared with all the other muscles. Huerta-Montauti et al. [9] found that the bottom sirloin flap and outside skirt were more tender than the inside skirt, but tenderness liking, flavor liking, and overall liking did not differ between these three muscles when papain was either applied alone or in conjunction with blade tenderization [9]. In the current study, the inside round cap was the least tender and the least juicy compared with all the other muscles, which resulted in low overall liking, suggesting this muscle may not be a suitable option for fajita production, even with enhancement.

Jeremiah et al. [22] assessed the palatability of several major beef muscles, including four common muscles to the current study: the outside skirt, inside round cap, bottom sirloin flap, and flank steak. All the muscles, except the outside skirt were considered slightly tough initially according to trained panelists, while the outside skirt was neither tough nor tender. Of all 33 muscles evaluated by [22], the outside skirt had the most intense flavor but consequently had the second least desirable flavor. The other common muscles had slightly intense flavor with slightly desirable flavor. Of the common muscles between the two studies, the inside round cap had the greatest overall palatability, followed by the flank steak, outside skirt, and bottom sirloin flap. Granted, Jeremiah et al. [22] roasted samples and used trained panelists compared with the untrained panelists used in the current study. In addition, no enhancement was used by [22]. However, their results are essentially reversed from the current findings. Belew et al. [23] determined Warner-Bratzler shear force values of 40 beef muscles, including all five used in the current study. According to their results, the outside skirt was the most tender of all the muscles evaluated, which they attributed to the high fat content of the diaphragm muscle. The bottom sirloin flap was the second most tender of the remaining common muscles, which was scored considerably more tender in the present findings. However, the second most tender according to [23] was the inside round cap, which contradicts the current findings, followed by the flank steak and inside skirt. Belew et al. [23] cooked samples as steaks, which aligns more closely to the cooking methods in the current study, but no enhancement was used.

Both types of enhancement improved all palatability traits regardless of the muscle. Previous findings have shown that the incorporation of various non-meat ingredients increased tenderness, juiciness, and flavor [2,3,5,6,24]. In alignment with the current findings, enhanced beef is typically juicier than non-enhanced beef [2,6]. Enhancement can increase muscle pH and decrease free water, increasing moisture retention [5,24]. Morrow et al. [10] found that enhanced flank steaks had a greater cooked moisture percentage than non-enhanced samples, and consumers scored those enhanced samples as juicier. Flavor liking and overall liking increased in the current study due to enhancement, again aligning with previous results [5,6,24]. According to previous results, sodium chloride can increase saltiness and enhance beef flavor intensity [25]; however, others have speculated that enhancement ingredients, particularly salt, could mask other flavors, including beef intensity [26,27]. Nonetheless, saltiness did not appear to be overwhelming or detrimental in the current results, as flavor liking improved with enhancement. Likewise, Morrow et al. [10] had consumers score flank steaks for saltiness using "Just about right" scales, where 50 was "Just about right". Their brine solution had a similar salt concentration as the current study. Scores above 50 indicated that the samples were too salty, and scores below 50 suggested the samples were not salty enough. Saltiness was similar between various methods of enhancement (tumbling, injection, or combination), closest to "Just about right" and greater than the non-enhanced flank steak samples, but not to the extent that the enhanced samples were too salty.

5. Conclusions

Muscle and enhancement independently influenced tenderness, juiciness, flavor liking, and overall liking. The MSA grade only impacted tenderness but had no effect on other palatability traits. The bottom sirloin flap was liked most overall compared with all the other muscles, while the inside round cap was liked less but did not differ from the inside skirt or flank steak. The samples enhanced with the "clean label" ingredient, sodium bicarbonate, were the most tender and juicy, the samples enhanced with phosphate were intermediate, and the non-enhanced control samples were the least tender and juicy, regardless of muscle or MSA grade. Despite differences in tenderness and juiciness between the enhanced samples, flavor and overall liking were similar between the clean and phosphate-enhanced samples, and both were liked more than the control samples. Enhancement was necessary for acceptable eating quality of the muscles evaluated in this study; however, the inside round cap may not be a suitable fajita option due to significantly reduced tenderness and juiciness compared with the other muscles. Flavor and overall liking were not different between the two enhancement formulations, but tenderness and juiciness varied. These results indicate that a "clean label" enhanced product is possible without compromising cooking yield or consumer satisfaction; in fact, the reverse was true. This may be an important finding for markets in which consumers are sensitive to ingredient labeling and desire a "natural" product.

Author Contributions: N.H., R.P., A.G., and M.M. contributed to the conceptualization and experimental design of the study; N.H., C.B., and A.G. undertook the study; A.G. conducted the formal data analysis and presentation of the data; A.G. wrote the original manuscript draft; reviews and editing of the completed manuscript were conducted by N.H. and R.P.; project administration was the responsibility of M.M. and A.G. All authors have read and agreed to the published version of the manuscript.

Funding: This research was funded through a research gift from Teys Australia.

Conflicts of Interest: The authors declare no conflicts of interest.

References

1. Polkinghorne, R.; Thompson, J.M.; Watson, R.; Gee, A.; Porter, M. Evolution of the Meat Standards Australia (MSA) beef grading system. *Aust. J. Exp. Agric.* **2008**, *48*, 1351–1359. [CrossRef]

2. Garmyn, A.J.; Garcia, L.G.; Spivey, K.S.; Polkinghorne, R.J.; Miller, M.F. Consumer assessment of beef muscles from Australian and US production systems with or without enhancement. *Meat Muscle Biol.* **2019**, (in press). [CrossRef]

3. Lees, A.; Konarska, M.; Tarr, G.; Polkinghorne, R.; McGilchrist, P. Influence of kiwifruit extract infusion on consumer sensory outcomes of striploin (M. *longissimus lumborum*) and outside flat (M. *biceps femoris*) from beef carcasses. *Foods* **2019**, *8*, 332. [CrossRef] [PubMed]

4. Baublits, R.T.; Pohlman, F.W.; Brown, A.H.; Johnson, Z.B. Effects of sodium chloride, phosphate type and concentration, and pump rate on beef biceps femoris quality and sensory characteristics. *Meat Sci.* **2005**, *70*, 205–214. [CrossRef] [PubMed]

5. Baublits, R.T.; Pohlman, F.W.; Brown, A.H.; Yancey, E.J.; Johnson, Z.B. Impact of muscle type and sodium chloride concentration on the quality, sensory, and instrumental color characteristics of solution enhanced whole-muscle beef. *Meat Sci.* **2006**, *72*, 704–712. [CrossRef] [PubMed]

6. Hardcastle, N.C.; Garmyn, A.J.; Legako, J.F.; Brashears, M.M.; Miller, M.F. The effect of finishing diet on consumer perception of enhanced and non-enhanced Honduran beef. *Meat Muscle Biol.* **2018**, *2*, 277–295. [CrossRef]

7. Molina, M.E.; Johnson, D.D.; West, R.L.; Gwartney, B.L. Enhancing palatability traits in beef chuck muscles. *Meat Sci.* **2005**, *71*, 52–61. [CrossRef]

8. Vote, D.J.; Platter, W.J.; Tatum, J.D.; Schmidt, G.R.; Belk, K.E.; Smith, G.C.; Speer, N.C. Injection of beef strip loins with solutions containing sodium tripolyphosphate, sodium lactate, and sodium chloride to enhance palatability. *J. Anim. Sci.* **2000**, *78*, 952–957. [CrossRef] [PubMed]

9. Huerta-Montauti, D.; Miller, R.K.; Schuehle Pfeiffer, C.E.; Pfeiffer, K.D.; Nicholson, K.L.; Osburn, W.N.; Savell, J.W. Identifying muscle and processing combinations suitable for use as beef for fajitas. *Meat Sci.* **2008**, *80*, 259–271. [CrossRef] [PubMed]

10. Morrow, S.J.; Garmyn, A.J.; Hardcastle, N.C.; Brooks, J.C.; Miller, M.F. The effects of enhancement strategies of beef flanks on composition and consumer palatability characteristics. *Meat Muscle Biol.* **2019**, *3*, 457–466. [CrossRef]

11. Sindelar, J.; Schilling, W.; Campbell, J. Clean label product ingredients … their role in moving the meat industry forward. In Proceedings of the Presented at the Reciprocal Meat Conference, Kansas City, MO, USA, 24–27 June 2018.

12. Sheard, P.R.; Tali, A. Injection of salt, tripolyphosphate and bicarbonate marinade solutions to improve the yield and tenderness of cooked porkloin. *Meat Sci.* **2004**, *68*, 305–311. [CrossRef] [PubMed]

13. Sen, A.R.; Naveena, B.M.; Muthukumar, M.; Babji, Y.; Murthy, T.R.K. Effect of chilling, polyphosphate and bicarbonate on quality characteristics of broiler breast meat. *Br. Poult. Sci.* **2005**, *46*, 451–456. [CrossRef]

14. Hardcastle, N.C.; Garmyn, A.J.; Miller, M.F. Effect of enhancement on three beef muscles with phosphate or alternative functional ingredients on the eating quality of Australian beef. In Proceedings of the Reciprocal Meat Conference, Fort Collins, CO, USA, 23–26 June 2019.

15. Watson, R.; Gee, A.; Polkinghorne, R.; Porter, M. Consumer assessment of eating quality–Development of protocols for MSA testing. *Aust. J. Exp. Agric.* **2008**, *48*, 1360–1367. [CrossRef]

16. Meat Standards Australia (MSA). Completing Your MSA Beef Vendor Declaration. Available online: https://producer.msagrading.com.au/Learning/S4P2 (accessed on 1 November 2019).

17. AUS-MEAT. Handbook of Australian Beef Processing. *The AUS-MEAT Language. Version 6.* Available online: https://www.ausmeat.com.au/WebDocuments/Producer_HAP_Beef_Small.pdf (accessed on 1 November 2019).

18. AOAC. *Official Methods of Analysis*, 18th ed.; AOAC International: Arlington, VA, USA, 2006.

19. Williams, J.B. Marination: Processing technology. In *Handbook of Meat and Meat Processing*, 2nd ed.; Hui, Y.H., Ed.; CRC Press: Boca Raton, FL, USA, 2012; pp. 495–504.

20. World Population Review. Lubbock, Texas Population 2020. Available online: http://worldpopulationreview.com/us-cities/lubbock-population/ (accessed on 22 January 2020).

21. Mohan, C. *Buffers. A Guide for the Preparation and Use of Buffers in Biological Systems*; EMD Bioscience: San Diego, CA, USA, 2006; pp. 20–21.

22. Jeremiah, L.E.; Gibson, L.L.; Aalhus, J.L.; Dugan, M.E.R. Assessment of palatability attributes of the major beef muscles. *Meat Sci.* **2003**, *65*, 949–958. [CrossRef]

23. Belew, J.B.; Brooks, J.C.; McKenna, D.R.; Savell, J.W. Warner-Bratzler shear evaluations of 40 bovine muscles. *Meat Sci.* **2003**, *64*, 507–512. [CrossRef]

24. Robbins, K.; Jensen, J.; Ryan, K.J.; Homco-Ryan, C.; McKeith, F.K.; Brewer, M.S. Consumer attitudes towards beef and acceptability of enhanced beef. *Meat Sci.* **2003**, *64*, 721–729. [CrossRef]

25. Stetzer, A.J.; Cadwallader, K.; Singh, T.K.; McKeith, F.K.; Brewer, M.S. Effect of enhancement and ageing on flavor and volatile compounds in various beef muscles. *Meat Sci.* **2008**, *79*, 13–19. [CrossRef] [PubMed]

26. Rose, M.N.; Garmyn, A.J.; Hilton, G.G.; Morgan, J.B.; VanOverbeke, D.L. Comparison of tenderness, palatability, and retail caselife of enhanced cow subprimals with nonenhanced cow and United States Department of Agriculture Select subprimals1. *J. Anim. Sci.* **2010**, *88*, 3683–3692. [CrossRef] [PubMed]

27. Wicklund, S.E.; Homco-Ryan, C.; Ryan, K.J.; McKeith, F.K.; Mcfarlane, B.J.; Brewer, M.S. Aging and enhancement effects on quality characteristics of beef strip steaks. *J. Food Sci.* **2005**, *70*, 242–248. [CrossRef]

Article

Lamb Age has Little Impact on Eating Quality

Claire E. Payne [1,2,3,*], **Liselotte Pannier** [1,2], **Fiona Anderson** [1,2], **David W. Pethick** [1,2] **and Graham E. Gardner** [1,2]

1   Engineering and Education, College of Science, Health, Murdoch University, Perth 6150, Australia;
     L.Pannier@murodch.edu.au (L.P.); F.Anderson@murdoch.edu.au (F.A.);
     D.Pethick@murdoch.edu.au (D.W.P.); G.Gardner@murdoch.edu.au (G.E.G.)
2   Australian Cooperative Research Centre for Sheep Industry Innovation, Armidale, New South Wales 2351,
     Australia
3   Department of Primary Industries and Regional Development, Perth 6151, Australia
*   Correspondence: claire.payne@murdoch.edu.au

Received: 15 January 2020; Accepted: 11 February 2020; Published: 13 February 2020

Abstract: There is an industry wide perception that new season lamb has better eating quality than old season lamb. This study aims to identify differences in consumer eating quality scores between two age classes in lamb. Consumer eating quality scores from eight cuts across the carcass were evaluated from new season (NS; $n = 120$; average age = 240 days) and old season lambs (OS; $n = 121$; average age = 328 days), sourced from four different flocks. Cuts were grilled (loin, topside, outside, knuckle and rump) or roasted (leg, shoulder, rack) and scored by untrained consumers for tenderness, juiciness, liking of flavour and overall liking. There was no difference in eating quality scores between the two age classes for the loin, leg, shoulder and rack. This was similarly shown in the topside with the exception of juiciness scores where NS lambs were higher than OS lambs. There was also a lack of age difference in the outside with the exception of flock 3 where NS lambs scored higher than OS lambs for all sensory traits. Across all sensory traits, OS lambs received on average 2.8 scores lower for the knuckle and 3.1 scores lower for the rump compared to NS lambs. These results show little difference in eating quality between NS and OS lamb, and highlight the potential to develop high quality OS or "autumn lamb" products, with a similar premium price at retail as NS lambs.

Keywords: lamb age; eating quality; consumer; sensory evaluation; sheepmeat

---

1. Introduction

Consumer purchasing behaviour when buying meat is driven by eating quality. Eating quality is heavily influenced by animal age, with increased age resulting in decreased eating quality [1,2]. This is mostly due to increases in collagen concentration and crosslinking that occur as animals mature, which decreases muscle tenderness [2–4]. Due to its negative impact on eating quality animal age is factored into price grids and grading systems at slaughter.

Sheep age in Australia is currently identified using dentition, with lambs classified as having no permanent incisors in wear, yearlings or hoggets as having 1–2 erupted permanent incisors in wear, and mutton as having 2 or more erupted permanent incisors [5,6]. Within the lamb classification, there are two sub-categories, new season (NS) lamb (approx. 5–8 months old) and old season (OS) lamb (approx. 10–12 months old). Studies that describe eating quality differences between lamb and mutton show that consumers generally prefer lamb [1]. Thompson et al. [2] reported lamb samples receiving significantly higher consumer scores for tenderness, juiciness, liking of flavour and overall liking compared to mutton samples. Although eating quality differences between lamb and mutton are well described, there is little evidence showing clear eating quality differences within younger animals, particularly across multiple cuts. Pannier et al. [7] compared consumer eating quality scores of the loin and topside cuts between lambs (average age: 335 days) and yearlings (average age: 685 days). There was little to no difference in eating quality scores for the loin between the age groups, however the lambs received higher scores than yearlings for the topside. This study suggests that animal age has a greater impact on eating quality in low quality cuts, for example the topside, than high quality cuts such as the loin. This is possibly attributed to differences in muscle function and characteristics, for example collagen solubility, between low quality and high quality cuts.

The impact of animal age on eating quality of cuts other than the loin and topside has yet to be described, particularly within young animals. Despite this lack of evidence, retail marketing of NS lamb, or "young spring lamb" as a high quality product, has resulted in an industry wide perception that NS lamb has better eating quality than OS lamb. Therefore NS lamb receives a premium price at market compared to OS lamb [8]. It is currently unknown if cuts from OS lamb have the same or worsened quality compared to NS lamb, and therefore if the reduced price at market is justified.

This study aims to further explore the relationship between animal age and eating quality across multiple cuts from NS and OS lambs. It is hypothesised that NS and OS lambs will not differ in their consumer eating quality scores for high quality cuts, however the low quality cuts from OS lambs will receive lower scores than those from NS lambs.

2. Materials and Methods

2.1. Experimental Design and Slaughter Details

Animals were obtained from four commercial sheep flocks across South Australia and Victoria with each flock described in Table 1. NS and OS lambs were balanced across each flock (total $n = 241$), and lambs from both age classes (NS and OS) within each flock genetically linked by similar sires and dams. Both age groups on each property were kept in the same paddock 3 weeks prior to slaughter. Lambs were weighed prior to being transported to a commercial processing plant, where they were held in lairage overnight and slaughtered the following day. All carcasses were subjected to a medium-voltage electrical stimulation [9] and trimmed according to AUS-MEAT specifications [10]. Carcasses were chilled overnight (3–4 °C) before sampling.

Table 1. Location, number, breed, average age and kill group of new season and old season lambs within each flock.

Flock	Location	Kill Group	*n*	New Season Sire Type × Dam Breed	New Season Average Age	*n*	Old Season Sire Type × Dam Breed	Old Season Average Age
1	Lochabar, South Australia	1	30	White Suffolk × White Suffolk	209 [a]	30	White Suffolk × Merino	298 [b]
2	Avenue Range, South Australia	2	30	Poll Dorsett × Border Leicester Merino	252 [a]	30	Border Leicester × Merino	308 [b]
3	Struan, South Australia	3	30	Poll Dorsett × Border Leicester Merino	250 [a]	30	Border Leicester × Merino	350 [b]
4	Greta, Victoria	3	30	Border Leicester composites	252 [a]	31	Border Leicester composites	357 [b]
Total			120			121		

Values within rows followed by different letters are significantly different at $p < 0.05$.

2.2. Sample Collection and Measurements

Within 1 h of slaughter, hot carcass weight (HCWT) and GR tissue depth (GR) (measured at the girth rib site, which is 11cm from the midline to the lateral surface of the 12th rib, using a standard GR knife) were measured on all carcasses. Ultimate pH was measured at 24 h post-slaughter on the left portion of the *M. longissimus thoracis* et *lumborum* (LTL) (TPS WP-80 v5.2). From the carcass saddle region the left short loin (AUS-MEAT 4880) [10] up to the 12th rib was removed. Eye muscle area (EMA) (mm^2) was determined by measuring the width and depth of the exposed LTL surface at the 12th rib using a digital calliper. C-site fat depth (mm) was measured 45 mm from the spine along the LTL at the 12th rib on the exposed surface with a digital calliper. From 176 carcasses, 8 different cuts were removed for eating quality analysis. From the remaining 64 carcasses, 2 cuts were removed and used as links, which served as starter samples during the eating quality sessions to prime the palate. Cuts are further described in Table 2. From the total cuts collected, 813 grill cuts were eaten by 1355 consumers in 23 eating quality sessions, and 528 roast cuts were eaten by 880 consumers in 15 eating quality sessions. All grill cuts were sliced into 5 steaks of 15 mm thickness and trimmed for subcutaneous fat and epimysium (silver skin). The leg and shoulder cuts once removed from the carcass were netted using meat netting. The rack cuts were frenched, and the cap fat was removed. All cuts were vacuum-packed and stored at 2 °C for 5 days of aging before being frozen at −20 °C. Cuts were thawed in a 4 °C fridge 24 h prior to sensory testing.

Table 2. Number and type of cuts collected and eaten by consumers according to cooking method.

Cut	AUS-MEAT Code	Cooking Method	*n* Tested
Knuckle	5072	Grill	163
Loin	5150	Grill	163
Rump	5074	Grill	163
Outside	5075	Grill	163
Topside	5077	Grill	161
Leg	4830	Roast	176
Shoulder	5050	Roast	176
Rack	4748	Roast	176

The remaining loin from each carcass was removed for fresh colour, intramuscular fat (IMF) and shear force (SF) measurements. Fresh eye muscle colour was measured on all carcasses approximately 49 h post-mortem on a section of the LL, which was exposed to air for 30–60 min. Lightness (*L* *), redness (*a* *) and yellowness (*b* *) were measured using a Minolta Chromameter, D65 illuminant with a 2° standard observer and 8 mm aperture, as per Pearce et al. [9]. IMF (expressed as a percentage) measurements were taken from a 40 g LL muscle sample. Samples were ground, freeze-dried in a Coolsafe 95-15 Pro (Scanvac, Lillerød, Denmark) and analysed using near-infrared technology (Technicon InfrAlyser 450 (19 wavelengths)) to estimate chemical fat content, as described by Perry et al. [11]. An additional 65 g of loin muscle was collected for shear force testing. Samples were vacuum-packed, aged for 5 days at 1 °C and then frozen at −20 °C until subsequent testing. Packaged frozen samples were cooked in a water bath at 71 °C for 35 min and then cooled in running water for 30 min after cooking. Six cores (approximately 3–4 cm long, 1 cm^2) from each loin sample were cut and Warner–Bratzler shear force (WBSF) was measured on each core sample using a Lloyd texture analyser with a Warner–Bratzler shear blade fitted [12]. Laboratory processing of loin samples and measurement of WBSF was performed at the University of New England Meat Science Department (Armidale, New South Wales, Australia).

2.3. Sensory Testing

The sensory testing protocol for grilled lamb cuts is outlined in Thompson et al [13]. The grilled cuts were cooked on a Silex grill (S-tronic steaker, Silex, Hamburg, Germany), with the top plate set to a temperature of 185 °C and the bottom plate set to 190 °C. Steaks were removed from the grill with an approximate internal temperature of 65 °C, measured with a heat resistance thermometer, rested for 1.5 min and halved before serving.

Leg and shoulder roast cuts were trimmed into a 15 cm × 15 cm block. Cuts were rolled and secured using butchers string prior to cooking. Roast cuts were cooked in an Electrolux 10 tray dry oven and set to a temperature of 160 °C. To achieve an internal temperature of 65 °C, roasts were removed from the oven at an internal temperature of 60 °C and rested for 10 min. Roasts were then sliced into 4 mm samples. Ten suitable slices from each cut were selected for consumer testing. Any external fat and connective tissue seams were removed and slices were trimmed to approximately 50 mm wide × 50 mm long × 4 mm thick. The 10 consumer samples were placed in steel pans which were maintained at a temperature of 50 °C until serving.

Untrained consumers were recruited by an independent company and screened to include individuals aged between 18 and 70 years old. Consumers scored each sample for tenderness, juiciness, liking of flavour and overall liking on a scale of 0 to 100, where a score of 0 indicates a tough, dry, unliked sample, while a score of 100 indicates a very tender, juicy, liked sample. Overall liking scores were used to determine a classification of high quality or low quality for each cut, based on a cut off of 65, which represents a 4-star Meat Standard Australiarating. Each consumer started with a common starter sample that was removed from the link cuts described in Section 2.2. Consumers then tested six consecutive samples over a period of 1 h allocated using a Latin square design [13]. An individual tasting session consisted of 60 consumers that either participated in a grill session or roast session. Overall, 36 cuts were tested within each tasting session, and each cut was sampled by 10 different consumers so that each cut received 10 consumer responses. The study was conducted in accordance with the Declaration of Helsinki, and the protocol was approved by the Ethics Committee of Murdoch University on 11 October 2018 (2018/129).

2.4. Statistical Analysis

Linear mixed effects models in SAS (SAS Version 9.1, SAS Institute, Cary, NC, USA) were used to analyse all consumer scores for overall liking, tenderness, juiciness and liking of flavour. The base model for each sensory trait included fixed effects for cut (loin, topside, outside, rump, knuckle, rack, shoulder, leg), age class (new season, old season) and flock within kill group. Animal identification and consumer identification within each consumer session were included as random terms. In addition to the base model, HCWT and GR tissue depth were analysed individually as covariates. All relevant first-order interactions between fixed effects were tested and non-significant ($p > 0.05$) terms were removed in a stepwise manner. In addition, HCWT, IMF%, SF, GR tissue depth, C-site fat depth, EMA, L *, a * and b * were also analysed as dependant variables to determine the phenotypic variability of each trait between the NS and OS lambs. These models included fixed effects for age class and flock within each kill group. Relevant first-order interactions between fixed effects were included and non-significant ($p > 0.05$) terms were removed in a stepwise manner.

3. Results

3.1. Effect of Lamb Age on Consumer Sensory Scores

Consumer scores for tenderness, juiciness, liking of flavour and overall liking varied between NS and OS lambs within some cuts and some flocks (Tables 3 and 4).

Table 3. Numerator degrees of freedom and F-values for the effects of the base linear mixed effects models of the tenderness, juiciness, flavour and overall liking scores.

Effect	NDF	Tenderness	Juiciness	Liking of Flavour	Overall Liking
Cut	7	284.6 **	155.6 **	117.0 **	169.8 **
Age class	1	2.1	2.7	1.1	1.1
Flock (Kill group)	2	7.7 **	11.8 **	11.6 **	9.7 **
Cut * Age class	7	2.7 *	4.1 **	2.7 *	ns
Flock (Kill group) * Age class	3	3.0 *	1.5	ns	ns
Cut * Flock (Kill group)	21	6.4 **	8.0 **	3.4 **	5.1 **
Cut * Flock (Kill group) * Age class	24	1.8 *	1.6 *	ns	ns

NDF: numerator degrees of freedom; *: $p < 0.05$: **: $p < 0.001$.

Table 4. Sensory properties of cuts from new season (NS) and old season (OS) lambs within each flock (least square means ± S.E).

Cut	Flock	Tenderness		Juiciness		Liking of Flavour		Overall Liking	
		NS	OS	NS	OS	NS	OS	NS	OS
Knuckle	1	72.8 ± 1.8	68.7 ± 1.8	68.9 ± 1.7 [a]	63.9 ± 1.8 [b]	69.3 ± 1.6	66.5 ± 1.7	70.6 ± 1.7	67.6 ± 1.8
	2	67.1 ± 1.8	64.1 ± 1.8	67.1 ± 1.7	64.1 ± 1.7	67.8 ± 1.6	64.6 ± 1.6	68.0 ± 1.7	65.8 ± 1.7
	3	72.2 ± 1.8 [a]	64.0 ± 1.8 [b]	65.8 ± 1.7	62.0 ± 1.7	67.7 ± 1.7	63.3 ± 1.6	70.0 ± 1.7 [a]	63.4 ± 1.7 [b]
	4	65.1 ± 1.8	66.9 ± 1.8	64.5 ± 1.7	68.8 ± 1.7	67.3 ± 1.7	67.1 ± 1.7	67.4 ± 1.7	68.5 ± 1.7
Loin	1	64.2 ± 1.8	68.2 ± 1.8	59.1 ± 1.7	62.2 ± 1.8	63.5 ± 1.6	66.6 ± 1.7	63.4 ± 1.7	66.9 ± 1.7
	2	64.3 ± 1.8	67.6 ± 1.8	62.8 ± 1.7	62.8 ± 1.7	65.4 ± 1.6	65.8 ± 1.6	65.6 ± 1.7	66.5 ± 1.7
	3	62.9 ± 1.8	63.1 ± 1.8	61.2 ± 1.7	57.2 ± 1.8	62.7 ± 1.7	62.8 ± 1.7	63.3 ± 1.7	62.5 ± 1.8
	4	64.8 ± 1.8	60.2 ± 1.8	60.8 ± 1.7	59.8 ± 1.7	64.8 ± 1.7	61.3 ± 1.6	65.2 ± 1.7	62.0 ± 1.7
Rump	1	74.4 ± 1.8	72.6 ± 1.8	74.1 ± 1.7 [a]	69.3 ± 1.8 [b]	71.8 ± 1.6	69.1 ± 1.7	74.2 ± 1.7	71.4 ± 1.7
	2	70.2 ± 1.7	67.1 ± 1.8	71.1 ± 1.7 [a]	64.9 ± 1.7 [b]	70.3 ± 1.6	66.2 ± 1. 6	70.9 ± 1.7 [a]	66.2 ± 1.7 [b]
	3	65.8 ± 1.7 [a]	58.8 ± 1.8 [b]	60.2 ± 1.7 [a]	53.7 ± 1.7 [b]	61.1 ± 1.7	58.4 ± 1.7	61.8 ± 1.7	58.4 ± 1.7
	4	61.5 ± 1.8	62.8 ± 1.8	63.5 ± 1.7	62.6 ± 1.7	63.7 ± 1.7	63.7 ± 1.7	64.1 ± 1.7	64.4 ± 1.7
Outside	1	57.0 ± 1.7	61.4 ± 1.8	64.3 ± 1.7	65.7 ± 1.8	62.6 ± 1.6	63.2 ± 1.7	61.4 ± 1.7	63.9 ± 1.8
	2	53.4 ± 1.8	51.8 ± 1.8	58.3 ± 1.7	56.2 ± 1.7	59.1 ± 1.7	57.8 ± 1.6	57.9 ± 1.7	56.3 ± 1.7
	3	57.3 ± 1.8 [a]	47.4 ± 1.8 [b]	60.5 ± 1.7 [a]	52.5 ± 1.7 [b]	59.1 ± 1.7 [a]	53.6 ± 1.7 [b]	59.7 ± 1.7 [a]	51.9 ± 1.7 [b]
	4	56.1 ± 1.7	53.0 ± 1.8	60.1 ± 1.7	59.3 ± 1.7	61.8 ± 1.6	58.9 ± 1.6	60.8 ± 1.7	59.0 ± 1.7
Topside	1	48.7 ± 1.8	49.9 ± 1.8	55.2 ± 1.7	52.9 ± 1.8	54.4 ± 1.6	56.4 ± 1.7	52.8 ± 1.7	54.9 ± 1.8
	2	46.9 ± 1.8	45.5 ± 1.8	55.8 ± 1.7 [a]	50.9 ± 1.7 [b]	55.6 ± 1.7	54.2 ± 1.6	52.8 ± 1.7	51.9 ± 1.7
	3	42.7 ± 1.8	38.2 ± 1.8	43.7 ± 1.7	42.3 ± 1.7	48.9 ± 1.7	47.0 ± 1.7	46.3 ± 1.7	44.0 ± 1.7
	4	43.9 ± 1.8	42.1 ± 1.8	50.4 ± 1.7	50.4 ± 1.7	52.3 ± 1.7	53.3 ± 1.7	50.3 ± 1.7	50.5 ± 1.7
Rack	1	71.4 ± 1.8	73.7 ± 1.8	65.6 ± 1.7	69.4 ± 1.7	67.2 ± 1.6	70.4 ± 1.6	67.8 ± 1.7	71.1 ± 1.7
	2	67.3 ± 1.8	68.6 ± 1.8	60.5 ± 1.7	62.7 ± 1.7	64.1 ± 1.6	66.1 ± 1.6	64.8 ± 1.7	67.1 ± 1.7
	3	69.5 ± 1.7	71.0 ± 1.7	63.3 ± 1.6	65.0 ± 1.6	64.3 ± 1.6	64.8 ± 1.6	66.1 ± 1.6	66.7 ± 1.6
	4	69.5 ± 1.7	67.5 ± 1.7	65.4 ± 1.6	64.0 ± 1.6	67.5 ± 1.6	67.6 ± 1.6	68.6 ± 1.6	67.2 ± 1.6
Leg	1	55.3 ± 1.8	58.0 ± 1.8	51.5 ± 1.7	51.7 ± 1.7	57.1 ± 1.6	59.1 ± 1.6	57.1 ± 1.7	58.1 ± 1.7
	2	50.0 ± 1.7	51.6 ± 1.8	46.9 ± 1.7	48.0 ± 1.7	57.0 ± 1.6	54.1 ± 1.6	54.4 ± 1.7	52.3 ± 1.7
	3	54.7 ± 1.7	51.2 ± 1.7	50.1 ± 1.6	46.9 ± 1.6	54.0 ± 1.6	52.7 ± 1.6	54.9 ± 1.6	52.1 ± 1.6
	4	57.8 ± 1.8	58.7 ± 1.7	53.4 ± 1.6	55.0 ± 1.6	58.1 ± 1.6	60.7 ± 1.6	58.5 ± 1.6	60.5 ± 1.6
Shoulder	1	59.7 ± 1.8 [a]	65.0 ± 1.8 [b]	57.2 ± 1.7 [a]	62.1 ± 1.7 [b]	59.2 ± 1.6 [a]	64.4 ± 1.6 [b]	58.5 ± 1.7 [a]	65.3 ± 1.7 [b]
	2	67.2 ± 1.8	66.7 ± 1.8	62.9 ± 1.7	64.8 ± 1.7	62.9 ± 1.6	61.3 ± 1.6	64.9 ± 1.7	63.2 ± 1.7
	3	64.7 ± 1.7	60.5 ± 1.7	58.8 ± 1.6	56.5 ± 1.6	56.2 ± 1.6	56.5 ± 1.6	57.6 ± 1.6	56.7 ± 1.6
	4	64.1 ± 1.7	62.6 ± 1.7	62.2 ± 1.6	61.2 ± 1.6	63.9 ± 1.6	62.2 ± 1.6	63.9 ± 1.6	62.7 ± 1.6

Values within rows for each sensory trait followed by different letters are significantly different at $p < 0.05$.

Within the grill cuts tested, the loin had no difference in scores between NS and OS lambs for all eating quality traits across all flocks. This was similarly seen in the topside with the exception of juiciness scores in flock 2, where NS lambs scored 4.9 scores higher than OS lambs. Within knuckle, rump and outside cuts, there was at least one flock where NS lambs scored higher than OS lambs ($p < 0.05$, Table 4). Within these cuts, these differences ranged between 7–10 points for tenderness, 5–8 points for juiciness and 5–8 points for overall liking. These differences most often occurred within flock 3. Liking of flavour scores did not differ between NS and OS lambs within any of the grill cuts with the exception of the outside from flock 2 where NS lambs scored 5.7 scores higher than OS lambs.

Within the roast cuts the rack and leg had no difference in scores between NS and OS lambs for all sensory traits across all flocks. There was also no age class effect within the shoulder cut for all sensory traits in flocks 2, 3 and 4. There was an age class effect seen in flock 1, however the effect was opposite to all other cuts with OS lambs scoring 5.3, 4.9, 5.2 and 6.8 sensory scores higher for tenderness, juiciness, liking of flavour and overall liking, respectively compared to the NS lambs ($p < 0.05$, Table 4).

These differences described remained mostly unchanged when carcass measurements were included. When corrected for HCWT the effect of age on juiciness scores within flock 1 for the knuckle and rump disappeared, however the magnitude of difference remained the same. There was no change in differences between sensory scores for NS and OS lambs when corrected for GR fat depth.

3.2. Carcass Data and Instrumental Meat Quality Measurement

Significant differences in carcass characteristics were found between NS and OS lambs. OS lambs were 82.5 days older and 1.55 kg heavier, had 1.6 mm more GR tissue depth, 0.57% more IMF%, 0.78 mm^2 more EMA and 0.3 more redness (a *) in the loin than NS lambs ($p < 0.05$ for all measurements, Table 5). These differences were not consistent across flocks, with only flock 3 consistently aligning with this overall trend. In this case the differences between OS and NS lambs also tended to be greatest, with 5.5 kg heavier carcasses, 3.5 mm more GR tissue depth, 1.4 mm^2 EMA and 1.2% more IMF% ($p < 0.05$ for all measurements, Table 5). Carcass measurements including C-site fat depth, shear force, lightness, yellowness and ultimate pH did not differ between age classes overall; however, some differences did exist within individual flocks. OS lambs from flock 2 and 3 had significantly more C-site fat depth than NS lambs with differences of 1.2 mm and 0.8 mm, respectively ($p < 0.05$). For shear force, only flock 1 displayed a significant difference between age classes with OS lambs having 0.6 kgF less than NS lambs. OS lambs from flock 2 showed more yellowness in the loin with a b * difference of 1.2, while OS lambs from flock 3 had lower ultimate pH with a difference of 0.1 from NS lambs.

Table 5. Carcass characteristics, colour parameters and instrumental measures of new season and old season lambs from each flock (predicted means ± S.E.), and the differences between new season and old season lambs for these measurements.

	Flock	New Season	Old Season	Difference
Hot carcass weight (kg)	1	22.4 ± 0.5	21.75 ± 0.5	0.6
	2	23.04 ± 0.5	22.84 ± 0.5	0.2
	3	21.43 ± 0.5 [a]	26.94 ± 0.5 [b]	−5.5
	4	28.61 ± 0.5 [a]	30.07 ± 0.5 [b]	−1.5
Girth Rib site tissue depth (mm)	1	10.02 ± 0.7	11.64 ± 0.8	−1.6
	2	16.03 ± 0.7	17.02 ± 0.7	−1
	3	14.98 ± 0.8 [a]	18.52 ± 0.8 [b]	−3.5
	4	20.1 ± 0.7	20.28 ± 0.7	−0.2
C-site fat depth (mm)	1	2.52 ± 0.3	2.09 ± 0.3	0.4
	2	2.87 ± 0.3 [a]	4.12 ± 0.3 [b]	−1.2
	3	2.84 ± 0.3 [a]	3.65 ± 0.3 [b]	−0.8
	4	5.1 ± 0.3	4.53 ± 0.3	0.6

Table 5. *Cont.*

	Flock	New Season	Old Season	Difference
Eye muscle area (mm^2)	1	13.93 ± 0.4 [a]	16.1 ± 0.4 [b]	−2.2
	2	15.3 ± 0.4 [a]	13.88 ± 0.4 [b]	1.4
	3	15.37 ± 0.4 [a]	16.78 ± 0.4 [b]	−1.4
	4	17.74 ± 0.4	18.69 ± 0.4	−0.9
Intramuscular fat (%)	1	3.31 ± 0.2 [a]	4.08 ± 0.2 [b]	−0.8
	2	4.36 ± 0.2	4.84 ± 0.2	−0.5
	3	4.18 ± 0.2 [a]	5.42 ± 0.2 [b]	−1.2
	4	5.22 ± 0.2	5.06 ± 0.2	0.2
Shear force at day 5 (KgF)	1	4.43 ± 0.2 [a]	3.83 ± 0.2 [b]	0.6
	2	3.72 ± 0.2	3.51 ± 0.2	0.2
	3	4.15 ± 0.2	3.84 ± 0.2	0.3
	4	3.56 ± 0.2	3.62 ± 0.2	−0.1
Lightness (L*)	1	34.17 ± 0.8	32.66 ± 0.9	1.5
	2	34.83 ± 0.8	33.94 ± 0.8	0.9
	3	35.49 ± 0.8	35.98 ± 0.8	−0.5
	4	33.36 ± 0.8	33.73 ± 0.8	−0.4
Redness (a*)	1	4.82 ± 0.2	4.76 ± 0.2	0.1
	2	5.53 ± 0.2 [a]	6.3 ± 0.2 [b]	−0.8
	3	5.48 ± 0.2 [a]	6.13 ± 0.2 [b]	−0.7
	4	6.28 ± 0.2	6.06 ± 0.2	0.2
Yellowness (b*)	1	16.73 ± 0.3	16.66 ± 0.3	0.1
	2	16.45 ± 0.3 [a]	17.64 ± 0.3 [b]	−1.2
	3	16.19 ± 0.3	16.93 ± 0.3	−0.7
	4	18.25 ± 0.3	17.78 ± 0.3	0.5
Ultimate pH	1	5.61 ± 0.02	5.58 ± 0.02	0.03
	2	5.64 ± 0.02	5.65 ± 0.02	−0.01
	3	5.8 ± 0.02 [a]	5.73 ± 0.02 [b]	0.1
	4	5.73 ± 0.02	5.76 ± 0.02	−0.03

Values within rows followed by different letters are significantly different at $p < 0.05$.

4. Discussion

4.1. Effect of Lamb Age on Eating Quality

There was no consistent effect of age class on consumer sensory scores across all flocks. In partial support of our hypothesis, there was no age class effect for some high quality cuts (loin, shoulder and rack), yet for the knuckle and rump, NS lambs had higher sensory scores than OS lambs. Contrary to our expectations for the low quality cuts (outside, topside and leg) there was little to no difference in consumer scores between age classes.

Most cuts across all flocks showed no difference in consumer scores between NS and OS lambs. This could be attributed to the small differences in age (approximately 3.5 months). This aligns well with other studies that showed no difference in sensory scores of the loin and topside cuts between animals aged 8.5 and 20 months old [1], or between lambs of two dentition categories: lambs that had milk teeth intact, and lambs that had lost milk teeth or had permanent teeth that just erupted [14]. These results suggest that eating quality is not impacted by animal age within lambs, but rather within more mature animals.

The little to no difference in eating quality could also be due to differing genetics between the age classes within each flock. NS lambs are designed to be fast growing for early turn off and are often sired by terminal rams, and terminal or maternal dams. While OS lambs are slower growing and late maturing, often sired by maternal rams and merino dams [15], as reflected in the lambs sourced for this study. It was shown that terminal sired lambs scored lower for both loin and topside tenderness, juiciness, liking of flavour and overall liking in comparison to maternal sired lambs [16]. In the loin, dam breeds also differed in sensory scores with lambs from merino dams scoring 1.9, 1.2, 1.5 and 1.4 scores higher than maternal merino dams for tenderness, juiciness, liking of flavour and overall liking, respectively [16]. It is possible that OS lambs are genetically superior for eating quality and therefore counteract any differences associated with age.

In contrast to most NS and OS cuts that received similar consumer scores, there was a significant age class effect for the knuckle and rump cuts. This was not expected due to their ranking as high quality cuts based on the consumer sensory scores. There are no previous studies comparing consumer scores across sheep ages for these cuts, however is was assumed that cuts deemed as high quality based on consumer scores would reflect the same results as the loin in other studies, as this is a consistently high quality and high scoring cut. Furthermore, Pannier et al. [7] showed a clear variation in the impact of age between a high quality and low quality cut, further reinforcing the notion that cuts with high scores were affected by animal age. However, these results suggest that the impact of animal age on eating quality cannot simply be explained by a ranking of high or low quality based on consumer scores. Differences between age classes in the knuckle and rump may be due to the muscle characteristics of these cuts, such as collagen solubility, rather than the overall quality.

As animal age increases, collagen solubility and tenderness are shown to decrease [4,17,18]. Decreases in collagen solubility vary between cuts, possibly explaining the decreased eating quality only seen in the knuckle and rump cuts of OS lambs. Cross et al. [19] compared collagen concentrations of topside, outside and knuckle cuts within 6 month old and 12 month old lambs. Soluble collagen was lower in the older animals, decreasing the most in the knuckle (5%), followed by the outside and topside (3.3 and 0.5%). In cattle, Girard et al. [20] also found greater decreases in collagen solubility from 6 to 12 months of age for the rump (18.3%) compared to the outside (3.3%). These studies found that with increased animal age, collagen solubility decreases the most rapidly in the knuckle and rump cuts, which therefore may contribute to rapid decreases in tenderness.

There was no difference in sensory scores between NS and OS lambs for the roast cuts. This is most likely due to the cooking method. Cooking temperature and time has significant effects on the heat solubility of collagen [21]. As temperature and time increases connective tissue becomes softer [22,23]. This is attributed to the conversion of collagen to gelatine at high temperatures [24]. Thus, differences seen in sensory scores between NS and OS lambs in the grilled leg cuts (knuckle and rump), which are attributed to decreases in collagen solubility, are mitigated or slowed by the roasting process.

4.2. Carcas Data and Instrumental Meat Quality Measures

NS lambs were lighter and leaner when compared to OS lambs, which is not surprising due to the difference in maturity of the two age classes. Older animals that are more mature are often heavier and fatter than younger animals [25]. OS lambs also had higher IMF% values which is in agreement with other reports of IMF% increasing with increased animal age [1,16,26]. Shear force values did not differ between the age classes, which aligns well with the lack of difference in tenderness scores between NS and OS lambs. These results are also in agreement with previous studies that showed no age class effect on loin or topside shear force in lambs of various ages [1,14,27,28]. It is well described that decreases in tenderness that occur with age are due to increases in collagen crosslinks [4,17,18,29]. More crosslinking increases the heat resistance of collagen during the cooking process [4], which in turn results in tougher meat. The lack of shear force and tenderness difference between NS and OS lambs supports the notion that collagen crosslinking does not seem to occur in sheep until at least 12 months old, and potentially older for some cuts.

OS lambs were redder compared to NS lambs, yet still considered acceptable for consumers (L * values of 34 and a * values higher that 14.5 are considered acceptable to consumers) [30]. This is in agreement with Kim et al. [28] and Pethick et al. [1] who demonstrated redder meat in older animals, suggesting increased oxidative fibres and aerobicity as animals age [31].

5. Conclusions

This study has shown little to no differences in eating quality between NS and OS lambs across several major cuts of the carcass. This is in contrast to the current perception that NS lambs, or "young spring lambs", have better quality than OS, carry over lambs. OS lambs should therefore receive the same premium price as NS lambs at retail. This study highlights the potential to develop high quality OS or "autumn lamb" products, with the greatest opportunity in roast cuts. This study also demonstrates that in lambs, cut type is more significant in the impact of eating quality than animal age. Differences in eating quality was only seen within two particular cuts, knuckle and rump, and is attributed to decreased rates of collagen solubility; therefore future work should involve identifying changes in collagen solubility across cuts.

Author Contributions: Conceptualization, D.W.P. and L.P.; methodology, C.E.P., D.W.P., G.E.G. and L.P.; software, G.E.G.; validation, C.E.P., D.W.P., F.A., G.E.G. and L.P.; formal analysis, C.E.P. and L.P.; investigation, C.E.P., F.A. and L.P.; resources, D.W.P., G.E.G. and L.P.; data curation, C.E.P.; writing—original draft preparation, C.E.P.; writing—review and editing, C.E.P., D.W.P., F.A., G.E.G. and L.P.; visualization, C.E.P., D.W.P., F.A., G.E.G. and L.P.; supervision, D.W.P., F.A., G.E.G. and L.P.; project administration, C.E.P., D.W.P. and L.P.; funding acquisition, D.W.P. and L.P. All authors have read and agreed to the published version of the manuscript.

Funding: This research was funded by Meat and Livestock Australia, grant number L.EQT.1810.

Acknowledgments: The authors gratefully acknowledge the producers who supplied their lambs and the many people who assisted with site supervision, sample collection and measurement, as well as abattoir staff. For the South Australia site: J. Hocking, E. Winslow, R. Withers and J. Peterse. For the Western Australian sites: M. Boyce and A. Williams. The authors also gratefully acknowledge the Australian Cooperative Research Centre for Sheep Industry Innovation for their financial support and contribution to the project.

Conflicts of Interest: The authors declare no conflict of interest. The funders had no role in the design of the study; in the collection, analyses, or interpretation of data; in the writing of the manuscript, or in the decision to publish the results.

References

1. Pethick, D.; Hopkins, D.; D'Souza, D.; Thompson, J.; Walker, P. Effects of animal age on the eating quality of sheep meat. *Aust. J. Exp. Agric.* **2005**, *45*, 491–498. [CrossRef]
2. Thompson, J.; Hopkins, D.; D'Souza, D.; Walker, P.; Baud, S.; Pethick, D. The impact of processing on sensory and objective measurements of sheep meat eating quality. *Aust. J. Exp. Agric.* **2005**, *45*, 561–573. [CrossRef]
3. Jeremiah, L.; Tong, A.; Gibson, L. The influence of lamb chronological age, slaughter weight and gender on carcass and meat quality. *Food Res. Int.* **1998**, *31*, 227–242. [CrossRef]
4. Young, O.; Braggins, T. Tenderness of ovine semimembranosus: Is collagen concentration or solubility the critical factor? *Meat Sci.* **1993**, *35*, 213–222. [CrossRef]
5. Sheep Producers Australia. Australia's New Definition of Lamb—What You Need to Know—Sheep Producers Australia. Available online: http://sheepproducers.com.au/lamb-definition/ (accessed on 18 August 2019).
6. Meat and Livestock Australia. Age/Dentition: Solutions to Feedback. Available online: https://solutionstofeedback.mla.com.au/sheep/agedentition/ (accessed on 18 August 2019).
7. Pannier, L.; Gardner, G.; Pethick, D. Effect of Merino sheep age on consumer sensory scores, carcass and instrumental meat quality measurements. *Anim. Prod. Sci.* **2018**, *59*, 1349–1359. [CrossRef]
8. Meat and Livestock Australia. Market Reports & Prices. Available online: https://www.mla.com.au/prices-markets/market-reports-prices/ (accessed on 1 February 2020).
9. Pearce, K.; Van De Ven, R.; Mudford, C.; Warner, R.; Hocking-Edwards, J.; Jacob, R.; Pethick, D.; Hopkins, D. Case studies demonstrating the benefits on pH and temperature decline of optimising medium-voltage electrical stimulation of lamb carcasses. *Anim. Prod. Sci.* **2010**, *50*, 1107–1114. [CrossRef]
10. HAM. *Handbook of Australian Sheepmeat Processing*; AUS-MEAT Limited: Murarrie, Australia, 2018.
11. Perry, D.; Shorthose, W.; Ferguson, D.; Thompson, J. Methods used in the CRC program for the determination of carcass yield and beef quality. *Aust. J. Exp. Agric.* **2001**, *41*, 953–957. [CrossRef]
12. Hopkins, D.L.; Toohey, E.S.; Warner, R.D.; Kerr, M.; Van de Ven, R. Measuring the shear force of lamb meat cooked from frozen samples: Comparison of two laboratories. *Anim. Prod. Sci.* **2010**, *50*, 382–385. [CrossRef]

13. Thompson, J.; Gee, A.; Hopkins, D.; Pethick, D.; Baud, S.; O'Halloran, W. Development of a sensory protocol for testing palatability of sheep meats. *Aust. J. Exp. Agric.* **2005**, *45*, 469–476. [CrossRef]

14. Wiese, S.; Pethick, D.; Milton, J.; Davidson, R.; McIntyre, B.; D'souza, D. Effect of teeth eruption on growth performance and meat quality of sheep. *Aust. J. Exp. Agric.* **2005**, *45*, 509–515. [CrossRef]

15. Robertson, S.; Friend, M.; Sargeant, K. Flexibility in Livestock Systems is Important for Risk Management in Variable Climates. Available online: Evergraze.com.au/library-content/flexibility-in-livestock-systems-is-important-for-risk-management-in-variable-climates/ (accessed on 1 February 2020).

16. Pannier, L.; Gardner, G.; Pearce, K.; McDonagh, M.; Ball, A.; Jacob, R.; Pethick, D. Associations of sire estimated breeding values and objective meat quality measurements with sensory scores in Australian lamb. *Meat Sci.* **2014**, *96*, 1076–1087. [CrossRef] [PubMed]

17. Weston, A.R. Review: The Role of Collagen in Meat Tenderness. *Prof. Anim. Sci.* **2002**, *18*, 107–111. [CrossRef]

18. Berge, P. Comparison of Muscle Composition and Meat Quality Traits in Diverse Commercial Lamb Types. *J. Muscle Foods* **2003**, *14*, 281–300. [CrossRef]

19. Cross, H.R.; Smith, G.C.; Carpenter, Z.L. Palatability of individual muscles from ovine leg steaks as related to chemical and histological traits. *J. Food Sci.* **1972**, *37*, 282–285. [CrossRef]

20. Girard, I.; Aalhus, J.; Basarab, J.; Larsen, I.; Bruce, H. Modification of muscle inherent properties through age at slaughter, growth promotants and breed crosses. *Can. J. Anim. Sci.* **2011**, *91*, 635–648. [CrossRef]

21. Vasanthi, C.; Venkataramanujam, V.; Dushyanthan, K. Effect of cooking temperature and time on the physico-chemical, histological and sensory properties of female carabeef (buffalo) meat. *Meat Sci.* **2007**, *76*, 274–280. [CrossRef] [PubMed]

22. Winegarden, M.; Lowe, B.; Kastelic, J.; Kline, E.A.; Plagge, A.R.; Shearer, P. Physical changes of connective tissues of beef during heating. *Food Res.* **1952**, *17*, 172–184. [CrossRef]

23. Ritchey, S.; Cover, S.; Hostetler, R. Collagen content and its relation to tenderness of connective tissue in 2 beef muscles. *Food Technol.* **1963**, *17*, 194.

24. Lawrie, R.A. *Lawrie's Meat Science*, 6th ed.; Woodhead Publishing Limited: Cambridge, UK, 1998.

25. Wood, J.; MacFie, H.; Pomeroy, R.; Twinn, D. Carcass composition in four sheep breeds: The importance of type of breed and stage of maturity. *Anim. Sci.* **1980**, *30*, 135–152. [CrossRef]

26. McPhee, M.; Hopkins, D.; Pethick, D. Intramuscular fat levels in sheep muscle during growth. *Aust. J. Exp. Agric.* **2008**, *48*, 904–909. [CrossRef]

27. Hopkins, D.; Stanley, D.; Martin, L.; Toohey, E.; Gilmour, A.R. Genotype and age effects on sheep meat production 3. Meat quality. *Aust. J. Exp. Agric.* **2007**, *47*, 1155–1164. [CrossRef]

28. Kim, Y.H.B.; Stuart, A.; Black, C.; Rosenvold, K. Effect of lamb age and retail packaging types on the quality of long-term chilled lamb loins. *Meat Sci.* **2012**, *90*, 962–966. [CrossRef] [PubMed]

29. Bailey, A.; Shimokomaki, M. Age related changes in the reducible cross-links of collagen. *FEBS Lett.* **1971**, *16*, 86–88. [CrossRef]

30. Khliji, S.; Van de Ven, R.; Lamb, T.; Lanza, M.; Hopkins, D. Relationship between consumer ranking of lamb colour and objective measures of colour. *Meat Sci.* **2010**, *85*, 224–229. [CrossRef]

31. Brandstetter, A.M.; Picard, B.; Geay, Y. Muscle fibre characteristics in four muscles of growing male cattle: II. Effect of castration and feeding level. *Livest. Prod. Sci.* **1998**, *53*, 25–36. [CrossRef]

Article

Attitudinal Determinants of Beef Consumption in Venezuela: A Retrospective Survey

Lilia Arenas de Moreno [1], Nancy Jerez-Timaure [2,*], Jonathan Valerio Hernández [3], Nelson Huerta-Leidenz [4] and Argenis Rodas-González [5]

1   Facultad de Agronomía, Instituto de Investigaciones Agronómicas, Universidad del Zulia, Maracaibo 15205, Venezuela; lilia.arenas@gmail.com
2   Instituto de Ciencia Animal, Facultad de Ciencias Veterinarias, Universidad Austral de Chile, Valdivia 5090000, Chile
3   Universidad Autónoma Chapingo, Km.38.5 Carretera México Texcoco, Chapingo 56230, Mexico; jevh_93@hotmail.com
4   Department of Animal and Food Sciences, Texas Tech University, Box 42141, Lubbock, TX 79409-2141, USA; nelson.huerta@ttu.edu
5   Department of Animal Science, Faculty of Agricultural & Food Sciences, University of Manitoba, Winnipeg, MB R3T 2N2, Canada; Argenis.RodasGonzalez@umanitoba.ca
*   Correspondence: nancy.jerez@uach.cl; Tel.: +56-953-567-344

Received: 15 January 2020; Accepted: 13 February 2020; Published: 16 February 2020

Abstract: Consumer surveys were conducted in the Western, Central, and Eastern regions of Venezuela to determine buying expectations, motivations, needs, perceptions, and preferences of beef consumers, and their acceptance of domestic (and foreign) beef, as affected by different intrinsic and extrinsic factors. Data (n = 693) were gathered by face-to-face interviews on the way out of fresh markets, butcher stores, supermarkets, and, in some cases, at home by using a 45-question structured questionnaire. Responses were subjected to factorial analysis of correspondence (FA) and hierarchical cluster analysis. From the FA, the first two factors explain 74% of the common variance. Factor 1 comprises intrinsic attributes such as color, smell, tenderness, flavor, juiciness, and freshness; while Factor 2 contains extrinsic attributes, mostly related to the origin. The FA profiling data showed that it is possible to concentrate on the traits that consumers usually use as a criterion to perceive beef quality, and to purchase beef. Using cluster analysis, four groups of consumers were mainly distinguished by region, intrinsic attributes, and credence attributes related to production system, aging, traceability, and hygiene. Results from this study will be helpful in designing strategies for recovering and enhancing the future, domestic beef demand.

Keywords: eating quality; beef; tenderness; flavor; meat purchase decision-making; meat buying criteria

---

1. Introduction

Meat is regarded as the most valuable livestock product [1]. Its consumption remains relatively steady in the developed world; however, in developing countries, its annual per capita consumption has doubled since 1980 [1]. This is not the case in Venezuela, where there is currently a strong contraction in the demand for beef due to rampant hyperinflation and a drastic loss of purchasing power [2–4]. The Venezuelan beef market has three main marketing channels: "Traditional" (local butcher shops that represents 60 percent of the market), which offers beef and beef products of different qualities, depending on location and the surrounding community's economic circumstance; "Modern" (supermarkets and medium-sized grocery stores, selling packaged, higher quality that represents 30 percent of the market); and "Industrial" representing 10 percent of the market and is comprised of beef renderers and packers [3,4]. Higher oil prices during the first decade of the millennium allowed for subsidies to import beef products and (or) slaughter cattle from USA, Brazil, Argentina, and Nicaragua; however, in 2003, Venezuela banned all US beef and beef products because of Bovine Spongiform Encefalopathy regulatory concerns [3]. Domestic beef production, imports, and per capita consumption from 2000 to 2019 have been estimated from different private sources [4]. The beef per capita consumption that experienced a sustained rise during the period 2005 to 2011 and reached a historic peak of 24 kg in 2011, fell dramatically to 7 kg (a 70% decrease) in 2018 [3]. Additionally, a lower domestic production (ca. 180,000 MT) and no imports were estimated for 2019, given the deteriorated macroeconomic and market conditions [4]. The population of Venezuela could reach 28.7 million inhabitants by July 2020, a reduction of 7.42% from the population (ca. 31 million) estimated in 2018 [5]. This anticipated population contraction could be explained by the publicly known refugee crisis in the country [6].

Because of the chaotic economic conditions and agri-food situation in Venezuela [2] any consumer survey conducted regarding the current consumers' habits, needs, quality requirements, and (or) preferences would be of little value for medium- and long-term planning purposes. However, it is clear that, before the ongoing situation of high food insecurity in the country [2], Venezuelans had a high preference for beef as center of the plate [7] and, given the expected axiomatic income elasticity of its demand [8], it can be hypothesized that it may eventually return to its former level if the purchasing power and economic conditions are recovered.

While beef demand is income- and price-sensitive, the purchase decision-making process is also affected by a set of attitudinal and credence factors. Organoleptic attributes perceived by consumers can add or subtract most value to a product, because this set of sensory traits largely determines the acceptance. Perceptions about domestic (and foreign) beef by final consumers are also heavily influenced by their varied socio-cultural background. Perceptions of intrinsic fresh meat attributes (e.g., nutritional value and organoleptic quality) and concerns (ethical quality, food safety, provenance, aging process, feeding, and production technologies, etc.) that influence the purchasing decision-making are poorly documented in Venezuela. Formal surveys designed to discern demand determinants (habits, traditions, preferences, and needs) are not easy to access. This is due to the subject of confidential marketing research for retail corporations or trade associations (e.g., Mercedes Hércules and Asociados, C.A. report for the Venezuelan Beef Council-CONVECAR), cited by Jerez-Timaure et al. [7].

Retrospective data regarding attitudinal responses of beef Venezuelan consumers during the first decade of this century could be helpful in designing strategies for recovering and enhancing the future domestic demand. Hence, the present study aims to (a) learn about attitudinal responses expressed by Venezuelan beef consumers at point of purchase or at home, during the 2007–2008 period, and (b) discover, with their opinions, unmet needs, and opportunities in order to create and capture costumers' preferences, values, and loyalty to Venezuelan fresh beef products.

2. Materials and Methods

2.1. Questionnaire Design Mode of Data Collection and Data Classification

The survey's method was qualitative and consisted in the application of a structured, questionnaire containing 45 questions divided into five sections that included: (1) Socio-demographic characteristics of the participants (Section I, seven questions); (2) Habits in beef consumption (Section II, four questions); (3) Criteria for assessing the quality of raw and cooked meat (intrinsic quality attributes (Section III, 10 questions); (4) Criteria for the evaluation of extrinsic quality attributes (Section IV, 14 questions) and, (5) Motivations for the purchase and/or consumption of beef (Section V, 10 questions).

The questionnaire was administered verbally, in person (face-to-face) using the traditional paper-and-pencil mode to randomly selected households in a total of 693 individuals representing three most populated regions of Venezuela, as follows: (a) Western region with 181 consumers from three cities (Maracaibo, San Francisco, and Cabimas); (b) Central region with 327 consumers from the Caracas,

Maracay, and Valencia; and Eastern region with 186 individuals from Barcelona, Puerto La Cruz and Lecherias. Data were collected between 2007 and 2008. It is worth noting that the domestic beef production in 2007 and 2008 was 490,000 MT and 400,000 MT, respectively, whereas the beef imports for the same years were 290,000 MT and 380,000 MT, respectively [3]. It is also noteworthy that the Venezuelan population for 2007 and 2008 was 27.2 million and 27.7 million inhabitants, respectively [9]. The personalized interview method was used at the exit of wet markets, butcheries, supermarkets and, in some cases, at home. This method was considered the most appropriate because of the length of the questionnaire [10]. To ensure the participation of consumers involved in the purchase of food (meats) and groceries, the selection of participants was conditioned on two aspects: (a) to be a beef consumer, (b) to be responsible for the family's food purchasing-decisions [10,11].

2.1.1. Demographic Characteristics and Beef Consumption Attitudes/Habits

Questions and options to responses related to the socio-demographic characteristics of the participants and habits in beef consumption (Sections I and II) follow: (a) Age (Under 30 years old/30–39 years/40–49 years/50–59 years/60 years. or older/No response); (b) Gender (Male/Female); (c) Civil status (Single/Married/Other/No response); (d) Educational level (Elementary/High School/College/Advanced degree/No response); (e) Occupation (Student/Full-time housewife/Working housewife/Worker/Independent Professional/Full-time employee/Part-time employee/Informal businessman Business owner/Other); (f) Family size (1 Member/2 Members-couple/3 Members/4 Members/More than 4 members); (g) Number of children at home (0/1/2/3/more than 3/No response); (h) Preference for beef (I love it/like it very much/I like it/It does not matter to me/It is not the one I prefer/only if I have no other option/I don't like it.); (i) Frequency of preparation of meals with beef (Daily/once every 2 or 3 days/Once a week/Once or twice per month/Rarely); (j) Frequency of eating beef as a center of the plate (Daily/once every 2 or 3 days/Once a week/Once or twice per month/Rarely); and (k) Changes in the frequency of consumption due to information received about diet/health issues (Increased consumption/There has been no change/Decreased temporarily but then returned to usual/Decreased consumption/Stop consuming).

2.1.2. Criteria for Assessing the Quality of Raw and Cooked Beef

Questions corresponding to Sections III and IV of the questionnaires focused on aspects related to the criteria for beef quality assessment, which have also been considered in previous research [8–12]. Participants indicated their degree of agreement with the above-mentioned questions/statements using the five-point Likert scale: Definitely yes (5 points); Yes (4 points); Indifferent/Don't know (3 points); No (2 points); Definitely no (1 point).

Intrinsic Attributes

Questions/statements to learn about consumers' perceptions on intrinsic attributes were: (1) "Beef tenderness is important", (2) "The color in raw beef is important", (3) "The smell of raw beef is important", (4) "The amount of fat present in raw beef is important", (5) "Preference to buy a leaner beef", (6) "Freshness (appearance/conservation) is important", (7) "A highly marbled beef is indicative of good quality" [after ensuring understanding of marbling (i.e., white flecks of intramuscular fat within the lean sections of the beef cut) and noting a chart with color pictures], (8) "The juiciness of cooked beef is important", (9) "Good flavor is important", and (10) "Beef with no fat (leaner beef) tastes better" (i.e., more flavorful).

Extrinsic Attributes

Questions/statements to learn about consumers' perceptions on extrinsic attributes follows: (1) "Hygiene is a very important decision factor", (2) "The type of cut is important", (3) "The shape and size of the cut are important for cooking", (4) "The beef aging process is important", (5) "Are you concerned or would like to know how animals for beef production are fed?", (6) "Are you concerned that animals intended for meat production will be treated with hormones and/or antibiotics to accelerate their growth?", (7) "If you are given the option to buy branded beef, even if it was more expensive, would you be willing to pay for it?", (8) "If you knew the region where the beef was produced, even if it was more expensive, would you be willing to pay for it?", (9) "If you knew the breed of the animal that produced the meat, even if it were more expensive, would you still be willing to pay it?", (10) "If beef imported from the USA or Canada -and certified by its country of origin-were regularly offered, would they have a greater consumption than the domestic one?", (11) "If beef imported from Colombia -and certified by its country of origin-were regularly offered, would it have a greater consumption than the domestic one?", (12) "If beef imported from other Latin American countries -and certified by its country of origin-were regularly offered, would they have a greater consumption than the domestic one?", (13) "When purchasing packed meat, do you pay attention for (check out) the packing date?", and (14) "If you had the opportunity to buy at the same price imported beef or beef produced in Venezuela, would you prefer domestic beef?".

2.1.3. Consumer's Buying/Consumption Motivations and Perceptions about Freshness

The set of questions/statements to learn on buying/consumption motivations (section V) follows: (1) "Would you pay a higher price for beef if you're guaranteed that it will be very tasty (juicy, tender, nicer taste and appearance, altogether?)", (2) "Would you be willing to pay more for a well-marbled beef?", (3) "Safety for my family is the most important aspect", (4) "Would you pay a better price if you are guaranteed that the beef is very safe for your family?", (5) "Beef with no fat or leaner is good for the human health", (6) "Would you be willing to pay more for a certified beef free of hormones, antibiotics or other additives?", (7) "Would you be willing to pay more for beef certified as Natural or Organic?", (8) "Do you trust the experience of the butcher and his advice at the time of buying beef?", and (9) "Would you be willing to pay a higher price for beef that allows you to prepare it in an easier and faster way (e.g., seasoned, portioned products, ready to cook). Additionally, participants indicated freshness preferences for buying and/or consuming beef (10) by choosing one of the following seven options: 7 = "Hot beef" knowing that it is derived from today's slaughtered cattle, 6 = "Very fresh" (beef without refrigeration), 5 = "Refrigerated (cold) meat, with good appearance and color in the showcase", 4 = "Indifferent/Does not know", 3 = "Aged beef with good color", 2 = "Well-aged beef, even if it has lost some color", and 1 = "Frozen beef".

2.2. Statistical Analyses

All analyses were performed using the software R [12] statistical package. Preliminary and then a Confirmatory Factorial Analysis (FA) of Correspondence [13], and a hierarchical clustering technique were used. The FA helps to determine the combination of attribute levels that are important for consumers, while the cluster technique allows to clustering consumers on several categories, in hierarchical manners. The correlation matrix suggested a high level of collinearity in the group of variables that formed the matrix; Bartlett's test assessed the homogeneity of variances ($\alpha = 0.05$), while Univariate (Shapiro-Wilk) and Multivariate (Mardia) normality tests proved that all variables did not have a normal distribution. The FA is part of the multivariate dependency method and does not require assumption tests such as normality; hence, preliminary analyses determined that FA was worthwhile for this data. Frequencies' distribution and Chi-square analysis for demographics (section I) and beef consumption habits (section II) by region were performed.

The interviewees were classified by means of a cluster analysis of the hierarchical type in order to create exclusive homogeneous classes; with maximum divergence among them, using the k-means method. Frequency analysis was conducted for each cluster or group created. The differences between the groups and the characteristics of the participants were analyzed by Chi-square with the data provided by the frequency analysis.

3. Results

3.1. Demographics Characteristics and Beef Consumption Habits

Descriptive parameters of demographic characteristics of the participants are shown in Table 1 and differences were established with contingency tables and Chi-square analysis among cities. By analyzing the whole sample structure, most of the consumers were women (67.15%). Regarding occupation, 25.65% expressed that they were in full-time employment, and 20.3% were housewives without formal work. Fifty-two percent of the consumers were in the most productive age (30 to 49 years) and 52.89% of them were graduates from college (Table 1). Fifty-nine percent of the interviewed belonged to a family group of more than 3 people with 2 children or fewer (86%). Significant differences among cities were detected for all variables ($p < 0.05$) but civil status ($p = 0.14$). These findings indicate a difference in the distribution frequencies of almost all demographic traits within the region.

In general terms, the Central region sample is mostly comprised by women; additionally, it is the region that concentrated most of the full-time housewives and the highest educational level percentage (college or advanced degrees). Consumers from the Central region comprise the capital city (Caracas) and two highly populated cites (Maracay and Valencia).

Table 2 shows the consumer´s beef consumption habits of the whole sample and stratified by region. The majority (93.8%) of the interviewed consumers had a high preference for beef, as indicated by their three most frequent responses ("I love it", "I like it a lot" and "I like it"). It was also observed that most of the interviewees prepared meals with beef as center of the plate every 2 or 3 days (63.0% and 56.9%, respectively); and this trend was very similar across regions; however, the distribution frequency for each question was different among regions ($p < 0.001$).

Table 1. Demographic characteristics for the whole sample and by region.

Questions	Optional Responses	Total, %	Central Region (n = 327)	Western Region (n = 181)	Eastern Region (n = 186)	p-Value
Age	≤29 years old	15.27	6.8	5.3	3.2	0.004
	30–39 years old	25.79	10.4	8.4	7.1	
	40–49 years Old	26.66	12.4	5.8	8.5	
	50–59 years Old	20.17	9.8	4.6	5.8	
	≥60 years	9.37	5.6	1.7	2.0	
	No response	2.74	2.2	0.3	0.3	
Gender	Male	32.85	13.3	8.6	11.0	0.01
	Female	67.15	33.9	17.4	15.9	
Civil Status	Single	29.97	15.4	8.2	6.3	0.14
	Married	58.50	26.9	14.0	17.6	
	Other	11.10	4.5	3.7	2.9	
	No response	0.43	0.3	0.1	0.0	
Educational level	Elementary	9.51	5.5	3.7	0.3	<0.001
	High School	36.02	18.9	11.8	5.3	
	College	46.69	18.6	8.8	19.3	
	Advanced degree	6.20	2.9	1.4	1.9	
	No response	1.59	1.3	0.3	0.0	
Occupation	Student	6.20	3.0	2.2	1.0	<0.001
	Full-time housewife	20.32	12.7	5.9	1.7	
	Working housewife	9.22	1.9	2.4	4.9	
	Worker	4.47	2.4	1.3	0.7	
	Independent Professional	16.28	6.5	2.4	7.3	
	Full-time employee	25.65	10.7	6.9	8.1	
	Part-time employee	3.75	0.3	1.4	2.0	
	Informal businessman	8.07	5.5	2.3	0.3	
	Business owner	1.59	0.6	0.4	0.6	
	Others	4.47	3.6	0.7	0.1	
Family Size	1 member	4.47	2.0	0.6	1.9	<0.001
	2 members-couple	11.82	4.9	2.6	4.3	
	3 members	24.78	8.1	6.2	10.5	
	4 members	24.78	12.0	5.8	7.1	
	More than 4 members	34.15	20.2	11.0	3.0	
Number of children at home	0	36.17	19.6	8.2	8.4	<0.001
	1	28.10	9.7	8.4	10.1	
	2	21.90	9.7	5.8	6.5	
	3	8.21	5.3	1.6	1.3	
	More than 3	4.90	2.7	1.7	0.4	
	No response	0.72	0.1	0.4	0.1	

Table 2. Beef consumption habits for the whole sample and by region.

Consumption Habits	Optional Responses	Total, %	Central Region ($n = 327$)	Western Region ($n = 181$)	Eastern Region ($n = 186$)	p-Value
Preference for beef	I love it	15.13	7.9	4.9	2.3	<0.001
	I like it very much	25.22	5.0	5.9	14.3	
	I like it	37.90	21.5	8.5	8.4	
	It does not matter to me/It is not the one I prefer	9.37	5.2	3.2	1.0	
	It is not the one I prefer	9.22	6.2	2.6	0.4	
	Only if I have no other option	2.74	1.6	0.9	0.3	
	I do not like it	0.43	0.1	0.1	0.1	
Frequency of preparations od meals with beef	Daily	9.37	5.3	3.3	0.7	<0.001
	Every 2 or 3 days	62.97	26.8	14.7	21.5	
	Once a week	19.74	11.1	5.5	3.6	
	1 or 2 times a month	3.60	2.4	0.6	0.6	
	Rarely	4.32	1.4	2.4	0.4	
Frequency of eating beef as a center of the plate	Daily	7.78	4.8	2.7	0.3	<0.001
	Every 2 or 3 days	56.92	27.1	13.5	16.3	
	Once a week	26.22	11.1	6.1	9.1	
	1 or 2 times a month	5.76	3.2	2.0	0.6	
	Rarely	3.31	1.0	1.7	0.6	
Changes in the frequency of consumption due to information received about diet/health issues	Increased consumption	2.02	1.6	0.3	0.1	<0.001
	There has been no change	61.53	32.3	13.0	16.3	
	Decreased but then returned to usual	13.26	1.3	4.3	7.6	
	Decreased consumption	22.19	11.8	7.6	2.7	
	Stop consuming	1.01	0.1	0.9	0.0	

3.2. Factorial Analysis of Correspondence FA

A preliminary FA including all variables/questions from the survey was performed to exclude those variables with little weight on each factor. Four variables/questions with a Kaiser-Meyer-Olkin (KMO) value less than 0.6 were excluded from further analysis, these were: marbling (as an intrinsic trait), butcher trustworthiness, freshness preferences for buying and/or consuming beef and preference for ready-to-eat meat. The confirmatory FA analysis showed that the most significant variables/questions were simplified into three (3) factors according to the R-square value; the first two FAs gathered the traits explaining most of the total variability, and the first two factors explain 74% of the common variance (Table 3).

Table 3. Results of the confirmatory factorial analysis for questions addressing intrinsic attributes, extrinsic attributes, and consumer purchasing motivations.

Question/Variable	FA1	FA2	FA3
Intrinsic Attributes			
Beef tenderness is important	**0.52**	−0.06	−0.04
The smell of raw beef is important	**0.61**	−0.06	−0.07
Flavor is important	**0.58**	0.10	−0.13
The color in raw beef is important	**0.61**	−0.09	−0.06
Leaner beef taste better	0.00	0.22	−0.02
The amount of fat is important	0.05	0.12	−0.09
Juiciness is important	**0.41**	0.20	−0.14
Preferences for leaner beef	0.10	0.06	0.02
Freshness (appearance/conservation) is important	**0.59**	0.01	−0.17
Extrinsic Attributes			
Hygiene is a very important	**0.59**	−0.05	−0.10
Type of cut is important	0.30	0.37	−0.16
Shape and size of cut is important	0.17	0.35	−0.15
Beef aging is important	−0.21	**0.48**	0.01
Animal feeding is important	0.23	0.38	−0.03
Use of hormones/antibiotics is important	0.30	0.36	0.05
Beef brand is important	0.01	**0.53**	0.10
Traceability is important	−0.07	**0.67**	0.09
Breed information is important	−0.12	**0.66**	0.05
Beef imported from the USA or Canada would have a greater consumption than the domestic one?	−0.12	0.33	**0.66**
Beef imported from Colombia would have a greater consumption than the domestic one?	−0.05	0.02	**0.86**
Beef imported from other Latin American countries would have a greater consumption than the domestic one?	−0.06	0.12	**0.78**
Label information is important	0.11	0.26	−0.12
Domestic beef is better than the imported one	0.23	0.09	−0.34
Consumer's Purchasing Motivations			
Would you pay more for a tasty beef?	0.25	**0.44**	0.04
Safety for my family is the most important aspect	**0.45**	0.16	−0.12
Would you pay a better price if you are guaranteed that the beef is very safe for your family?	0.41	0.29	−0.04
Leaner beef is healthier	0.24	0.26	−0.12
Would you be willing to pay more for a beef certified as free of hormones, antibiotics or other additives?	0.31	**0.40**	0.05
Would you be willing to pay more for beef certified as Natural or Organic?	0.28	0.36	0.17
Would you be willing to pay more for a well-marbled beef?	−0.21	0.37	0.06
Variance by factor (%)	40	35	26
Cumulative variance (%)	40	74	100
R-square	0.82	0.81	0.84

The root means square of the residuals (RMSR) is 0.07. Bartlett's K-squared = 2844.4, df = 33, $p < 0.001$. Numbers highlighted in bold are the highest standarized factors loading (≥ 0.40).

Projections of main variables/questions answered by Venezuelan consumers on intrinsic and intrinsic attributes and buying/consumption motivations (in the FA1 and FA2 axes) are shown in Figure 1. The first factor (FA1) is mainly defined by the importance of "intrinsic attributes" such as tenderness, smell, flavor, color, juiciness, and freshness. The second factor (FA2) shows that extrinsic attributes like origin (traceability, breed information) and beef brand are important. The third factor (FA3) (not shown in Figure 1) highlights the consumer perception that imported beef is better than its domestic counterpart as an extrinsic attribute of beef quality.

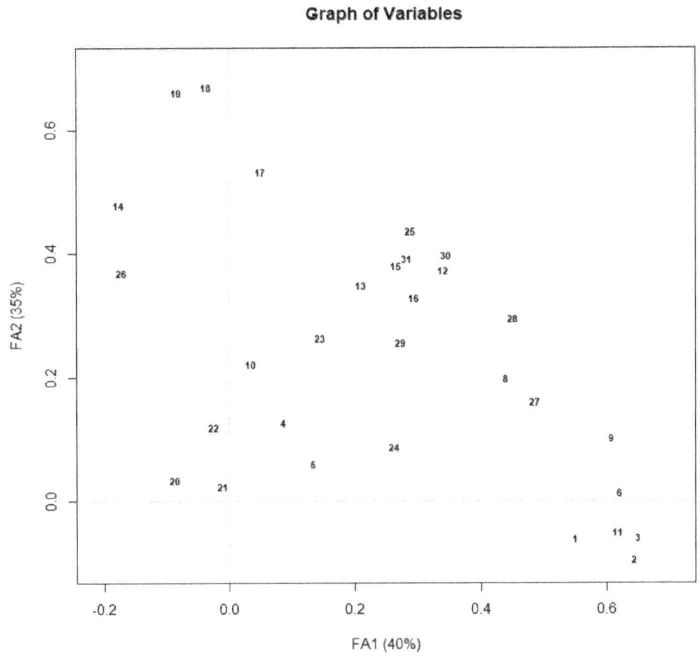

Figure 1. Projection of main variables/questions about consumers' perceptions of intrinsic and intrinsic attributes and consumers buying/consumption motivations in the FA1 and FA2 axes. Variables/question are shown as numbers as follow: Intrinsic attributes: Beef tenderness is important (1); The color in raw beef is important (2); The smell of raw beef is important (3); The amount of fat in raw beef is important (4); Leaner beef taste better (5); Freshness is important (6); A highly marbled beef is indicative of good quality (7); The juiciness is important (8); Good flavor is important (9); Leaner beef taste better (10). Extrinsic attributes: Hygiene is very important (11); The type of cut is important (12); The shape and size of the cut are important (13); Beef aging is important (14); Animal feeding is important (15); Use of hormones/antibiotics is important (16); Beef brand is important (17); Traceability is important (18); Breed information is important (19); Beef imported from the USA or Canada would have a greater consumption than the domestic one? (20); Beef imported from Colombia would have a greater consumption than the domestic one? (21); Beef imported from other Latin American countries would have a greater consumption than the domestic one? (22); Label information is important (23); Domestic beef is better than the imported one (24). Consumer's purchasing motivations: Would you pay a higher price for tasty beef (25); Would you be willing to pay more for a well-marbled beef? (26); Safety for my family is the most important aspect (27); Would you pay more for safe beef? (28); Leaner beef is good for human health (29); Would you be willing to pay more for a certified beef free of hormones, antibiotics or other additives? (30); Would you be willing to pay more for beef certified as Natural or Organic? (31); Do you trust the experience of the butcher (32); Freshness preferences for buying and/or consuming beef (33); Would you be willing to pay a higher price for beef that allows you to prepare it in an easier and faster way (34).

In Figure 2, intrinsic (FA1) and extrinsic attributes (FA2) are represented by the two-dimensional space defined by these factors or axes, by region. Preferences and motivations for purchasing beef based on perceived freshness were different ($p < 0.05$), and more noticeable between consumers from Western and Eastern regions. However, consumers from the Central region showed wider preferences and motivations that cover those expressed by counterparts in both Western and Eastern regions. Eastern consumers were more concentrated and gave more importance to origin and production traits (breed type, animal feeding, use of hormones). Additionally, they recognized the importance of aging to improve beef quality; therefore, they were willing to pay more for branded beef, traceability, and tastier meat. Consumers from the Western region gave more importance to intrinsic beef attributes (tenderness, flavor, juiciness, smell, color, and freshness), while consumers in the Central region showed a broader spectrum of preferences and motivations to buy beef.

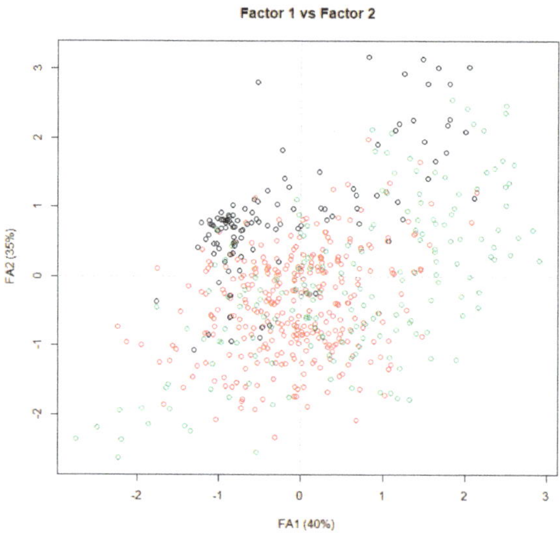

Figure 2. Projection of the Central (green), Western (black), and Eastern (red) regions for the first and second factors (FA1 and FA2, respectively).

3.3. Cluster Analysis Characteristics of Different Types of Consumers

Cluster analysis allowed identifying the conformation of four groups of consumers that showed to have a small as possible of internal variants, the advantages of this method are using variance analysis to check the distance between clusters. Figure 3 shows the hierarchical clustering of the four groups of consumers and the Euclidean distance among them. For instance, G1 (showed by red color) and G4 (purple color) were the closest groups, whereas G2 (green color) and G3 (blue color) resulted to be the most divergent ones.

The frequencies for each cluster (group of consumers) variables/question that resulted with $p < 0.05$ in the Chi-square test and resulted with the highest eigenvalue in the FA analysis, are depicted in Table 4 (intrinsic attributes), Table 5 (extrinsic attributes), and Table 6 (consumer´s buying/consumption motivations and perceptions.

Table 4. Intrinsic attributes with significant differences ($p < 0.05$ *) in the groups of consumers identified by cluster analysis.

Question/Variable	G 1 ($n = 138$)		G 2 ($n = 390$)		G 3 ($n = 138$)		G 4 ($n = 27$)	
	n	%	n	%	n	%	n	%
Beef Tenderness Is Important								
1	0	0.00	1	0.26	0	0.00	0	0.0
2	0	0.00	6	1.54	2	1.45	0	0.0
3	0	0.00	8	2.05	3	2.17	0	0.0
4	137	99.28	204	52.31	24	17.39	20	74.1
5	1	0.72	171	43.85	109	78.99	7	25.9
Color of Raw Beef Is Important								
1	0	0.00	0	0.00	0	0.00	0	0.0
2	0	0.00	6	1.54	0	0.00	0	0.0
3	0	0.00	10	2.56	1	0.72	0	0.0
4	133	96.38	207	53.08	23	16.67	14	51.9
5	5	3.62	167	42.82	114	82.61	13	48.1
Smell of Raw Beef Is Important								
1	0	0.00	1	0.26	0	0.00	0	0.0
2	0	0.00	2	0.51	0	0.00	0	0.0
3	0	0.00	5	1.28	1	0.72	0	0.0
4	117	84.78	197	50.51	20	14.49	9	33.3
5	21	15.22	185	47.44	117	84.78	18	66.7
Freshness (Appearance/Conservation) Is Important								
1	1	0.72	0	0.00	0	0.00	0	0.0
2	0	0.00	2	0.51	0	0.00	0	0.0
3	0	0.00	0	0.00	1	0.72	0	0.0
4	135	97.83	307	78.72	22	15.94	9	33.3
5	2	1.45	81	20.77	115	83.33	18	66.7
Juiciness of Cooked Beef Is Important								
1	0	0.00	1	0.26	0	0.00	0	0.0
2	0	0.00	15	3.85	5	3.62	0	0.0
3	0	0.00	19	4.87	2	1.45	0	0.0
4	134	97.10	308	78.97	47	34.06	14	51.9
5	4	2.9	47	12.05	84	60.87	13	48.10
Good flavor is Important								
1	0	0.00	1	0.26	0	0.00	0	0.0
2	0	0.00	2	0.51	0	0.00	0	0.0
3	0	0.00	5	1.28	0	0.00	0	0.0
4	131	94.93	285	73.08	27	19.57	2	7.4
5	7	5.07	97	24.87	111	80.43	25	92.6

Participants indicated their degree of agreement using the 5-point Likert scale: 1: Definitely no; 2: No; 3: Indifferent/Don't know; 4: Yes; 5: Definitely yes. * obtained from Chi-squared test.

Table 5. Extrinsic attributes with significant differences ($p < 0.05$ *) in the groups of consumers identified by cluster analysis.

Question/Variable	G 1 ($n = 138$)		G 2 ($n = 390$)		G 3 ($n = 138$)		G 4 ($n = 27$)	
	n	%	n	%	n	%	n	%
Hygiene Is a Very Important Factor								
1	0	0.00	0	0.00	0	0.00	0	0.0
2	0	0.00	1	0.26	0	0.00	0	0.0
3	0	0.00	0	0.00	0	0.00	0	0.0
4	123	89.13	179	45.90	7	5.07	2	7.4
5	15	10.87	210	53.85	131	94.93	25	92.6
The Beef Aging Process Is Important								
1	0	0.00	11	2.82	12	8.70	6	22.2
2	1	0.72	86	22.05	24	17.39	11	40.7
3	4	2.90	136	34.87	27	19.57	9	33.3
4	131	94.93	148	37.95	41	29.71	1	3.7
5	2	1.45	9	2.31	34	24.64	0	0.0
Are you Concerned or Would Like to Know How Animals for Beef Production are Fed?								
1	0	0.00	2	0.51	1	0.72	5	18.5
2	0	0.00	43	11.03	6	4.35	11	40.7
3	1	0.72	24	6.15	5	3.62	3	11.1
4	132	95.65	285	73.08	57	41.30	5	18.5
5	5	3.62	36	9.23	69	50.00	3	11.1
Are You Concerned that Animals Intended for Meat Production Are Treated with Hormones and/or Antibiotics to Accelerate Their Growth?								
1	0	0.00	4	1.03	2	1.45	0	0.0
2	0	0.00	35	8.97	5	3.62	0	0.0
3	1	0.72	23	5.90	3	2.17	1	3.7
4	128	92.75	280	71.79	53	38.41	18	66.7
5	9	6.52	48	12.31	75	54.35	8	29.6
If You Are Given the Option to Buy Branded Beef, Even If it Was More Expensive, Would You be Willing to Pay for It?								
1	0	0.00	2	0.51	7	5.07	2	7.4
2	1	0.72	101	25.90	26	18.84	7	25.9
3	1	0.72	26	6.67	10	7.25	8	29.6
4	135	97.83	246	63.08	41	29.71	7	25.9
5	1	0.72	15	3.85	54	39.13	3	11.1
If You Knew the Region Where the Beef Was Produced, Even If It Was More Expensive, Would You be Willing to Pay for It?								
1	0	0.00	3	0.77	8	5.80	2	7.4
2	1	0.72	168	43.08	35	25.36	13	48.1
3	3	2.17	50	12.82	7	5.07	10	37.0
4	133	96.38	163	41.79	55	39.86	1	3.7
5	1	0.72	6	1.54	33	23.91	1	3.7
If You Knew the Breed of the Animal that Produced the Meat, Even If It Were More Expensive, Would You Still be Willing to Pay It?								
1	0	0.00	2	0.51	12	8.70	3	11.1
2	1	0.72	161	41.28	32	23.19	12	44.4
3	4	2.90	57	14.62	11	7.97	10	37.0
4	131	94.93	162	41.54	56	40.58	1	3.7
5	2	1.45	8	2.05	27	19.57	1	3.7
If you had the opportunity to buy at the same price imported beef vs. beef produced in Venezuela, would you willing to pay for it?								
1	0	0.00	1	0.26	0	0.00	0	0.0
2	16	11.59	38	9.74	7	5.07	5	18.5
3	7	5.07	36	9.23	5	3.62	10	37.0
4	113	81.88	284	72.82	38	27.54	12	44.4
5	2	1.45	31	7.95	88	63.77	0	0.0

Participants indicated their degree of agreement using the five-point Likert scale: 1: Definitely no; 2: No; 3: Indifferent/Don't know; 4: Yes; 5: Definitely yes. * obtained from Chi-squared test.

Table 6. Consumer's buying/consumption motivations and perceptions with significant differences (p < 0.05 *) in the groups of consumers identified by cluster analysis.

Variable	G 1 (n = 138)		G 2 (n = 390)		G 3 (n = 138)		G 4 (n = 27)	
	n	%	n	%	n	%	n	%
Would you Pay a Higher Price for Beef If You Are Guaranteed that It Will be Very Tasty (Juicier, Tender, Nicer Taste and Appearance, Altogether?								
1	0	0.00	3	0.77	2	1.45	0	0.0
2	0	0.00	46	11.79	5	3.62	0	0.0
3	1	0.72	23	5.90	3	2.17	0	0.0
4	135	97.83	295	75.64	53	38.41	22	81.5
5	2	1.45	23	5.90	75	54.35	5	18.5
Safety for My Family Is the Most Important Aspect								
1	0	0.00	0	0.00	0	0.00	0	0.0
2	0	0.00	17	4.36	0	0.00	0	0.0
3	0	0.00	10	2.56	0	0.00	0	0.0
4	135	97.83	272	69.74	22	15.94	10	37.0
5	3	2.17	91	23.33	116	84.06	17	63.0
Would You be Willing to Pay More for a Certified Beef Free of Hormones, Antibiotics or Other Additives?								
1	0	0.00	0	0.00	0	0.00	0	0.0
2	0	0.00	36	9.23	12	8.70	0	0.0
3	0	0.00	21	5.38	6	4.35	0	0.0
4	133	96.38	302	77.44	42	30.43	20	74.1
5	5	3.62	31	7.95	78	56.52	7	25.9
Would You be Willing to Pay More for Beef Certified as Natural or Organic?								
1	0	0.00	2	0.51	1	0.72	0	0.0
2	10	7.25	63	16.15	19	13.77	0	0.0
3	1	0.72	35	8.97	10	7.25	0	0.0
4	123	89.13	262	67.18	39	28.26	17	63.0
5	4	2.90	28	7.18	69	50.00	10	37.0

Participants indicated their degree of agreement using the five-point Likert scale: 1: Definitely no; 2: No; 3: Indifferent/Don't know; 4: Yes; 5: Definitely yes. *obtained from Chi-squared test.

It is worthwhile to highlight that most of the intrinsic attributes were very important for all consumers interviewed; however, consumers from G3 group were more sensitive to these attributes than those from G1 group. All consumers grouped in G1 were from the Eastern region (n = 138) and represent 20% of the total data. Consumers from G1 assigned greater importance to extrinsic factors such as hygiene and beef aging; however, they also would like to know about the origin, feeding practices and breed of the animals; hence, they are willing to pay more for higher quality, freedom of hormones, and certified organic beef. The G2 group (n = 390) represents 56.3% of the total consumers, mostly from the central region (76.77%), who were more sensitive to intrinsic attributes, mainly freshness, tenderness, flavor, and color and less worry about extrinsic attributes like origin, breed and feeding, among others. The G3 group considers as very important, the beef intrinsic quality traits (tenderness, color, smell, freshness, and juiciness); and represents 20% of the total data (n = 138), most of them (85%) from the Western region. Lastly, G4 was the smallest group comprised of only 24 consumers, all from the Western region (n = 27). Consumers in G4 did not give importance to extrinsic factors such as beef aging, origin, breed, or feeding practice but were willing to pay more for safe, good-quality beef; they also preferred domestic beef rather than the imported ones.

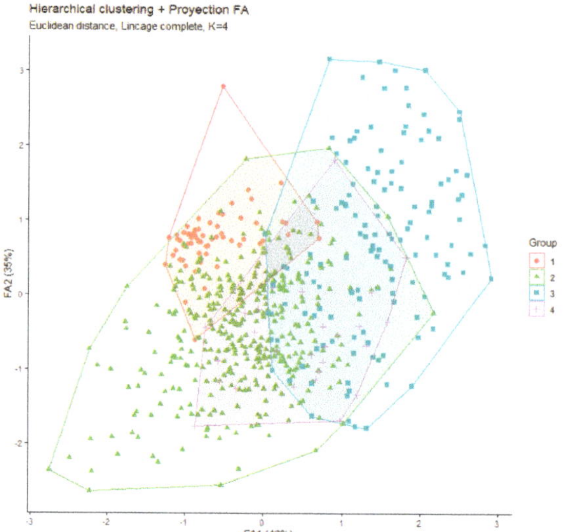

Figure 3. Hierarchical clustering and group of consumers projected in the FA1 and FA2 axes.

4. Discussions

Consumers are the last segment of the beef chain, and having their expectations met is an important part of their satisfaction and shopping behavior. It is, therefore, important to understand the factors affecting consumer behavior [14]. On the other hand, consumer perception is generally dynamic, and its development needs to be monitored continuously as one of the information resources for decision-making in all the food chain. Villalobos et al. [15] stated that Chilean consumers can modify their choices and habits, prioritizing quality attribute differentiators over economic aspects at the time of purchase.

Few studies have evaluated the demand determinants based on beef quality attributes in developing countries. Castillo and Carpio [16] conducted a survey of 574 households in Ecuador and reported that they have a positive interest in all credence attributes (i.e., sanitary control, aging, animal welfare and traceability). In Brazil, Giacomazzi et al. [17] reported that the most valued attributes for beef consumers are appearance, price, and type of cut; while brand and certifications had little relevance as purchase-decision criteria.

To get a broader representation of Venezuelan consumers, surveys were performed in the regions that concentrate most of the population, including Caracas, the capital city of the country. From the total sample of consumers ($n = 693$), most of them (67.1%) were women. This finding concurs with those of Segovia et al. [18], Pierce-Colfer et al. [19], Mahbubi et al. [20] and Castillo and Carpio [16], supporting the argument that women dominate the decision-making for beef consumption in their family.

Educational status affects food consumption patterns, which reflects more nutrition and dietetic knowledge, and health-food perceptions, as stated by Bhurosy and Jeewon [21]. For Krystallis and Arvanitoyanis [10], age and education are the most important socio-demographic characteristics that influence consumers' attitudes towards meat. Older consumers associate meat purchases to the risk concept, mainly due to perceived diet-health issues.

Most of the consumers interviewed herein have families comprised of four or more members (Table 1). Yee et al. [22] stated that smaller families tend to be well-nourished and healthier because they have a broader opportunity to access sources of animal protein. Bonny et al. [23] highlighted other demographic factors such as income, gender, employment status, occupation, city or region, number of children and adults at home; while the frequency of beef consumption and degree of doneness had little effect on beef preferences.

Participants in this survey showed a relatively high preference for beef. The majority (93.8%) responded that they like beef or like it very much, and more than 50% of the responders consumed beef at least every 2 or 3 days. High frequencies of beef consumption have also been reported by the majority (>50%) of responders in other Latin American countries like Mexico (i.e., 75%, by Vilaboa-Arroniz et al. [24], and 95% by Ngapo et al. [25]), Ecuador (79%, by Castillo and Carpio [16]), Costa Rica (91%, by CORFOGA, [26]), and Chile (82%, [27]). Additionally, in Australia, Ardeshiri and Rose [28] reported from a survey of 1002 Australian residents that the majority of respondents (51.8%) purchase beef once a week.

Intrinsic and extrinsic beef attributes are important determinants of consumer´s demand for beef; however, the relative importance of these factors varies from country to country [16] or from city to city (in the same country), as it was detected in the present study. Intrinsic cues relate to physical aspects of the product (e.g., color, shape, appearance, etc.) whereas extrinsic ones relate to the product but are not physically part of it (brand, origin, store, packaging, type of cut, production information, etc.).

From the preliminary FA, four variables/questions with a low KMO value were excluded from further FA analysis, namely marbling (as an intrinsic trait and willingness), butcher trustworthiness, freshness preferences for buying and/or consuming beef, and preference for ready-to-eat meat. These results showed that these attributes were not as important for Venezuelan consumers.

Among the intrinsic attributes, marbling, and color stand out as the most important beef quality traits in other surveys [29]. Marbling is well known by consumers in countries with long-standing appreciation for beef quality, and it is frequently used as an intrinsic clue (and consumers are willing to pay for highly marbled beef); particularly in those countries with well-established beef carcass quality systems or meat standards (i.e., USA, Canada, Australia, Japan) [30]. Additionally, it is widely accepted that the use of early-maturing *Bos taurus* genetics, castration and grain feeding favors deposition of intramuscular fat [31]. However, in Venezuela, most of the cattle are genetically composed by *Bos indicus* types (mainly Brahman straight-breeds and crossbreds) because of their adaptability to harsh, tropical conditions. Brahman-influenced breed-types have been characterized in many studies as having lower intramuscular fat content (and hence, lower marbling scores) as compared to *Bos taurus* breed types [31]. Additionally, the beef cattle in Tropical America are mostly raised and fattened on pastures with little supplementation (i.e., lower energy diets), which contributes to a poor marbling performance. Moreover, the predominant slaughter cattle in Venezuela are intact males whose carcasses are leaner (with lower marbling scores) than castrates and cull females. For instance, Jerez and Huerta [32] described beef carcasses derived from grass-fed bulls with low marbling scores (fluctuating between "Traces" and "Slight" amounts). This observation can explain why most Venezuelan consumers are not familiar with marbling or do not recognize it as an important quality trait/clue.

Preference for frozen vs. refrigerated beef or ready-to-eat beef was irrelevant. Indifference (or lack of understanding) to ready-to-eat meals may be explained for the entrenched custom of many Venezuelans to prefer to eat at home. On the other hand, extrinsic attributes such as butcher trustworthiness were disregarded by consumers. This may be due to the growth of self-service store (without personalized attention) in the cities under study. It should be noted that many of the surveys were carried out at the exit of this type of store. Additionally, there are many anecdotes of bad experiences dealing with butchers, which leads to a bad opinion of these merchants. Previous experiences could modify the quality expectations of meat products [29].

The confirmatory FA revealed that intrinsic attributes such as tenderness, color, juiciness, smell, flavor, and freshness were deemed important by most consumers; however, cluster analysis revealed that intrinsic attributes, such as tenderness, color, flavor, and smell were more important for G3 than for the G1 group of consumers. The brilliant red color is desirable for most consumers and it determines the purchasing decision, probably because consumers perceive discoloration as an indicator of spoilage [33]. In other countries, consumers also relate red–purple color with freshness and brown color with a lack of freshness [11,14,27].

Most of the consumers from this survey (G3, G2, and G4) highly regarded tenderness as an important attribute for beef preference, and interestingly, they acknowledged more the beneficial effect of aging on beef quality/tenderness than marbling. The latter perception is supported by science because Khan et al. [34] and many others demonstrated that beef aging positively affects the final texture, juiciness, and aroma developed through the major biochemical processes-proteolysis, lipolysis, and oxidation.

Several extrinsic factors related to origin (i.e., breed information, animal feeding, traceability, etc.) were important to Venezuelan consumers. According to the FA analysis, these attributes are regarded as some of the most important considerations for explaining differences among consumers. The G1 group was more sensitive to extrinsic attributes of beef; generally, consumers preoccupied about origin of beef also have a higher preference for healthier beef (free of hormones, additives or certified as natural or organic). Bernués et al. [35] pointed out that animal feeding and the provenance are the most important extrinsic attributes for European consumers and supported that origin of meat has also been associated with meat safety. In France, a survey with 625 consumers [36] revealed that consumers are willing to pay for meat products with guaranteed attributes such as labels, traceability, tenderness, and certifications. Meanwhile, in Latin American (Chile), Villalobos et al. [15] found that quality assurance was the most important attribute in the consumer's beef purchase decision process, followed by country of origin, production system, and price.

Perceptions of beef quality and motivations to buy beef have shown to be different among regions. However, from a marketing standpoint, results from the Central region—which included the most populated cities—indicate that its attitudinal pattern could be representative of the rest of the country. However, cluster analysis provides a more precise consumer segmentation and allowed to determine the demographics and preferences profile of each group. For instance, cluster analysis segregated consumers from the Eastern region in one group (G1); and this group had a particular interest in origin and traceability, perhaps because most (73%) of them had some college education and 55.07% were independent professionals or full-time employees; hence, they seem to be better informed about beef production. On the other hand, the majority of the G2 consumers were from the Central region, with a high-school educational level (41.5%), and most of them were full-time housewives. Most of the consumers from the Western region were concentrated in the G4 group, who dismissed extrinsic attributes over intrinsic ones.

5. Conclusions

Two factors could explain the higher proportion of heterogeneity in this sample of Venezuelan beef consumers. Intrinsic attributes such as tenderness, color, smell, flavor, freshness, and juiciness; as well as extrinsic attributes such as aging, hygiene, origin, breed, and animal feeding information were important for Venezuelan consumers.

The FA analysis of the profiling data showed a distinct location for each region in the multivariate space and four groups of consumers were defined by cluster analysis, which demonstrated that the relative importance that consumers are giving to the different beef attributes, as determinants of its purchase and consumption, varies widely between regions, and, in turn, depends on the educational level, occupation and other sociodemographic characteristics.

This reality constitutes a great opportunity for consumer-driven product development and further market segmentation and indicates that consumers need to be well informed and educated on quality and food safety matters, by means of trustworthy sources of information.

Author Contributions: L.A.d.M. and N.H.-L. designed the consumers' questionnaire and survey. L.A.d.M. was responsible for the selection, induction, and training of personnel who conducted the interviews and collected field data that support this research. L.A.d.M., N.J.-T., and J.V.H. performed statistical analyses, tabulated survey results and wrote the result and discussion section. L.A.d.M., N.H.-L., A.R.-G., and N.J.-T. reviewed the literature and interpreted, designed, and revised the structure of the manuscript. All authors searched and reviewed the literature, discussed the contents of the manuscript, and approved submission. All authors have read and agreed to the published version of the manuscript.

Funding: This research was funded by the Scientific and Humanistic Development Council of the Universidad del Zulia (CONDES-LUZ), grant number CC0976-07 (Research Program "Alternative Systems of Production. Marketing and Biotechnology of Meats").

Acknowledgments: We thank Carlos Carpio (Texas Tech University), Eugenio Mendoza (Universidad del Zulia), and Martin O'Connor, for their valuable comments and suggestions which helped improved the manuscript.

Conflicts of Interest: The authors declare no conflict of interest.

References

1. Food and Agriculture Organization of the United Nations (FAO). Meat & Meat Products, Agriculture and Consumer Protection Department, Animal Production and Health. Available online: http://www.fao.org/ag/againfo/themes/en/meat/home.html (accessed on 8 December 2019).
2. Gutiérrez, B.A. La situación agroalimentaria en Venezuela: Hacia una nueva estrategia. *Foro.* 2019, 3, pp. 31–52. Available online: https://www.revistaforo.com/2019/0305-04 (accessed on 8 December 2019).
3. FAS-USDA. Foreign Agricultural Service, United States Department of Agriculture, GAIN Report No. VE1804. Venezuela: Livestock and Products Annual. Annual Report 2018. Available online: https://apps.fas.usda.gov/newgainapi/api/report/downloadreportbyfilename?filename=Livestock%20and%20Products%20Annual_Caracas_Venezuela_8-13-2018.pdf (accessed on 7 February 2020).
4. FAS-USDA. Foreign Agricultural Service, United States Department of Agriculture, GAIN Report No. VE2019-0001. Venezuela: Livestock and Products Annual. Available online: https://www.fas.usda.gov/data/venezuela-livestock-and-products-annual-1 (accessed on 9 December 2019).
5. Central Intelligence Agency (CIA), United States of America. The World Factbook. Available online: https://www.cia.gov/library/publications/the-world-factbook/fields/335rank.html#VE (accessed on 5 February 2020).
6. O'Neil, S.K. A Venezuelan refugee crisis. In *Preparing for the Next Foreign Policy Crisis: What the United States Should Do*; Stares, P.B., Ed.; Council on Foreign Relations: New York, NY, USA, 2019; pp. 77–90.
7. Jerez-Timaure, N.; Arenas de Moreno, L.; Giuffrida de Mendoza, M. *Que Buena es la Carne Bovina Venezolana*; Landaeta de Jiménez, M., Ed.; Fundación Bengoa para la Alimentación y Nutrición: Caracas, Venezuela, 2010.
8. Huerta-Leidenz, N.; Valdez-Muñoz, A.; Lopez, F.; Scott, T.; Howard, S.T.; Dixon, C.; Belk, K.E. Retail display case merchandising's consist and price-elasticity of demand for U.S. beef and pork variety meats sold in Mexican grocery stores. *Rev. Fac. Agron. (LUZ)* **2017**, *34*, 216–235.
9. Instituto Nacional de Estadística (INE). Ministerio del Poder Popular de Planificación. República Bolivariana de Venezuela. Demograficos. Available online: http://www.ine.gov.ve/index.php?option=com_content&view=category&id=98&Itemid=51 (accessed on 6 February 2020).
10. Krystallis, A.; Arvanitoyannis, I.S. Investigating the concept of meat quality from consumer's perspective: The case of Greece. *Meat Sci.* **2006**, *72*, 164–176. [CrossRef] [PubMed]
11. Bernués, A.; Olaizola, A.; Corcoran, K. Labelling information demanded by European consumers and relationships with purchasing motives, quality and safety of meat. *Meat Sci.* **2003**, *65*, 1095–1106. [CrossRef]
12. Core Team. R: A language and Environment for Statistical Computing. R Foundation for Statistical Computing. Available online: https://www.R-project.org/ (accessed on 12 January 2019).
13. Loret-Segura, S.; Ferreres-Traver, A.; Hernandez-Baeza, A.; Tomas-Marco, I. El análisis factorial exploratorio de los ítems: Una guía práctica, revisada y actualizada. *An. Psicol.* **2014**, *30*, 1151–1169. [CrossRef]
14. Fonti-i-Furnols, M.; Gerrero, L. Consumer preference, behavior and perception about meat and meat products: An overview. *Meat Sci.* **2014**, *98*, 361–371. [CrossRef]
15. Villalobos, P.; Padilla, C.; Ponce, C.; Rojas, A. Beef consumer preferences in Chile: Importance of quality attribute differentiators on the purchase decision. *Chil J. Agric. Res.* **2009**, *70*, 85–94. [CrossRef]
16. Castillo, M.J.; Carpio, C.E. Demand for high-quality beef attributes in developing countries: The case of Ecuador. *J. Agric. Appl. Econ.* **2019**, *51*, 568–590. [CrossRef]
17. Giacomazzi, C.M.; Talamini, E.; Kindlein, L. Relevance of brands and beef quality differentials for the consumer at the time of purchase. *R. Bras. Zootec.* **2017**, *46*, 354–365. [CrossRef]

18. Segovia, E.; Contreras, D.; Marcano, D.; Pirela, R.; Albornoz, A. Conducta del consumidor de carne bovina según clase socioeconómica en el municipio Maracaibo, Estado Zulia, Venezuela. *Agroalimentaria* **2005**, *11*, 113–121.

19. Pierce-Colfer, C.P.; Achdiawan, R.; Roshetko, J.M.; Mulyoutami, E.; Yuliani, E.L.; Mulyana, A.; Moeliono, M.; Adnan, H.E. The balance of power in household decision-making: Encouraging news on gender in southern Sulawesi. *World Dev.* **2015**, *76*, 147–164. [CrossRef]

20. Mahbubi, A.; Uchiyama, T.; Hatanaka, K. Capturing consumer value and clustering customer preferences in the Indonesian halal beef market. *Meat Sci.* **2019**, *156*, 23–32. [CrossRef] [PubMed]

21. Bhurosy, T.; Jeewon, R. Overweight and obesity epidemic in developing countries: A problem with diet, physical activity, or socioeconomic status? *Sci. World J.* **2014**, *2014*, 1–7. [CrossRef] [PubMed]

22. Yee, A.Z.H.; Lwin, M.O.; Ho, S.S. The influence of parental practices on child promotive and preventive food consumption behaviors: A systematic review and meta-analysis. *Int. J. Behav. Nutr. Phys. Act.* **2017**, *14*, 1–14. [CrossRef] [PubMed]

23. Bonny, S.P.F.; Gardner, G.E.; Pethick, D.W.; Allen, P.; Legrand, I.; Wierzbicki, J.; Farmer, L.J.; Polkinghorne, R.J.; Hocquette, J.F. Untrained consumer assessment of the eating quality of European beef: 2. Demographic factors have only minor effects on consumer scores and willingness to pay. *Animal* **2017**, *11*, 1399–1411. [CrossRef]

24. Vilaboa-Arroniz, J.; Díaz-Rivera, P.; Ruiz-Rosado, O.; Platas-Rosado, D.; González-Muñoz, S.; Juarez-Lagunes, F. Beef consumption patterns in the Papaloapan region, Veracruz, México. *Agric. Soc. Desarro* **2009**, *6*, 145–159.

25. Ngapo, T.M.; Braña Varela, D.; Rubio Lozano, M.S. Mexican consumers at the point of meat purchase. Beef choice. *Meat Sci.* **2017**, *134*, 34–43. [CrossRef]

26. Corporación de Fomento Ganadero (CORFOGA). Estudio C.U.A.S Consumo, Uso y Actitudes: Carnes. Eds. AW-Investigación de Mercados. San Jose, Costa Rica. 2006. Available online: https://www.corfoga.org/carnes-de-costa-rica/ (accessed on 6 February 2019).

27. Schnettler, B.; Clesia, M.; Candia, A.; Llancanpán, F.; Sepúlveda, J.; Denegri, M.; Miranda, H.; Sepúlveda, N. The Importance of colour, fat content and freshness in the purchase of beef in Temuco, La Araucanía region, Chile. *Rev. Cien. FCV-LUZ* **2010**, *20*, 623–632.

28. Ardeshiri, A.; Rose, J.M. How Australian consumer value intrinsic and extrinsic attributes of beef products. *Food. Qual. Prefer* **2018**, *65*, 146–163. [CrossRef]

29. Henchion, M.M.; McCarthy, M.; Resconi, V.C. Beef quality attributes: A Systematic review of consumers perspectives. *Meat Sci.* **2017**, *128*, 1–7. [CrossRef]

30. Polkinghorne, R.J.; Thompson, J.M. Meat standards and grading: A world view. *Meat Sci.* **2016**, *86*, 227–235. [CrossRef]

31. Byers, F.M.; Cross, H.R.; Schelling, G.T. Integrated nutrition, genetics, and growth management programs for lean beef production. In *Designing Foods: Animal Product Options in the Marketplace*; National Academy Press: Washington, DC, USA, 1988; pp. 283–291.

32. Jerez-Timaure, N.; Huerta-Leidenz, N. Effects of breed type and supplementation during grazing on carcass traits and meat quality of bulls fattened on improved savannah. *Livest. Sci.* **2009**, *121*, 219–226. [CrossRef]

33. Manchini, R.A.; Ramanathan, R. Effects of postmortem storage time on color and mitochondria in beef. *Meat Sci.* **2014**, *98*, 65–70. [CrossRef] [PubMed]

34. Khan, M.I.; Jung, S.; Nam, K.C.; Jo, C. Postmortem aging of beef with a special reference to the dry aging. *Korean J. Food Sci. An.* **2016**, *36*, 159–169. [CrossRef] [PubMed]

35. Bernués, A.; Olaizola, A.; Corcora, K. Extrinsic attributes of red meat as indicators of quality in Europe: An application for market segmentation. *Food. Qual. Prefer* **2003**, *14*, 265–276. [CrossRef]

36. Ellies-Oury, M.P.; Lee, A.; Hocquette, J.F. Meat consumption-what French consumers feel about the quality of beef? *Ital. J. Anim. Sci.* **2019**, *18*, 646–656. [CrossRef]

 *foods*

Article

Extending Aging of Beef *Longissimus Lumborum* From 21 to 84 Days Postmortem Influences Consumer Eating Quality

Andrea Garmyn [1],*, Nicholas Hardcastle [1], Rod Polkinghorne [2], Loni Lucherk [3] and Mark Miller [1]

[1] Department of Animal and Food Sciences, Texas Tech University, Lubbock, TX 79409, USA; nicholas.hardcastle@ttu.edu (N.H.); mfmrraider@aol.com (M.M.)

[2] Birkenwood Pty. Ltd., 46 Church St, Hawthorn, VIC 3122, Australia; rod.polkinghorne@gmail.com

[3] Department of Agricultural Sciences, West Texas A&M University, Canyon, TX 79016, USA; llucherk@wtamu.edu

* Correspondence: andrea.garmyn@ttu.edu; Tel.: +1-806-834-6559

Received: 23 January 2020; Accepted: 13 February 2020; Published: 17 February 2020

Abstract: Our objective was to determine the effect of extending postmortem aging from 21 to 84 days on consumer eating quality of beef *longissimus lumborum*. Strip loins were collected from 108 carcasses. The *longissimus lumborum* muscle was isolated from strip loins and assigned to one of ten postmortem aging periods from 21 to 84 days (7-day increments) and balanced within four anatomical positions within the muscle. Consumer evaluations for tenderness, juiciness, flavor, and overall liking were conducted using untrained consumer sensory panels consisting of 1080 individual consumers, in accordance with the Meat Standards Australia protocols. These scores were then used to calculate an overall eating quality (MQ4) score. Postmortem aging had no effect ($P > 0.05$) on tenderness, but juiciness, flavor liking, overall liking, and MQ4 declined ($P < 0.05$) as aging period increased. Samples aged 21 to 42 days were most preferred having greater ($P < 0.05$) overall liking and greater ($P < 0.05$) MQ4 scores than samples aged 70 to 84 days postmortem. These results suggest that *longissimus lumborum* samples should not be wet-aged longer than 63 days to prevent potential negative eating experiences for consumers; however, altering storage conditions, specifically reducing temperature, could potentially allow for longer chilled storage without such negative effects on flavor and overall liking.

Keywords: beef; consumer; eating quality; extended postmortem aging; sensory testing

---

1. Introduction

The effects of postmortem aging of beef *longissimus* tenderness are relatively well documented [1,2] up to 28 days postmortem. However, the 2015 National Beef Tenderness Survey indicated average postfabrication storage or aging times for retail strip loin and foodservice top loin in the Unites States were 27.2 days and 34.6 days, respectively, but could range up to 101 days [3]. Moreover, the effects of longer term (>28 days) postmortem aging on consumer perception of eating quality are less well defined and more variable, especially in other palatability traits, such as flavor or juiciness.

Colle et al. [4] found increasing postmortem aging of *longissimus lumborum* steaks from 2 to 14 days improved tenderness, but no additional improvement was observed after 14 days. Also, consumer acceptability, juiciness, and flavor liking were similar from 2 through 63 days of postmortem aging [4]. Tenderness can be improved by aging strip loin subprimals for 56 days postmortem, but flavor intensity can decline with an increase in off-flavor intensity at this aging length [5]. However, Hughes et al. [6] reported an improvement in all consumer eating quality traits (tenderness, juiciness, flavor and overall liking) of beef *longissimus* from 2 to 12 weeks of postmortem aging, but no further improvement when extending postmortem aging to 20 weeks.

Given the potential variation in the postmortem aging length of beef, and especially *longissimus lumborum*, available at retail in the US, there is a need to better quantify the effect of extended postmortem aging on consumer eating quality of beef *longissimus lumborum*. To our knowledge, no studies have investigated extended aging of US beef *longissimus lumborum* through 12 weeks (84 days) of chilled vacuum-packaged storage. Therefore, our objective was to determine the effect of extending postmortem aging from 21 to 84 days on consumer eating quality. We hypothesize that tenderness will continue to improve throughout postmortem aging, but flavor may be negatively impacted as the potential for lipid oxidation and off-flavor development increases. Due to the strong reliance of flavor liking on overall liking, we believe overall liking or acceptance may also be negatively impacted at the longest aging periods.

2. Materials and Methods

2.1. Animals

A total of 108 carcasses were utilized in the current study. All cattle were Continental crossbred steers that were considered grain fed and were sourced from a single supplier. Hormone growth promotants (HGPs) were administered 160 days before harvest to all cattle in this study; specifically, all cattle were implanted with Revalor 200 (Merck Animal Health, DeSoto, KS, USA), which contains 200 mg trenbolone acetate and 20 mg estradiol, following manufacturer recommendations for administration.

2.2. Slaughter Procedure, Carcass Grading, and Subprimal Collection

Cattle were slaughtered on a single day at a commercial abattoir in Schuyler, NE. After the carcass chilling period, trained personnel collected carcass data including: marbling score, ossification using USDA skeletal maturity standards [7], 12th rib fat thickness, ribeye area, and HCW. In addition, ultimate pH was collected at the time of carcass grading using a handheld temperature-pH meter equipped with an intermediate junction pH sensor (TPS Model WP-90 with pH sensor part #111227, TPS Pty Ltd., Brendale, QLD, Australia).

Strip loins (Institutional Meat Purchase Specification #180) were collected from the right side of each carcass. Subprimals were vacuum packaged individually and were transported under refrigeration to Texas Tech University, Lubbock, TX. Upon arrival, subprimals were held in chilled storage at 2 °C until 21 days postmortem. Strip loins were fabricated in accordance with MSA protocols [8]. All external fat and connective tissue were removed from strip loins prior to steak fabrication. In addition, the *gluteus medius* was removed from the strip loin leaving only the *longissimus lumborum*. *Longissimus* muscles were fabricated into 2.5 cm steaks and were further processed into smaller pieces measuring approximately 5 × 5 cm. Steak pieces were wrapped in plastic and vacuum packaged as sets of five based on position within the strip loin. Four sets (4 anatomical positions–Anterior [A]1, Anterior [A]2, Posterior [P]3, Posterior [P]4) of five steak pieces were retained from each strip loin for subsequent consumer testing. One of ten postmortem aging periods (every 7 days from 21 days to 84 days) was assigned and balanced within each position to avoid any positional effect when examining postmortem aging. All consumer steaks were vacuum packaged, boxed, maintained in chilled storage (2.0 °C ± 1.0 °C), and frozen on their respective day based on a predetermined postmortem aging designation. Samples were stored at −20 °C until being thawed for consumer sensory testing.

2.3. Consumer Sensory Evaluation

Consumer testing was conducted according to MSA grill protocols [8]. Steak samples were thawed at 2–4 °C for 24 h prior to consumer evaluation. All steaks were cooked on a Silex clamshell grill (Model S-143K, Silex Grills Australia Pty Ltd., Marrickville, Australia) with a temperature set at 135 and 142 °C for the top and bottom plate, respectively. The Silex grill was preheated 30 min prior to the start of the panels. A strict and detailed time schedule was followed to ensure all steaks were prepared identically [9]. Each cooking round consisted of ten samples that were cooked at the same time on one grill. All steaks were cooked for 5 min and 45 s, followed by a 3-min rest period. After the rest period, each steak was cut in half into two equal size pieces and served to two separate predetermined consumer panelists.

The Texas Tech University Institutional Review Board approved procedures for use of human subjects for consumer panel evaluation of meat sensory attributes (IRB#: 2017-598). Consumer panels were conducted in the Texas Tech University Animal and Food Sciences Building. Consumer panelists (n = 1080) were recruited from Lubbock, Texas and the surrounding local communities. Panelists had to be regular red meat eaters aged 18 to 75 years old to be able to participate. Each consumer was monetarily compensated and were only allowed to be participate one time. Each session consisted of 20 people with three sessions being conducted on a given evening. Each session lasted approximately 60 min. Each panelist was seated at a numbered booth and was provided with a ballot, plastic utensils, a toothpick, unsalted crackers, a napkin, an empty cup, a water cup, and a cup with diluted apple juice (10% apple juice and 90% water). Each ballot consisted of a demographic questionnaire, seven sample ballots, and a post panel survey regarding beef purchasing habits. Before beginning each panel, consumers were given verbal instructions by Texas Tech personnel about the ballot and the process of testing samples. Panels were conducted in a large classroom that is equipped with standard fluorescent lighting overhead (i.e., no red filters were used) with tables that were divided into individual consumer booths.

Each consumer evaluated seven samples. One steak sample was included in the cooking order as a warm-up sample for consumers and to provide linkage across all testing nights. The link samples were always served in the first position, followed by six test samples served in predetermined, balanced order representing one of ten postmortem aging periods. A Latin-square design was utilized to balance the order and presentation of the samples, ensuring that each product was presented an equal number of times in the six test positions before and after every other product. Each sample had 10 consumer observations (i.e., five consumer steaks all being cut in half and served to two individuals each). Consumers scored palatability traits tenderness, juiciness, flavor liking, and overall liking on 100 mm line scales verbally anchored at 0 (not tender, not juicy, dislike extremely) and 100 (very tender, very juicy, like extremely). Consumers were asked to rate the overall quality or satisfaction of each sample as "unsatisfactory", "good everyday quality", "better than everyday quality", or "premium quality." The 10 individual scores for each trait were averaged to generate mean sensory scores for each palatability trait and satisfaction prior to analysis. A composite score (MQ4) was calculated using the following equation: tenderness × 0.3 + juiciness × 0.1 + flavor liking × 0.3 + overall liking × 0.3 [8]. Weightings for tenderness and flavor liking have been adjusted from the original weightings by [8] for a balanced contribution to the MQ4 value. The weightings give an indication of the relative importance of the four sensory attributes (tenderness, juiciness, flavor, overall satisfaction) to the final meat quality score.

2.4. Statistical Analysis

Data were analyzed in SAS using PROC GLIMMIX (version 9.4, SAS Inst. Inc., Cary, NC, USA). For consumer sensory analyses, postmortem aging period, position, and their interaction were included as fixed effects. Testing day was included as a random effect. Marbling score was included as a covariate, but was significant ($P < 0.05$) only for juiciness. Treatment least squares means were separated with the PDIFF option of SAS using a significance level of $P \leq 0.05$. Mean separation tests for all pairwise comparisons were performed using the PDIFF function, which requests that P-values for differences of all least squares means be produced. The PROC CORR of SAS was used to assess the relationship between consumer eating quality traits by generating Pearson correlation coefficients. The PROC FREQ of SAS was used to summarize consumer demographic information.

3. Results and Discussion

3.1. Carcass Traits

All carcasses were graded using USDA grading specifications. Carcass characterization can be found in Table 1. The average marbling score was representative of USDA Select, but ranged from USDA Standard to average Choice. However, very few carcasses (5.6%) had marbling scores representative of Choice carcasses (marbling score $\geq$ 400). As a result of the variation and the known relationship between eating quality and marbling score in the *longissimus lumborum* [10–12], marbling score was tested for inclusion as a covariate in the statistical analysis. As previously mentioned, marbling score was only required in the model for consumer juiciness scores ($P < 0.05$).

Table 1. Simple Statistics for Carcass Traits of Beef Cattle ($n = 108$).

Item	Mean	Standard Deviation	Minimum	Maximum
HCW, kg	364.8	32.8	285.0	436.6
Fat thickness, mm	13.0	4.7	1.3	29.2
LM area, cm ^{2}	94.7	11.8	66.1	125.2
Marbling score [1]	327.2	47.4	200	560
Ossification [2]	119.9	18.0	110	250
Ultimate pH	5.61	0.06	5.49	6.08

[1] Marbling: 200 = traces^{00}, 300 = slight^{00}, 400 = small^{00}, 500 = modest^{00} [7]; [2] Ossification: 100 = A^{00}, 200 = B^{00} [7].

3.2. Consumer Sensory

Demographic characteristics of participating consumers can be found in Table 2. Almost 70% of the participants were aged 20–49 years old, with a relatively even split between those three 10-year age brackets. Seventy-four percent of the population in Lubbock, TX is less than 50 years old [13,14], which aligns with participants in this study. We also believe this percentage is suitable according to the product studied. Participants were evenly distributed between male and female. Most participants (87.7%) identified with Caucasian/White or Hispanic as their ethnic origin, with a fairly even split between the two distinctions. For census purposes, persons who identify as Hispanic or Latino can identify as any race; however, in the latest census data available for Lubbock, TX, 35% reported themselves as Hispanic or Latino, while 65% reported themselves as not Hispanic or Latino [14]. The most common household size consisted of 2–3 adults, representing 74.2% of participants. Nearly half of the participants had no children living in their household. The level of education with the highest proportion of participants was for "some college/technical school" (39.2%), while high school and college graduates collectively accounted for another 45.5%. Additionally, the majority of consumers ate beef at least twice per week (76.9%). The most preferred degree of doneness was medium-rare, with medium and medium-well contributing another 49.1% collectively.

Consumer sensory outcomes can be found in Table 3. No interactions were detected ($P > 0.05$) between postmortem aging and position. Postmortem aging influenced ($P < 0.01$) juiciness, flavor liking, and overall liking, as well as the composite MQ4 score and satisfaction. Somewhat surprisingly, tenderness was not impacted by postmortem aging ($P = 0.29$); however, this could likely be explained by the minimum aging period in this study of 21 days postmortem, where a large portion of proteolysis has potentially already occurred at that stage [15–17]. With the exception of flavor liking, position affected ($P < 0.01$) all palatability traits, as well as the composite MQ4 score and satisfaction.

Table 2. Demographic characteristics of consumers (n = 1080) who participated in consumer sensory panels at Texas Tech University in Lubbock, TX.

Characteristic	Response	% of Consumers
Age Group	<20	7.2
	20–29	24.2
	30–39	24.9
	40–49	20.3
	50–59	12.6
	>60	10.7
Gender	Male	46.8
	Female	53.1
Occupation	Tradesperson	13.3
	Professional	24.8
	Administration	16.6
	Sales and Service	14.6
	Laborer	8.8
	Homemaker	3.1
	Student	10.0
	Unemployed/Retired	8.9
Ethnic Origin	African-American	10.0
	Asian	0.5
	Caucasian/White	41.3
	Hispanic	46.4
	Native American	0.5
	Other	1.3
Household Size (Adults)	1	13.3
	2	57.2
	3	17.0
	4	8.7
	5+	3.8
Household Size (Children)	0	46.6
	1	14.6
	2	23.1
	3	11.4
	4+	4.2
Annual Household Income	<$20,000	13.8
	$20,000–$50,000	32.1
	$50,001–$75,000	21.3
	$75,001–$100,000	16.6
	>$100,000	16.3
Level of Education	Non-High School Graduate	5.3
	High School Graduate	21.7
	Some College/Technical School	39.2
	College Graduate	23.8
	Post-College Graduate	10.0
Beef Consumption	Daily	10.0
	4–5 times a week	26.2
	2–3 times a week	40.7
	Weekly	15.1
	Every other week	5.1
	Monthly	3.0
Preferred Beef Degree of Doneness	Rare	2.9
	Medium-Rare	32.5
	Medium	24.9
	Medium-Well	24.2
	Well-Done	15.5

Table 3. The Effects of postmortem aging and position within the *longissimus lumborum* on eating quality scores as rated by consumers ($n = 1080$).

Postmortem Aging	Tenderness [1]	Juiciness [1]	Flavor Liking [1]	Overall Liking [1]	MQ4 [2]	Satisfaction [3]
21 days	66.1	60.8 [a]	61.1 [a]	62.6 [a]	63.0 [a]	3.37 [a]
28 days	67.6	59.8 [ab]	59.1 [ab]	61.1 [a]	62.3 [ab]	3.32 [ab]
35 days	66.5	60.8 [a]	59.1 [ab]	60.0 [ab]	61.8 [ab]	3.30 [ab]
42 days	66.0	58.0 [abc]	57.0 [bc]	59.3 [abc]	60.5 [abc]	3.18 [bc]
49 days	65.7	53.6 [d]	53.4 [cd]	55.8 [bcd]	57.9 [cd]	3.15 [c]
56 days	65.4	58.3 [abc]	53.6 [cd]	56.4 [bc]	58.5 [bcd]	3.21 [bc]
63 days	64.5	55.6 [cd]	51.7 [de]	55.0 [cd]	56.9 [cd]	3.10 [cd]
70 days	62.1	53.8 [d]	46.4 [fg]	50.2 [de]	53.0 [e]	2.97 [de]
77 days	65.2	55.2 [cd]	48.3 [ef]	51.9 [e]	55.0 [de]	3.07 [cde]
84 days	63.3	56.2 [bcd]	43.8 [g]	48.5 [e]	52.2 [e]	2.94 [e]
SEM [4]	1.9	1.8	2.0	2.0	1.8	0.07
P-value	0.29	<0.01	<0.01	<0.01	<0.01	<0.01
Position [5]						
A1	69.0 [a]	58.6 [ab]	54.1	57.3 [a]	60.0 [a]	3.22 [a]
A2	67.2 [a]	60.1 [a]	54.7	58.2 [ab]	60.0 [a]	3.24 [a]
P3	62.8 [b]	56.7 [b]	52.4	55.2 [bc]	56.7 [b]	3.13 [b]
P4	62.0 [b]	53.4 [bc]	52.1	53.6 [c]	55.6 [b]	3.06 [b]
SEM [4]	1.5	1.4	1.5	1.6	1.4	0.05
P-value	<0.01	<0.01	0.14	<0.01	<0.01	<0.01

[1] 0 mm = not tender, not juicy, dislike flavor extremely, dislike overall extremely; 100 mm = very tender, very juicy, like flavor extremely, like overall extremely; [2] MQ4 = tenderness × 0.3 + juiciness × 0.1 + flavor liking × 0.3 + overall liking × 0.3; [3] 2 = unsatisfactory, 3 = good everyday quality, 4 = better than everyday quality, 5 = premium quality; [4] Pooled (largest) SEM reported of least sqaures means; [5] A = anterior; P = posterior; [a-g] Within a column and main effect, least squares means without a common superscript differ ($P < 0.05$).

Juiciness generally decreased as postmortem aging increased, but several adjacent aging periods had similar ($P > 0.05$) juiciness scores. For example, samples aged 21 to 35 days had similar and greater juiciness than samples aged 63 days or longer. However, samples aged 42 and 56 days had similar juiciness as those samples aged 21 to 35 days. Flavor liking declined ($P < 0.01$) very clearly as aging period increased. Consumers generally did not differentiate between adjacent aging periods, and so samples were typically grouped together in two to three week aging bands before consumers would indicate flavor liking had declined. Consumers were not as discriminative against overall liking as they were flavor liking, but similar declining trends can be observed in the scores for those two traits. Samples aged 21 to 42 days were most preferred having greater ($P < 0.05$) overall liking and greater ($P < 0.05$) composite MQ4 scores than samples aged 70 to 84 days postmortem. Samples aged 49 to 63 days were essentially intermediate for overall liking and MQ4 score, but had similarities to samples aged both less than 49 days and greater than 63 days. Finally, when assessing satisfaction, a score below 3 indicates consumers scored the sample as "unsatisfactory". Despite statistical differences ($P < 0.05$), all samples aged 63 days or less would be classified as "good everyday quality". Consumers were more ($P < 0.05$) satisfied with samples aged up 35 days than samples aged 63 days or longer.

Gruber et al. [2] showed no improvement in tenderness, via decrease in Warner–Bratzler shear force (WBSF) values, beyond 21 days postmortem for *longissimus dorsi* muscle from upper two-thirds USDA Choice carcasses; however, WBSF values did continue to improve through 28 days postmortem for *longissimus dorsi* muscle from USDA Select carcasses, which more closely aligns with the quality grade of carcasses used in the current study. Those results suggest aging *longissimus* muscle beyond 21 days postmortem was only beneficial in carcasses with less marbling (Select). Hughes et al. [6] showed an improvement in eating quality traits (tenderness, juiciness, flavor and overall liking) of beef *longissimus* from 2 to 12 weeks of postmortem aging, but no further improvement at 20 weeks. Lipid oxidation increased throughout the postmortem storage period to levels slightly above acceptable for rancidity detection at 20 weeks, but MQ4 scores suggested the meat would still be acceptable through 20 weeks of storage as classified by consumers [6]. We would like to point out the storage temperature in that trial was maintained at $-1.0\,°C \pm 0.5\,°C$ [6], which was lower and slightly less variable than the current study. Lepper-Blilie et al. [18] also showed tenderness, according to trained panelists, improved linearly as postmortem aging increased from 14 days to 49 d days; however, there was no statistical improvement in tenderness beyond 21 days, which aligns with the current results. When extending aging from 2 to 63 days postmortem, Colle et al. [4] observed no improvement in consumer tenderness after 14 days postmortem, which again supports the current findings. However, acceptability, juiciness, and flavor liking did not differ between the various postmortem aging periods [4], which contradicts the current findings as flavor liking and overall liking decreased with increasing aging period in the present findings. We believe the discrepancy in results between these two studies could again be related to postmortem storage temperature, as previous work has shown increased storage temperature ($0\,°C$ vs. 5.0–$5.5\,°C$) of vacuum packaged beef can negatively affect shelf-life and palatability, especially when aging beef beyond 28 postmortem [19,20]. Colle et al. [4] stored vacuum-packaged muscle sections at $0\,°C$, which is again lower than the current study. Additionally, lipid oxidation increased with postmortem aging, but the authors did not believe their samples had reached the threshold for lipid oxidation based on TBARS values [4]. All samples, regardless of aging period, had less than 1 mg MDA/kg, explaining why they did not see any differences in consumer flavor scores because lipid oxidation had not reached a detectable level by consumers. Lipid oxidation was not evaluated in the current study.

Although postmortem aging can have a positive influence on meat tenderness [2,5,6], its impacts on beef flavor are inconsistent and less well defined, especially at extended postmortem aging periods. In one instance consumers liked the flavor of beef *longissimus lumborum* aged for 12 weeks more than the flavor of beef aged for 2 weeks but no further improvement to flavor liking scores was observed from aging beef for 20 weeks [6]. According to Brewer and Novakofski [1], postmortem aging up to 21 days had no influence on beef flavor; however, extended postmortem aging can promote the development of undesirable flavor characteristics and reduction of beef flavor intensity [5,18]. Lipid oxidation is limited by endogenous antioxidant mechanisms in living muscle [21], but the effectiveness of these antioxidants declines as postmortem aging time increases [22] which could results in increased lipid oxidation [6]. Since lipid oxidation has been linked with rancid flavor [23,24], this could explain why flavor liking decreased as postmortem aging increased in the current study. However, lipid oxidation was not measured in the current study, so we are only speculating that this could be responsible, in part, for the decline in flavor liking scores as postmortem aging increased.

As seen in Table 3, position also affected palatability traits. With the exception of flavor liking, position affected ($P < 0.01$) all palatability traits, as well as the composite MQ4 score and satisfaction. The two anterior-most portions (A1 and A2) were similar and were more tender ($P < 0.05$) than the two posterior-most portions (P3 and P4), which were also similar ($P > 0.05$). A tenderness gradient exists within the *longissimus* muscle [25,26] from the anterior to posterior ends of the strip loin. Muscle fiber angle is affected by steak position [25], and marbling can also vary depending on the anatomical position within the strip loin [27]. Both factors likely contribute to this tenderness gradient observed in the strip loin. The A2 position was juicier ($P < 0.05$) than the two posterior positions, but the A1 position did not differ ($P > 0.05$) in juiciness from any other position in the strip loin. Overall liking generally decreased from the anterior to posterior end of the *longissimus lumborum* muscle. The composite MQ4 score and satisfaction followed the same trend as tenderness, where the anterior portions received greater scores than the posterior portions. Despite statistical differences ($P < 0.05$), all samples would be classified as "good everyday quality." However, consumers were more ($P < 0.05$) satisfied with samples from the anterior portions of the *longissimus lumborum* than the posterior portions.

3.3. Correlations

To estimate the extent to which eating quality scores are linked to overall liking and satisfaction, correlation coefficients between palatability traits, MQ4, and satisfaction scores were determined (Table 4). Consumer overall liking was associated (r = 0.74; $P < 0.01$) with consumer tenderness and juiciness ratings, but most highly related with flavor liking (r = 0.93). Individual palatability traits were strongly correlated to each other (r $\geq$ 0.67), indicating that individual improvements of these traits could influence the perception of another trait. MQ4 was highly related ($P < 0.01$) to eating quality scores for tenderness, juiciness, flavor liking, and overall liking, as would be expected given it is a composited score of those traits. Satisfaction was positively linked ($P < 0.01$) to all eating quality traits, especially overall liking, and was highly correlated to MQ4 ($P < 0.01$).

Table 4. Pearson's correlation coefficients for the relationships between consumer sensory scores of *longissimus lumborum* aged 21 to 84 days postmortem.

Trait	Juiciness	Flavor Liking	Overall Liking	MQ4	Satisfaction
Tenderness	0.73 *	0.67 *	0.74 *	0.86 *	0.71 *
Juiciness		0.68 *	0.74 *	0.81 *	0.71 *
Flavor Liking			0.93 *	0.94 *	0.86 *
Overall Liking				0.96 *	0.88 *
MQ4					0.89 *

* Correlation coefficients were significant ($P < 0.01$).

4. Conclusions

With the exception of flavor liking, position affected all palatability traits, as well as the composite MQ4 score and satisfaction. Despite statistical differences, all samples would be classified as "good everyday quality" regardless of anatomical position within the strip loin. However, consumers were more satisfied with samples from the anterior portions of the *longissimus lumborum* than the posterior portions, likely as a result of greater tenderness from those samples. Postmortem aging influenced juiciness, flavor liking, and overall liking, as well as the composite MQ4 score and satisfaction, but not tenderness. Samples aged 21 to 42 days were most preferred having greater overall liking and greater MQ4 scores than samples aged 70 to 84 days postmortem. Overall liking was clearly driven by flavor liking, as demonstrated by the strongest relationship of the palatability traits. Despite statistical differences, all samples aged 63 days or less would be classified as "good everyday quality". Consumers were, however, more satisfied with samples aged up 35 days than samples aged 63 days or longer. These results suggest that *longissimus lumborum* samples should not be wet-aged longer than 63 days to prevent potential negative eating experiences for consumers; however, altering storage conditions, specifically reducing storage temperature, could potentially allow for longer chilled storage without such negative effects on flavor liking. Future research involving measurement of lipid oxidation should be conducted to confirm and help define the negative consumer response to flavor and overall liking with extended postmortem aging.

Author Contributions: R.P. and M.M. contributed to the conceptualization and experimental design of the study; N.H., L.L., and A.G. undertook the study; A.G. conducted the formal data analysis and presentation of the data; A.G. wrote the original manuscript draft; reviews and editing of the completed manuscript were conducted by M.M. and N.H.; project administration was the responsibility of M.M. and A.G. All authors have read and agreed to the published version of the manuscript.

Funding: This research was funded through a research gift from Teys Australia.

Conflicts of Interest: The authors declare no conflict of interest.

References

1. Brewer, S.; Novakofski, J. Consumer sensory evaluations of aging effects on beef quality. *J. Food Sci.* **2008**, *73*, S78–S82. [CrossRef] [PubMed]
2. Gruber, S.L.; Tatum, J.D.; Scanga, J.A.; Chapman, P.L.; Smith, G.C.; Belk, K.E. Effects of postmortem aging and USDA quality grade on Warner-Bratzler shear force values of seventeen individual beef muscles. *J. Anim. Sci.* **2006**, *84*, 3387–3396. [CrossRef] [PubMed]
3. Martinez, H.A.; Arnold, A.N.; Brooks, J.C.; Carr, C.C.; Gehring, K.B.; Griffin, D.B.; Hale, D.S.; Mafi, G.G.; Johnson, D.D.; Lorenzen, C.L.; et al. National Beef Tenderness Survey-2015: Palatability and Shear Force Assessments of Retail and Foodservice Beef. *Meat Muscle Biol.* **2017**, *1*, 138–148. [CrossRef]
4. Colle, M.J.; Richard, R.P.; Killinger, K.M.; Bohlscheid, J.C.; Gray, A.R.; Loucks, W.I.; Day, R.N.; Cochran, A.S.; Nasados, J.A.; Doumit, M.E. Influence of extended aging on beef quality characteristics and sensory perception of steaks from the gluteus medius and longissimus lumborum. *Meat Sci.* **2015**, *110*, 32–39. [CrossRef]
5. Juárez, M.; Larsen, I.L.; Gibson, L.L.; Robertson, W.M.; Dugan, M.E.R.; Aldai, N.; Aalhus, J.L. Extended ageing time and temperature effects on quality of sub-primal cuts of boxed beef. *Can. J. Anim. Sci.* **2010**, *90*, 361–370. [CrossRef]
6. Hughes, J.; McPhail, N.G.; Kearney, G.; Clarke, F.; Warner. R.D. Beef longissimus eating quality increases up to 20 weeks storage and is unrelated to meat colour at carcass grading. *Anim. Prod. Sci.* **2015**, *55*, 174–179. [CrossRef]
7. US Department of Agriculture. *United States Standards for Grades of Carcass Beef. Livestock and Seed Program*; Agricultural Marketing Service: Washington, DC, USA, 1997.
8. Watson, R.; Gee, A.; Polkinghorne, R.; Porter, M. Consumer assessment of eating quality-Development of protocols for MSA testing. *Aust. J. Exp. Agric.* **2008**, *48*, 1360–1367. [CrossRef]
9. Gee, A. *Protocol Book 4: For the Thawing Preparation, Cooking and Serving of Beef for MSA (Meat Standards Australia) Pathway Trials*; Meat and Livestock Australia: North Sydney, Australia, 2006.
10. Corbin, C.H.; O'Quinn, T.G.; Garmyn, A.J.; Legako, J.F.; Hunt, M.R.; Dinh, T.T.N.; Rathmann, R.J.; Brooks, J.C.; Miller, M.F. Sensory evaluation of tender beef strip loin steaks of varying marbling levels and quality treatments. *Meat Sci.* **2015**, *100*, 24–31. [CrossRef]
11. Platter, W.J.; Tatum, J.D.; Belk, K.E.; Koontz, S.R.; Chapman, P.L.; Smith, G.C. Effects of marbling and shear force on consumers' willingness to pay for beef strip loin steaks. *J. Anim. Sci.* **2005**, *83*, 890–899. [CrossRef]

12. Smith, G.C.; Carpenter, Z.L.; Cross, H.R.; Murphey, C.E.; Abraham, H.C.; Savell, J.W.; Davis, G.W.; Berry, B.W.; Parrish, F.C., Jr. Relationship of USDA marbling groups to palatability of cooked beef. *J. Food Qual.* **1985**, *7*, 289–308. [CrossRef]

13. United States Census Bureau. American Fact Finder. Available online: https://factfinder.census.gov/faces/tableservices/jsf/pages/productview.xhtml?src=bkmk (accessed on 22 January 2020).

14. World Population Review. Lubbock, Texas Population 2020. Available online: http://worldpopulationreview.com/us-cities/lubbock-population/ (accessed on 22 January 2020).

15. Dransfield, E.; Jones, R.C.D.; MacFie, H.J.H. Tenderising in *M. longissimus dorsi* of beef, veal, rabbit, lamb and pork. *Meat Sci.* **1981**, *5*, 139–147. [CrossRef]

16. Etherington, D.J.; Taylor, M.A.J.; Dransfield, E. Conditioning of meat from different species. Relationship between tenderizing and the levels of cathepsin B, cathepsin L, calpain I, calpain II, and β-glucuronidase. *Meat Sci.* **1987**, *20*, 1–18. [CrossRef]

17. Koohmaraie, M.; Whipple, G.; Kretchmar, D.H.; Crouse, J.D.; Mersmann, H.J. Postmortem proteolysis in longissimus muscle from beef, lamb and pork carcasses. *J. Anim. Sci.* **1991**, *69*, 617–624. [CrossRef] [PubMed]

18. Lepper-Blilie, A.N.; Berg, E.P.; Buchanan, D.S.; Berg, P.T. Effects of post-mortem aging time and type of aging on palatability of low marbled beef loins. *Meat Sci.* **2016**, *112*, 63–68. [CrossRef]

19. Carpenter, Z.L.; Beebe, S.D.; Smith, G.C.; Hoke, K.E.; Vanderzant, C. Quality characteristics of vacuum packaged beef as affected by postmortem chill, storage temperature, and storage interval. *J. Milk Food Technol.* **1976**, *9*, 592–599. [CrossRef]

20. Hur, S.J.; Park, G.B.; Joo, S.T. Effect of storage temperature on meat quality of muscle with different fiber type composition from Korean native cattle (Hanwoo). *J. Food Qual.* **2008**, *32*, 315–333. [CrossRef]

21. Sies, H. Oxidative stress: Introductory remarks. In *Oxidative Stress*; Sies, H., Ed.; Academic Press: Orlando, FL, USA, 1986; pp. 1–8.

22. Monahan, F.J. Oxidation of lipids in muscle foods: Fundamental and applied concerns. In *Antioxidants in Muscle Foods: Nutritional Strategies to Improve Aualit*; Decker, E.A., Faustman, C., Lopez-Bote, C.J., Eds.; John Wiley & Sons, Inc. Publication: New York, NY, USA, 2000; pp. 3–23.

23. Asghar, A.; Gray, J.I.; Buckley, D.J.; Pearson, A.M.; Booren, A.M. Perspectives on warmed-over flavor. *Food Technol.* **1988**, *42*, 102–108.

24. Mottram, D.S.; Edwards, R.A. The role of triglycerides and phospholipids in the aroma of cooked beef. *J. Sci. Food Agric.* **1983**, *34*, 517–522. [CrossRef]

25. Derington, A.J.; Brooks, J.C.; Garmyn, A.J.; Thompson, L.D.; Wester, D.B.; Miller, M.F. Relationships of slice shear force and Warner-Bratzler shear force of beef strip loin steaks as related to the tenderness gradient of the strip loin. *Meat Sci.* **2011**, *88*, 203–208. [CrossRef]

26. Wheeler, T.L.; Shackelford, S.D.; Koohmaraie, M. Beef longissimus slice shear force measurement among steak locations and institutions. *J. Anim. Sci.* **2007**, *85*, 2283–2289. [CrossRef]

27. Acheson, R.J.; Woerner, D.R.; Walenciak, C.E.; Colle, M.J.; Bass, P.D. Distribution of marbling throughout the M. longissimus thoracis et lumborum of beef carcasses using an instrument-grading system. *Meat Muscle Biol.* **2018**, *2*, 303–308. [CrossRef]

 *foods* | MDPI

Article

Beef Quality Preferences: Factors Driving Consumer Satisfaction

Chad Felderhoff [1], Conrad Lyford [1], Jaime Malaga [1], Rod Polkinghorne [2], Chance Brooks [3], Andrea Garmyn [3] and Mark Miller [3,*]

[1] Department of Agricultural and Applied Economics, Texas Tech University, Lubbock, TX 79409, USA; chad@muenstermilling.com (C.F.); Conrad.Lyford@ttu.edu (C.L.); Jaime.Malaga@ttu.edu (J.M.)

[2] Birkenwood Pty. Ltd., 46 Church St, Hawthorn, VIC 3122, Australia; rod.polkinghorne@gmail.com

[3] Department of Animal and Food Sciences, Texas Tech University, Lubbock, TX 79409, USA; chance.brooks@ttu.edu (C.B.); andrea.garmyn@ttu.edu (A.G.)

* Correspondence: mfmrraider@aol.com; Tel.: +1-806-742-2805

Received: 13 February 2020; Accepted: 2 March 2020; Published: 4 March 2020

Abstract: The current study was designed to broaden the understanding of the attributes impacting the sensory properties of beef when consumed. Using a survey of consumers from three different geographical regions in the United States (US), we determined the impacts of three attributes on overall satisfaction in several different ways. The two main statistical methods used were an Ordinary Least Squares (OLS) model and the Conditional Logit model. Perhaps the most important finding of this study was that flavor was the largest contributor to consumer satisfaction. This finding was consistent throughout all the models. In the base model, flavor represented 59% of the satisfaction rating. Additionally, results indicated domestic beef was preferred over Australian beef by US consumers. Another important finding of the study was the impact of the demographic variables of age, income, and gender on satisfaction. The older group generally placed more emphasis on tenderness, while younger people preferred juicier beef. Males were more responsive than females for all attributes, especially tenderness. Those with higher income were more responsive to tenderness for all quality levels, but the lower income group was more responsive to juiciness. Overall, flavor had the largest impact on consumers' satisfaction level in comparison to tenderness or juiciness.

Keywords: beef; consumer; demographics; eating quality; flavor; satisfaction; tenderness

---

1. Introduction

The beef market has always been under constant pressure of evolving preferences in areas such as taste, consistency, and healthfulness [1–3]. Differences in consumer lifestyles and their impacts on product selection are well recognized by processors and marketers of food [4,5]. Transformation in the demand structure presents many significant opportunities for beef producers and marketers. This changing demand structure signifies an important move from beef being marketed primarily as a homogenous commodity to a niche product [6].

Historically, the lack of ability by marketers to respond to the changing environment in the marketplace is exposed by the decrease in consumption. The United States Department of Agriculture (USDA) [7] has estimated that beef consumption along with other red meats has declined for the last several decades, based on per capita availability. This creates substantial concern within the beef production–marketing system. Changes in the pricing structure of competing meats alone cannot explain the shifts in beef demand [8].

Previous studies have identified beef characteristics that are perceived as desirable with the goal of increasing demand. In many cases researchers have focused on the tenderness attribute of beef in a retail environment [6,9–11]. The retail market is commonly believed to be the end market for beef;

hence, researchers typically focus their efforts on understanding the consumer purchase decision at the retail level. External factors present in the retail environment influence the consumers' perception of quality, suggesting that this environment may add an element of bias to studies [12]. The current study controls for these external factors by using an untrained sensory test outside of the retail or home environment to evaluate fundamental quality preferences in beef.

We look to answer many questions about consumers' preferences by using consumer choice modeling. The consumer choice model allows consumers to make decisions about products based on several key attributes [6]. The developed model evaluates the consumer rating of chosen attributes and their effects on consumer satisfaction by using a large scale study.

The overall focus of this study was to evaluate the extent to which United States (US) beef consumers vary in their preference. Multiple objectives arose, including determination of the impact of tenderness, juiciness, and flavor on consumers' overall satisfaction and what changes were required to increase consumer satisfaction to higher perceived quality levels. Moreover, we wanted to estimate the quality attributes that correspond with distinct levels of satisfaction, as well as evaluate the effect of the beef products' country of origin on consumer satisfaction. Our final objective was to determine if consumers' preference structures vary by demographic characteristics.

2. Materials and Methods

2.1. Sampling Methodology

Methods of gathering data in previous studies have varied widely, were generally limited to a smaller scope (less than 1000 consumers), and were less representative of the overall US population. Data in the current study were gathered by using the combination of a survey and untrained sensory tests in three US cities ($n = 1440$; 480/city). These data originated from the following diverse metropolitan areas across the US: (a) Phoenix, AZ, (b) Lubbock, TX and (c) Washington, DC/Baltimore, MD. These cities were chosen based on the results from a previous study conducted as representative of the overall US population [11]. Phoenix was chosen because of its diverse population and proximity to the western side of the US. Lubbock was used to represent the central region of the US and because Miller et al. [11] showed the preferences of beef consumers in Lubbock, TX, were not different from those of beef consumers in Dallas, TX. The Washington, DC/Baltimore area was chosen because of its diverse population and proximity to the east coast.

There were essentially two parts to the data collected in this study: survey data and untrained sensory results. However, this study focused on the demographic characteristics of the consumer along with the untrained sensory test results. The demographics were gathered from consumer responses to predetermined demographical questions on the survey.

The use of the untrained sensory test allowed for the gathering of measurements that lead to the discovery of the consumer's fundamental beef preference. In previous consumer studies, the cut (muscle) and origin (country and/or feeding system) were made known to the panelist [9,13]. This can present a problem because consumers have been shown to rank the cut of meat as the most important factor in determining beef quality [12]. The cut, along with other outside factors, can create biases in the consumers' perception of quality. Moreover, Ron et al. [14] found the consumers' perception of eating quality can be influenced by revealing quality-differentiated brand names and labeling claims to the consumer, particularly claims related to production practices. It was the above reasons that influenced us to use untrained "blind" sensory tests which control for specific external factors.

2.2. Product

One major variable in our study was the origin of the sample. The origin refers to the muscle being tested as well as the country and/or feeding system that the beef was generated from. Four subprimals were used in this study: outside round, *semimembranosus*; top sirloin butt, *gluteus medius*; tenderloin, *psoas major*; and strip loin, *longissimus lumborum*. Australian product was collected from cattle that were grass-fed, short fed grain for 70 days, or long fed with grain for 188 days. All grain-fed cattle had received a hormonal growth promotant (HGP) implant, whereas none of the grass-fed cattle were implanted. HGP usage was monitored and reported for Australian cattle when they were transferred to a Meat Standards Australia (MSA) licensed abattoir. Live animal information pertaining to diet (grass vs. grain) was made available through the animal identification system in Australia, but no data were collected on-farm.

Sixty carcasses (n = 20/feeding type) were selected in Australia at two Queensland abattoirs. Forty carcasses (grass and short fed grain) were obtained from one abattoir and 20 carcasses (long fed grain) were obtained from the second abattoir. US product was sourced from Select and High Choice USDA graded carcasses at a commercial abattoir in Nebraska. The US carcasses (n = 60; 30/quality grade) were selected to evenly represent each USDA grade, whereas the Australian carcasses were taken as a consecutive run from each category. US cattle were all commerically grain-fed. According to cattle feeding surveys, feedyard finishing rations in Nebraska consist predominately of corn, distillers grains, haw and straw, silage, and mineral [15]. Cattle in the Northern Plains (Nebraska, South Dakota, North Dakota) are on finishing rations an average of 137 d [15]. It was likely cattle had received at least one implant as 95% of cattle in this region reportedly receive some type of growth enhancing technology and over 80% of feedlots in the Northern Plains (including Nebraska) report using implants. However, exact duration of the finishing ration and actual HGP usage was not known in the current study due to their selection from a commercial abattoir. All carcasses were graded by a common MSA grader with the US carcasses also assessed by a senior USDA grader.

2.3. Preparation

Subprimals were fabricated into 2.5 cm steaks, and further processed into 2.5 × 5 × 5 cm steak pieces and vacuum packaged at 10 mbars of pressure (Cryovac barrier bag; moisture vapor transmission rate: 0.3 to 0.6 g/100 in^2/24 h; oxygen transmission rate: 1.5 to 3.5 cc/m^2/24 h; Sealed Air Food Care; Charlotte, NC, USA) as sets of five steak pieces in sequential order according to anatomical position. All samples were frozen (−10 °C) at 14 days postmortem, so that aging period did not differ between samples. The frozen steak pieces were sorted into a predetermined cook order. Steak preparation from the primal cuts, allocation to cooking order, and consumer allocation followed the MSA protocols [16].

Samples were thawed at 2–4 °C for 24 h prior to consumer panel evaluation. All samples were prepared on a Model S-143 K Silex clamshell grill (Silex Grills Australia Pty Ltd., Marrickville, Australia) with plate temperature set at 225 °C. A strict time schedule was used to ensure all steaks were prepared identically [16]. Ten sample steaks were prepared on the grill for each cooking round. All steaks were cooked for 5 min and 45 s. After a mandatory 3 min rest period, each steak was cut in half into two equally sized rectangular pieces and served to two separate preselected consumers.

2.4. Consumer Panels

The Texas Tech University Institutional Review Board approved procedures for use of human subjects for consumer panel evaluation of sensory attributes. The survey team provided a monetary incentive to local organizations for providing volunteers, which formed the consumer groups. The groups then met at a specified location to participate in the sensory test. Consumer panelists were only allowed to participate once.

The sensory test was administered on eight different nights to groups of sixty volunteers in each of the three cities. Consumers were asked to fill out a survey regarding demographics and prior beef preferences. The summary statistics for demographics of the population can be found in Table 1.

Table 1. Consumer demographics and responses to beef preference statements. Reported as percentages of consumers (*n* = 1440; 480/city).

	Overall	Lubbock	Washington DC	Phoenix
Age				
20–30	34.5	32.9	49.3	21.4
31–40	21.1	17.2	16.6	29.6
41–50	26.7	30.6	17.1	32.2
51–60	16.0	19.1	12.0	16.8
>60	1.7	0.2	5.0	0.0
Income				
<USD 20,000	12.0	16.0	12.7	7.2
USD 20,000–50,000	28.8	27.6	34.1	24.8
USD 51,000–75,000	24.5	26.3	20.2	26.8
USD 76,000–100,000	15.8	16.5	14.4	16.4
>USD 100,000	19.0	13.5	18.6	24.8
Gender				
Male	49.9	44.5	55.5	49.7
Female	50.1	55.5	44.5	50.3
Education				
Non-High School Graduate	2.7	1.6	1.6	2.6
High School Graduate	9.8	9.7	9.7	9.3
Some College/Technical School	28.1	32.0	32.0	31.4
College Graduate	36.6	35.4	35.4	34.6
Post Graduate	22.8	21.2	21.2	22.1
Preferred Doneness				
Blue	0.1	0.2	0.2	0.0
Rare	3.2	3.4	4.2	1.9
Medium Rare	26.2	27.6	29.3	21.4
Medium	30.2	32.0	28.7	29.6
Medium Well	30.0	31.5	24.5	33.5
Well Done	10.3	5.3	11.7	13.6
Statement				
I enjoy red meat. It's an important part of my diet.	46.7	56.3	44.6	39.1
I like red meat well enough. It's a regular part of my diet.	37.9	32.9	38.7	42.0
I do eat some red meat although, but it wouldn't worry me if I didn't.	12.4	8.8	13.5	15.0
I rarely/never eat red meat.	3.0	2.0	3.2	3.9
Regular Purchaser of Beef				
Yes	70.7	72.1	68.7	71.3
No	29.3	27.9	31.3	28.7
Grade of Beef Most Commonly Purchased [a]				
USDA Prime	14.5	9.0	18.9	15.8
USDA Choice	51.2	55.4	47.7	50.6
USDA Select	11.4	14.1	8.6	11.4
Other	22.9	21.5	24.9	22.2
Important Palatability Trait for Roasts				
Flavor	40.2	38.3	38.5	42.7
Tenderness	47.8	49.2	48.5	46.3
Juiciness	12.1	12.5	13.0	10.9
Important Palatability Trait for Steaks				
Flavor	39.4	38.2	40.9	38.6
Tenderness	46.8	45.9	46.7	47.3
Juiciness	13.8	15.8	12.4	14.1

[a] USDA = United States Department of Agriculture.

Each panelist was seated at a numbered consumer booth and provided a ballot, plastic utensils, toothpicks, a napkin, an expectorant cup, a cup of water, and palate cleansers to use between samples (unsalted crackers and a 10% apple juice, 90% water solution). Prior to the start of each panel, panelists were given verbal instructions about the ballot and the procedure for the testing of samples. Panelists were instructed to cut each sample using their utensils to a size representative of beef consumed in the home or restaurant. The panels were conducted in large rooms with tables that had been divided into individual sensory booths.

Eight notional products of differing quality were designated by the combination of four cuts and two USDA grades. Australian samples were allocated to the assumed closest USDA grade. Each consumer was served six of the eight products with three samples being drawn from Australian sourced product and three from US sourced product. The six products used were balanced to ensure that each product was tested an equal number of times in each US city. In addition, each of the six individual products were allocated by a Latin square design which balanced presentational order. Consumers were served a total of seven samples in a predetermined balanced order in accordance with a 6 × 6 Latin square. All consumers received a warm-up sample to orient them to the sample format and evaluation procedures. Data obtained from warm-up samples was excluded from the analysis, as these samples were not related to the trial. The warm-up samples were always served in the first position, followed by six test samples. This design provided a balance for frequency, order, and carryover effects [17]. All samples were identified with a unique identification code assigned by the MSA software [16]. Each sample was rated on a 100 mm continuous line scale for tenderness, juiciness, flavor, and overall liking. On the scale, 0 mm was verbally anchored as not tender, not juicy, dislike flavor extremely, and dislike overall extremely, and 100 mm was verbally anchored as very tender, very juicy, like flavor extremely, and like overall extremely. Additionally, consumers rated each sample as "unsatisfactory", "good everyday quality", "better than everyday quality", or "premium quality."

In this study, consumer satisfaction was measured for the cut of meat and was the dependent variable for all models in this study. There were two types of satisfaction. One was a discrete choice while the other was measured on a continuum. The continuum measurement was the consumer's perceived satisfaction on a 0 (worst) to 100 (best) scale. Discrete satisfaction levels were identified by the level of quality at which a consumer makes an acceptability decision, as follows: (2) "unsatisfactory"; (3) "good everyday quality"; (4) "better than everyday quality"; and (5) "premium quality." For the purposes of this study, many parameters were presented as having an impact on overall satisfaction. The attributes of focus for this study were tenderness, flavor, and juiciness, all of which were measured from 0 to 100. These attributes were initially assumed to have a diminishing marginal utility as it can generally be expected that as satisfaction rises additional increases in attributes have declining effects. Overall, we expected these attributes had a positive influence on satisfaction. In addition, four different sources of beef were used in this study: Australian grass-fed beef, Australian grain-fed beef, USDA Select, and USDA Choice. These were represented as dummy variables in the model.

2.5. Conceptual Framework

Given the panel nature of the data, there were two problem areas that could arise and decrease the accuracy of the results. One area was heteroskedascity, which was tested for by using White's test and corrected where found.

The second problem area was nonrepresentative samples. As is common in empirical marketing, it is important to eliminate nonrepresentative observations. Such observations typically reduce the effectiveness in estimating the model. The method chosen to correct for these observations was the Cook's Distance method. Cook's Distance (Di) is identified by the Equation (1) [18]:

$$D_i = \frac{\sum_{j=1}^{n} (y_j - y_j(i))^2}{(k+1)s_2}, i = 1, \dots, n \tag{1}$$

where: s = estimated root mean square error, y_j = regression estimate of the conditional mean $E(Y_j|x_{1j}, \dots, x_{kj})$, and $y_j(i)$ = regression estimate of the condition mean $E(Y_j|x_{1j}, \dots, x_{kj})$ with the ith data point $(y_i, x_{1i}, \dots, x_{kj})$ removed.

This produces a normalized measure of the influence of point i on all predicted mean values. A common rule of thumb is to treat any point i as an outlier when the value of (D_i) exceeds $\frac{4}{n-(k+1)}$ where n = number of observations and k = degrees of freedom. The use of Cook's Distance increases the efficiency of the model and increases R^2. The Cook's Distance procedure was used in the data for all models of this paper. The scatter plots that depict the relation of flavor to satisfaction show the effects of using Cook's Distance (Figures S1 and S2 for raw and cleaned, respectively). As one can see, the scatter plot narrows when the outliers were removed from the equation. The same can be seen for tenderness (Figures S3 and S4 for raw and cleaned, respectively) and juiciness (Figures S5 and S6 for raw and cleaned, respectively).

Marketers are continually looking for ways to better meet consumer preferences. An important key in understanding the preferences of the consumer is their experience while eating the product. From the evaluation of the consumers' reactions, marketers can then develop products with the appropriate attributes that can closely fit the preferences of consumers. In addition, the preference structure provides marketers with a method for making assumptions about populations that contain the same characteristics.

The food choice process is influenced by a large number of complex factors, including the person making the decision and the associated environment [12]. Product characteristics such as quality, price, and usefulness are among the common factors believed to influence the consumers' purchase decision. The relevance of product characteristics to the individual consumer lies in their ability to generate some response (positive or negative), relative to a consumer's perception of quality and ultimately the purchase decision [5]. Characteristics generating positive responses to quality perception are considered to add to utility gained by the consumption of the product.

The preference structure of the consumer is revealed by several methods. The basic concept for these methods is to measure the reaction of consumers to changing attribute levels. Commonly used methods include surveys, focus groups, feedback, and consumer trials. These methods are specifically designed to evaluate the interaction of the internal and external factors that influence the purchase decision. Internal factors refer to the product attributes while the external factors refer to consumer demographics and market characteristics such as labeling and stores. Attribute values are assigned by the consumers as a reflection of the satisfaction gained from the product. Factors such as age, income, and education level have been shown to influence the consumers' purchase decision [6,19,20]. In the current study, these factors were included in our model to capture their impact on satisfaction.

To better understand the impact of the above factors on satisfaction, we must first understand their interaction with consumer satisfaction. The interaction can be best explained by choice theory. Modern economic choice theory starts with the assumption that an individual's market behavior is generated by maximization of utility [21]. According to consumer choice theory, a person evaluates the amount of utility provided by each good and bases the purchase decision on the amount of utility to be gained from each good. In following the utility maximization theory, consumers look for products that maximize utility. Utility, as shown in Equation (2), is the combination of attributes possessed by a product.

$$U = f(x_1, x_2, x_3 \ldots, x_n) \tag{2}$$

U in this case is a function of attribute levels $(x_1, \ldots, x_n)$ and considered to be the utility or satisfaction gained from the meat. Our model focused on the attributes of the meat: tenderness, juiciness, and flavor. These variables add to satisfaction independently and when combined yield a utility.

2.6. Statistical Analysis

2.6.1. General Model

In an effort to study the effects of beef attributes on the consumer preference, we developed a model that accounts for chosen effects and attributes. Statistical analysis of the models was performed in SAS (SAS Inst. Inc., Cary, NC, USA). The natural logs of the independent variables were used in the general model, consistent with the expectation of diminishing marginal utility. Our base model for understanding the consumer preference structure is given in Equation (3):

$$Y = f(Tenderness, Juiciness, Flavor) \tag{3}$$

where Y is consumer satisfaction.

Two methods were used to evaluate the interaction of the attributes with consumer satisfaction. The reason for using two different methods was because the data presented two measurements of satisfaction. Each of the two measurements required a method that was specific to its characteristics.

2.6.2. Overall Satisfaction Model

The first method that we used to approach consumer satisfaction was an Ordinary Least Squares (OLS) model. This model was used in conjunction with the continuum satisfaction measurement. The random utility model provided a direct relationship between the satisfaction and the attributes. It was applied to the general model to determine the impact of tenderness, juiciness, and flavor on overall satisfaction. For this application our general model was fitted for OLS estimation (Equation (4)).

$$Y = \beta_0 + \beta_1 \ln Tenderness + \beta_2 \ln Juiciness + \beta_3 \ln Flavor + \varepsilon \tag{4}$$

Both the conditional logit and OLS model served specific purposes aimed at understanding the consumer preference structure.

To evaluate if the origin of beef had any impact on consumer satisfaction, a fixed effects model was used preliminarily to determine if there was a statistical difference in the origins. Once the statistical importance of the grouping had been determined, the base model was applied to each significant grouping.

Previous studies suggest that demographic characteristics have a substantial impact on beef preferences. The demographic characteristics of the consumers were initially dealt with by following the same procedure as Lusk and Fox [6], which involved dividing the demographical classes into two groups forming a high and low group. A dummy variable was assigned to the variable grouping in the common form of assigning the high grouping a value of one and the lower grouping a value of zero. The dummy variable was placed in the OLS regression creating a fixed effects model. If the dummy variable was statistically significant, it was assumed the two demographic categories had unique effects on the consumers' satisfaction.

To determine the attribute levels present, each discrete level was labeled as the attribute intensity model. This model identified statistically common expectations for quality levels that together result in satisfaction levels. This method was designed on the theory that the variance between the levels was changing. Therefore, the model cannot be simply evaluated as a simple linear regression.

2.6.3. Conditional Logit Model

The second method used was a conditional logit model. This model was used with the discrete levels of satisfaction. The discrete levels of satisfaction presented a unique problem to the model. The use of panel data allowed us to assume that the variances between the discrete levels were not uniform. Prior studies have used ordered probit, logit, and multinomial logit models to evaluate the equation. These models, however, do not account for the variance present between levels. The conditional logit model attempts to solve the issue with variance by breaking up the continuous discrete variable to account for the variance present between levels. A discrete choice can be evaluated by many statistical procedures, the most common being a form of logit modeling. The conditional logit model evaluated two consecutive discrete levels by evaluating both as a logit model. The value of one was given to the higher level and the value of zero was given to the lower level, thereby forming a simple logit model. The conditional logit model produced an odds ratio representing the probability of moving to the next discrete level with a one unit increase in the attribute. This model was specified similarly to the random utility model, but the dependent variable was in the form of a logit model of satisfaction states (Equation (5)).

$$Y = \beta_0 + \beta_1 Tenderness + \beta_2 Juiciness + \beta_3 Flavor + \varepsilon \tag{5}$$

This model told us the probability of increasing the satisfaction of the consumer by a change in the attribute structure. This was different from the other model in many ways. The conditional logit model provided a probability of increasing consumer satisfaction, but the random utility model demonstrated the overall satisfaction based on certain attributes; however, it does not describe how to increase satisfaction. The conditional logit model illustrated how the influence of each attribute changes satisfaction. Both the conditional logit and OLS model served specific purposes aimed at understanding the consumer preference structure.

3. Results and Discussion

Throughout the results the parameter estimates were considered to be the influence on consumer satisfaction unless otherwise noted. All values were positive unless a negative sign was present. Standard errors in the results were corrected for heteroskedasticity and were considered to be the robust standard errors.

3.1. Overall Satisfaction Model

The overall satisfaction model was the most common model used in this study. The model was adapted to examine the attribute makeup of satisfaction for origins, demographics, and discrete satisfaction levels. The following section explores the results from the various random utility models used.

The first objective and use of the random utility model was to determine the impact of the three palatability attributes on consumer satisfaction. In order to determine the impact of the attributes on satisfaction, the semilog base model was evaluated. This model produced a high R^2 and a moderate Residual Standard Deviation (RSD) (Table 2). By common convention, an RSD that is less than 10 is desired. Our RSD was larger than ten for some models, but it was never more than 12.

Table 2. Impacts of attributes on overall satisfaction using a random utility model.

Variables	Estimates	SE
Intercept	−105.28 *	1.551
Tenderness	23.99 *	0.751
Juiciness	15.17 *	0.782
Flavor	56.47 *	1.050

$n = 9357$; $R^2 = 0.77$; RSD $= 11.15$. * Denotes variables significant at $p < 0.05$.

The parameter estimates generated by the model were considered to be the impact on the continuum measurement of satisfaction, but since a semilog model was used, the estimates were transformed to represent their relationship with satisfaction. The transformation is explained as a one percent increase in any of the attributes results in a change in satisfaction. For example, a one percent increase in flavor resulted in the increase of overall satisfaction by 0.56 units. By the same transformation, a one percent increase in tenderness increased the overall satisfaction by 0.24 units, and a one percent increase in juiciness increased overall satisfaction by 0.15 units. A unit of satisfaction was represented by a one point change on the continuum scale.

The parameter estimates indicate that flavor had the largest impact on consumer satisfaction. It was the assumption of our model that all variables were exogenous; however, endogeneity of a variable could arise, creating potential problems. These three variables were the only measures used by the survey to measure consumer satisfaction, so we assumed that they were exogenous.

The level of importance to the consumer was depicted as (1) flavor, (2) tenderness, and (3) juiciness, with number one being the most important. The estimates suggested that flavor had a 58% greater ($p < 0.05$) impact on satisfaction than tenderness. The impact of tenderness on satisfaction was 43% greater ($p < 0.05$) than the impact of juiciness.

O'Quinn et al. [22] determined the relative contribution of each trait to overall liking by using multivariate regression, ultimately suggesting that flavor contributed the most (49.4%), followed by tenderness (43.4%), and juiciness (7.4%). They also reported that no single palatability trait was the most important, as beef palatability was dependent upon the acceptance of all three traits. Flavor had a greater contribution to beef overall liking [22], but not to the extent that we observed in the current findings. Previous work has shown strong relationships between beef flavor and overall acceptability or liking [23–25]. In fact, flavor was the most highly correlated trait to overall liking as opposed to juiciness or tenderness. The current results were not unexpected as the previous reports of beef eating quality for US consumers align with these coefficients for grain-fed beef [26,27] and grass-fed beef [28–30].

3.2. Impacts of Origin on Satisfaction

The next objective of the study was to determine if beef origin (the source country and cattle finishing system) had any impact on consumer preferences. For the preliminary determination, a fixed effects model was used. The results of the model showed that origin had an impact ($p < 0.05$) on consumer satisfaction (Table 3). The model was run using each origin as the base; therefore, the R^2 and the RSD are the same for each base. This was done to ensure that the rankings were consistent. USDA Select and USDA Choice had statistically different impacts on satisfaction, but the two Australian finishing systems (grass-fed or grain-fed) had similar impacts on satisfaction. More importantly, we found that Australian beef had a statistically different impact than the USDA cuts on satisfaction. The ranking produced by the model align with previous results [20,25,30,31] that consumers prefer the flavor of domestic beef, especially when compared to international grass-fed beef.

Table 3. Fixed effects model for origin impact on consumer satisfaction.

Base Origin	Variables	Estimates	SE	R^2 (RSD)
Australian Grass	Intercept	−105.73 *	1.548	0.78
	Tenderness	23.85 *	0.750	(11.47)
	Juiciness	15.18 *	0.784	
	Flavor	56.30 *	1.049	
	USDA Select	1.16 *	0.381	
	USDA Choice	2.09 *	0.378	
	Australian Grain	0.42	0.363	

* Denotes variables significant at $p < 0.05$.

Estimates of the pairwise model showed that US beef produced greater ($p < 0.05$) consumer satisfaction than the Australian grass-fed beef. One of the biggest differences in the origins was the consistently higher estimates for flavor in the US beef (Table 4). The parameter estimates revealed that the flavor of the USDA Choice was approximately 12% greater than Australian grass-fed and 3% greater than Australian grain-fed beef. Consumers were 2% more responsive to the flavor of Australian grain-fed beef than USDA Select, but responded more favorably to the tenderness of USDA Select.

The impact of origin was also evaluated with the attribute intensity model (Table 5). For USDA Choice beef and the Australian grain-fed beef, tenderness did not impact ($p > 0.05$) satisfaction at the "premium quality" level, but tenderness was a significant attribute affecting ($p < 0.05$) satisfaction of the USDA Select samples. Tenderness, juiciness, and flavor impacted ($p < 0.05$) satisfaction at all other levels across all origins. Results of this model showed that the flavor in Australian grass-fed beef was not as strong as the US beef at the "better than everyday quality" level, and Australian grain-fed beef was not as strong as the US beef especially at the "premium quality" level. At the premium level, USDA Choice flavor was 29% greater than Australian beef, while the USDA Select flavor was 42% higher than Australian grain-fed beef. For USDA Choice beef and the Australian grain-fed beef, tenderness did not impact ($p > 0.05$) satisfaction at the "premium quality" level, but tenderness was a significant attribute affecting ($p < 0.05$) satisfaction of the USDA Select samples.

Table 4. Comparisons of origins and attribute effects.

Origin	Variables	Estimates	SE	R^2 (RSD)	n
Australian Grass	Intercept	−103.00 *	3.15	0.78	1511
	Tenderness	27.12 *	1.70	(11.54)	
	Juiciness	14.76 *	1.95		
	Flavor	51.82 *	2.33		
Australian Grain	Intercept	−103.53 *	2.50	0.78	2998
	Tenderness	21.76 *	1.27	(11.50)	
	Juiciness	15.27 *	1.24		
	Flavor	57.23 *	1.70		
USDA Select	Intercept	−101.20 *	3.12	0.77	2410
	Tenderness	23.25 *	1.48	(11.78)	
	Juiciness	14.04 *	1.51		
	Flavor	56.06 *	2.30		
USDA Choice	Intercept	−112.26 *	1.97	0.79	2411
	Tenderness	24.91 *	1.02	(11.34)	
	Juiciness	16.83 *	1.05		
	Flavor	58.82 *	1.39		

* Denotes variables significant at $p < 0.05$.

We also used this model to understand the differences present in the levels between product origin (source country and finishing system). Results suggested that people were extremely decisive about what beef samples they deem "unsatisfactory." Attribute estimates for "unsatisfactory" and "good everyday quality" beef seemed to be fairly similar across the product origins. The results also showed that consumers were less decisive about what they deem "premium quality," indicated by large standard errors.

3.3. Impact of Demographics

Our last use of our random utility model was to determine the impact that demographic variables had on consumer satisfaction. The four demographic variables evaluated in this model were age, income, education, and gender, as each has been shown to impact consumer perception [4,19]. As seen in Table 6, education did not impact satisfaction ($p > 0.05$) in the overall model, but age, income, and gender influenced ($p < 0.05$) satisfaction.

A fixed effects model was used to determine the significance of the demographical groupings and to determine if any of the demographical effects carried over into the origins (Table 6). The only observed significance occurred in the age groups in the Australian beef and income groups in the USDA Choice. Because of the large scope of this study, these interactions were not investigated further.

Once we had determined that income, gender, and age impacted satisfaction, we applied the previous models. A pairwise comparison of the groupings was performed, followed by the attribute intensity model. These models allowed us to understand the reason for the differences between groups.

Table 5. Comparison of attribute intensity within different origins.

Variables	USDA Choice Beef				USDA Select Beef				Australian Grain-Fed Beef				Australian Grass-Fed Beef			
	Est.	SE	R^2 (RSD)	n	Est.	SE	R^2 (RSD)	n	Est.	SE	R^2 (RSD)	n	Est.	SE	R^2 (RSD)	n
Unsatisfactory																
Intercept	−38.65 *	2.974	0.58	424	−29.60 *	2.534	0.54	499	−37.53 *	2.313	0.59	701	−36.96 *	3.205	0.56	390
Tenderness	10.56 *	1.621	(9.87)		8.32 *	1.497	(9.09)		10.37 *	1.181	(8.88)		11.81 *	1.588	(8.85)	
Juiciness	7.39 *	1.743			6.51 *	1.502			5.59 *	1.284			7.63 *	1.812		
Flavor	29.12 *	1.699			26.04 *	2.083			29.88 *	1.316			25.59 *	1.609		
Good Everyday Quality																
Intercept	−109.88 *	11.570	0.60	987	−98.05 *	4.5013	0.62	1081	−107.59 *	4.248	0.62	1330	−105.48 *	5.716	0.60	652
Tenderness	18.40 *	2.542	(9.98)		19.07 *	1.745	(9.52)		17.63 *	1.909	(9.68)		23.89 *	2.515	(9.57)	
Juiciness	14.56 *	2.703			14.01 *	1.761			13.17 *	1.582			9.94 *	2.501		
Flavor	63.54 *	4.566			56.44 *	2.766			63.88 *	2.567			59.74 *	3.440		
Better than Everyday Quality																
Intercept	−152.52 *	13.663	0.62	667	−162.70 *	10.264	0.62	562	−134.02 *	18.178	0.56	705	−116.25 *	30.604	0.43	331
Tenderness	22.07 *	5.635	(7.13)		32.72 *	4.943	(6.79)		22.79 *	4.470	(7.95)		30.96 *	7.960	(9.58)	
Juiciness	11.12 *	3.795			11.50 *	2.412			15.36 *	3.557			15.43 *	6.914		
Flavor	89.98 *	9.158			84.06 *	5.320			74.79 *	9.217			56.43 *	17.274		
Premium Quality																
Intercept	−148.51 *	40.662	0.61	333	−179.27 *	16.164	0.82	268	−74.71 *	33.775	0.42	258	−216.42 *	18.378	0.81	136
Tenderness	32.83	21.819	(4.93)		30.77 *	9.330	(4.68)		16.76	13.319	(7.64)		28.95 *	12.413	(4.65)	
Juiciness	13.74 *	6.128			14.66 *	4.342			14.34 *	8.301			31.68 *	8.205		
Flavor	76.78 *	13.395			93.58 *	5.652			54.29 *	17.262			87.26 *	16.206		

* Denotes variables significant at $p < 0.05$. Est. denotes Estimates.

Table 6. Demographic effects on satisfaction.

Variables	Overall Model			USDA Choice Beef			USDA Select Beef			Australian Beef		
	Estimates	SE	R^2 (RSD)	Estimates	SE	R^2 (RSD)	Estimates	SE	R^2 (RSD)	Estimates	SE	R^2 (RSD)
Intercept	−116.65 *	1.256	0.84	−126.03	2.757	0.85	−113.85 *	2.529	0.82	−113.36 *	1.693	0.83
Tenderness	25.80 *	0.568	(9.66)	27.72 *	1.250	(9.16)	25.02 *	1.105	(9.84)	25.15 *	0.775	(9.77)
Juiciness	16.01 *	0.608		17.32 *	1.447		14.56 *	1.076		16.25 *	0.853	
Flavor	60.98 *	0.812		63.94 *	1.784		61.53 *	1.689		58.97 *	1.077	
Income	−0.78 *	0.218		−1.27 *	0.418		−0.69	0.436		−0.61	0.315	
Age	0.69 *	0.209		0.24	0.384		0.39	0.424		1.08 *	0.306	
Education	0.37	0.256		−0.63	0.481		−0.15	0.488		−0.29	0.382	
Gender	−0.52 *	0.202		−0.67	0.375		−0.63	0.406		−0.35	0.295	

* Denotes variables significant at $p < 0.05$.

3.3.1. Gender

The fixed effects model suggested that the females were going to be less responsive to changes in the attributes (Table 7). The pairwise comparison of gender indicated that males were more responsive to a change in all attributes than females. The most distinct difference was that males were 7% more responsive to a change in tenderness.

Table 7. Pairwise comparisons of gender on consumer satisfaction.

Variables	Male				Female			
	Estimates	SE	R^2 (RSD)	n	Estimates	SE	R^2 (RSD)	n
Intercept	−121.32 *	1.847	0.83	4650	−114.42 *	1.608	0.84	4580
Tenderness	26.84 *	0.840	(9.42)		25.02 *	0.766	(9.92)	
Juiciness	16.29 *	0.914			15.80 *	0.817		
Flavor	61.77 *	1.239			60.42 *	1.069		

* Denotes variables significant at $p < 0.05$.

Bonny et al. [32] reported demographic differences of European consumer scores for beef eating quality. Notably, men scored beef samples more favorably than women in general, but this trend varied between countries (Ireland, Northern Ireland, and Poland) and between palatability traits. Only Irish and Northern Irish males scored beef juicier with greater flavor liking than females, while only Polish males rated tenderness higher than females. Kubberød et al. [33] also reported men score meat more favorably than females, which was likely attributed to their more positive attitude towards red meat.

The attribute intensity model showed that males and females disagree on what cut of meat should be "unsatisfactory." At the "unsatisfactory" level, males were 13% more responsive to tenderness while females were 11% more responsive to juiciness (Table 8). A noticeable difference also occurred at the "better than everyday quality" level, where males were 24% more responsive to juiciness while females were 12% more responsive to tenderness. Males and females seemed to agree on the attribute make-up for a "premium quality" cut of beef.

Table 8. Comparison of attribute intensity for gender by satisfaction level.

Variables	Male				Female			
	Estimates	SE	R^2 (RSD)	n	Estimates	SE	R^2 (RSD)	n
Unsatisfactory								
Intercept	−51.09 *	1.936	0.70	833	−49.63 *	1.574	0.71	962
Tenderness	13.65 *	0.895	(7.27)		11.92 *	0.824	(7.44)	
Juiciness	8.25 *	0.977			9.13 *	0.874		
Flavor	33.92 *	1.180			33.87 *	0.919		
Good Everyday Quality								
Intercept	−108.02 *	2.819	0.67	2063	−112.18 *	2.995	0.70	1924
Tenderness	21.67 *	1.129	(8.33)		22.33 *	1.175	(8.72)	
Juiciness	14.59 *	1.093			13.46 *	1.093		
Flavor	58.95 *	1.785			62.03 *	1.843		
Better than Everyday Quality								
Intercept	−158.18 *	7.659	0.66	1148	−149.91 *	7.194	0.63	1158
Tenderness	24.97 *	2.766	(6.72)		27.96 *	2.932	(7.30)	
Juiciness	16.54 *	1.993			12.51 *	1.893		
Flavor	84.47 *	4.681			81.00 *	3.645		
Premium Quality								
Intercept	−199.92 *	8.465	0.76	602	−188.59 *	17.188	0.75	535
Tenderness	39.71 *	5.822	(4.77)		39.68 *	9.549	(4.88)	
Juiciness	14.20 *	3.535			13.86 *	3.506		
Flavor	95.54 *	6.246			90.14 *	9.382		

* Denotes variables significant at $p < 0.05$.

3.3.2. Age

The next demographical variable presented was age. The fixed effects model suggested that older consumers were more responsive to changes in the attributes. The outcome of the pairwise comparison showed that the two age groupings were relatively close in the preferences for the attributes (Table 9). One distinct difference was that younger participants were more responsive to changes in juiciness by almost 6%.

Table 9. Pairwise comparisons of age groups for consumer satisfaction.

Variables	20–40 Years Old				41–60 Years Old			
	Estimates	SE	R^2 (RSD)	n	Estimates	SE	R^2 (RSD)	n
Intercept	−118.25 *	1.6826	0.83	5113	−116.54 *	1.716	0.84	4117
Tenderness	25.58 *	0.7528	(9.75)		26.34 *	0.871	(9.60)	
Juiciness	16.48 *	0.8511			15.52 *	0.874		
Flavor	61.11 *	1.1169			60.61 *	1.181		

* Denotes variables significant at $p < 0.05$.

Bonny et al. [32] observed a small negative relationship between age and tenderness in France and Poland, and with age and juiciness in Ireland, Northern Ireland, and Poland. However, tenderness scores increased with age in Northern Ireland. Age can also influence red meat purchase intent of Canadian consumers [34]. Quagrainie et al. [34] reported that people aged 30 years old or younger were more likely to purchase meat products (specifically high quality pork, high quality beef, or ground beef) than older consumers.

The next area evaluated was the attribute intensity for each discrete level in the two age groups. Both groups were commonly decisive as to what cuts of meat were "unsatisfactory." Once again, as the satisfaction level increased the standard error increased, suggesting that the consumers were uncertain about the personal classification of "premium quality."

The older group had a greater response to the flavor attribute than the younger age group at all levels. This was most evident at the "good everyday quality" and "premium quality" levels (Table 10). At these levels, the older group was 7% more responsive to changes in flavor. There was less variation present in the older group's estimates suggesting they have developed their quality expectations more thoroughly. Inconsistencies between the age groups were evident at the "better than everyday quality" level, where older consumers were 38% more responsive to tenderness, but younger consumers were 18% more responsive to juiciness. The younger panelists were more responsive to juiciness in the top three levels by a minimum of 16%.

Table 10. Comparison of attribute intensity for age groups by satisfaction level.

Variables	20–40 Years Old				41–60 Years Old			
	Estimates	SE	R^2 (RSD)	n	Estimates	SE	R^2 (RSD)	n
Unsatisfactory								
Intercept	−50.18 *	1.577	0.71	964	−50.42 *	1.903	0.70	831
Tenderness	12.90 *	0.800	(7.25)		12.45 *	0.926	(7.48)	
Juiciness	8.85 *	0.868			8.60 *	0.971		
Flavor	33.28 *	0.949			34.62 *	1.095		
Good Everyday Quality								
Intercept	−108.81 *	2.901	0.67	2264	−112.10 *	2.884	0.71	1723
Tenderness	21.56 *	1.104	(8.67)		22.39 *	1.223	(8.30)	
Juiciness	15.39 *	1.105			12.41 *	1.089		
Flavor	58.87 *	1.708			62.82 *	1.926		

Table 10. *Cont.*

	20–40 Years Old				41–60 Years Old			
Variables	Estimates	SE	R^2 (RSD)	*n*	Estimates	SE	R^2 (RSD)	*n*
Better than Everyday Quality								
Intercept	−147.16 *	7.104	0.64	1296	−166.06 *	6.972	0.66	1010
Tenderness	23.67 *	2.489	(7.12)		32.65 *	3.439	(6.86)	
Juiciness	15.29 *	1.842			12.60 *	2.041		
Flavor	81.15 *	3.960			84.69 *	3.837		
Premium Quality								
Intercept	−194.23 *	13.979	0.77	584	−195.41 *	11.673	0.75	553
Tenderness	41.33 *	9.280	(4.88)		38.28 *	6.771	(4.76)	
Juiciness	15.52 *	3.523			13.06 *	3.338		
Flavor	89.77 *	10.136			95.72 *	6.127		

* Denotes variables significant at $p < 0.05$.

3.3.3. Income

The last demographic examined was income. The fixed effects model suggested that the lower income group was more responsive to a change in attributes. The results from the pairwise comparison illustrated that higher income consumers were more responsive to a change in tenderness, by nearly 11%, while the lower income group was more responsive to juiciness and flavor (Table 11).

Table 11. Pairwise comparisons of income groups on consumer satisfaction.

	USD 0–50,000				USD 50,000 and above			
Variables	Estimates	SE	R^2 (RSD)	*n*	Estimates	SE	R^2 (RSD)	*n*
Intercept	−116.38 *	1.536	0.82	3762	−117.96 *	1.985	0.85	5468
Tenderness	24.26 *	0.732	(10.17)		26.96 *	0.897	(9.32)	
Juiciness	16.08 *	0.745			15.85 *	1.037		
Flavor	62.12 *	1.017			60.16 *	1.330		

* Denotes variables significant at $p < 0.05$.

Bonny et al. [32] reported income level with respect to country had very little effect on consumer scores for beef eating quality. Only juiciness was influenced by income level [35]. Likewise, Hwang et al. [35] reported demographic characteristics had minimal impact on beef sensory scores given by Korean and Australia consumers.

The attribute intensity model showed there were inconsistencies about the attributes that qualify as an "unsatisfactory" cut of beef (Table 12). The higher income group was 15% more responsive to tenderness at this level, but the lower income group was 10% more responsive to juiciness. The biggest factor separating the higher income consumers from the lower income consumers seemed to be the difference in tenderness preferences at all levels. The lower income group was more responsive to flavor at all levels except for the "premium quality" level. The lower income group was 10% more responsive to juiciness at the "premium quality" level. The variation for the lower income group was greater, suggesting that the higher income group had a more defined perception of satisfaction.

Table 12. Comparison of attribute intensity for income groups by satisfaction level.

Variables	USD 0–50,000				USD 50,000 and above			
	Estimates	SE	R^2 (RSD)	n	Estimates	SE	R^2 (RSD)	n
Unsatisfactory								
Intercept	−49.27 *	1.973	0.70	670	−50.88 *	1.549	0.72	1125
Tenderness	11.63 *	1.003	(7.82)		13.34 *	0.742	(7.07)	
Juiciness	9.31 *	1.112			8.38 *	0.790		
Flavor	33.98 *	1.222			33.85 *	0.884		
Good Everyday Quality								
Intercept	−108.57 *	3.453	0.66	1656	−110.69 *	2.502	0.71	2331
Tenderness	21.08 *	1.278	(9.27)		22.67 *	1.050	(7.93)	
Juiciness	13.34 *	1.392			14.22 *	0.913		
Flavor	61.44 *	2.118			59.84 *	1.569		
Better than Everyday Quality								
Intercept	−151.60 *	8.210	0.62	961	−154.53 *	6.639	0.67	1345
Tenderness	25.36 *	3.301	(7.40)		27.09 *	2.493	(6.73)	
Juiciness	14.07 *	2.226			14.55 *	1.748		
Flavor	83.20 *	4.609			82.17 *	3.733		
Premium Quality								
Intercept	−188.06 *	13.819	0.79	472	−203.17 *	11.300	0.74	553
Tenderness	39.96 *	9.022	(5.15)		40.58 *	6.437	(4.56)	
Juiciness	15.27 *	4.260			13.73 *	2.961		
Flavor	88.31 *	10.890			96.69 *	5.248		

* Denotes variables significant at $p < 0.05$.

3.4. Attribute Intensity Model

In an effort to determine the impacts of the attributes at each satisfaction level, we applied the base model to each discrete satisfaction level. To ensure the integrity of this model, we had to first guarantee that the satisfaction levels were statistically different from each other. A fixed effects model was used to show that the levels were indeed different. The fixed effects model showed that the discrete levels follow their ordinal ranking as expected. The pairwise comparison of all the levels allowed us to determine the difference in the levels, implying that a change in attributes at the "premium quality" level had a 63% greater impact on satisfaction than the same change at the "unsatisfactory" level. This discrepancy gap became narrower as satisfaction level elevated. For example, the change in parameter estimates between "premium quality" and "better than everyday quality" was only 2%.

The base model was applied to each discrete level for the attribute intensity model (Table 13). This model also showed how the attributes' impacts vary between levels. Results showed that once again the flavor attribute garnered the largest parameter estimate in each of the levels. Juiciness appeared to be the attribute more responsible for change between "premium quality" and "better than everyday quality." The parameter estimate for juiciness increased by 25% between these two levels, while the other attributes decreased a small amount, suggesting diminishing marginal utility was present.

Results indicated that consumers were not decisive when classifying beef as "premium quality," as shown through the increase in standard error from "better than everyday quality" to "premium quality." On the other hand, consumers do not have a problem classifying beef as "unsatisfactory," as indicated by the smaller standard error and greater difference in parameter estimates at these levels. Our model demonstrated that the parameter estimates changed about 52% between "unsatisfactory" and "everyday quality" levels. This was the largest difference between levels found in our model.

Table 13. Comparison of attribute levels of corresponding satisfaction levels.

Satisfaction Level	Variables	Estimates	SE	R² (RSD)	n
Unsatisfactory	Intercept	−35.75 *	1.343	0.57	2018
	Tenderness	10.48 *	0.727	(9.78)	
	Juiciness	6.37 *	0.774		
	Flavor	27.92 *	0.850		
Good Everyday	Intercept	−104.42 *	3.392	0.61	4063
Quality	Tenderness	19.26 *	1.118	(9.73)	
	Juiciness	13.20 *	1.019		
	Flavor	60.65 *	1.658		
Better than	Intercept	−140.97 *	10.179	0.56	2273
Everyday Quality	Tenderness	25.71 *	2.929	(7.78)	
	Juiciness	13.44 *	1.973		
	Flavor	77.58 *	5.909		
Premium Quality	Intercept	−140.95 *	18.876	0.65	997
	Tenderness	25.60 *	8.529	(5.91)	
	Juiciness	17.92 *	4.683		
	Flavor	75.93 *	11.416		

* Denotes variables significant at $p < 0.05$.

3.5. Conditional Logit Model

The conditional logit model was used to determine the probability of changing discrete levels with an attribute change. This model was designed to compare two consecutive levels as a logit model. The higher level was given a value of one, while the lower level was given a value of zero. The point estimate produced by the model indicated the probability of changing satisfaction levels with a one unit change in the attribute. The point estimate was the exponentiated value of the parameter estimates. The tables included in this section show the parameter estimates along with the point estimates.

This model was first applied to the overall model. Previously in the conceptual model we assumed that the variables would exhibit a diminishing marginal utility. The results shown in Table 14 illustrate the diminishing marginal utility of flavor, as indicated by the decreasing point estimate as the satisfaction level increased.

Table 14. Conditional logit model.

Transition	Variables	Estimates	SE	Point Estimates
Level 2 to Level 3 [a]	Intercept	−4.43 *	0.135	
	Tenderness	0.04 *	0.002	1.037
	Juiciness	0.01 *	0.002	1.004
	Flavor	0.08 *	0.002	1.079
Level 3 to Level 4 [a]	Intercept	−8.10 *	0.207	
	Tenderness	0.04 *	0.002	1.042
	Juiciness	0.01 *	0.002	1.013
	Flavor	0.06 *	0.002	1.059
Level 4 to Level 5 [a]	Intercept	−11.86 *	0.458	
	Tenderness	0.06 *	0.005	1.065
	Juiciness	0.01 *	0.003	1.015
	Flavor	0.05 *	0.005	1.055

* Denotes variables significant at $p < 0.05$. [a] Level 2 = Unsatisfactory; Level 3 = Good everyday quality; Level 4 = Better than everyday quality; Level 5 = Premium quality.

Table 14 shows the consumers exhibited a linear preference structure between levels, which translated to an equal distribution of attributes when changing between levels. For example, the probability for a change from "better than everyday quality" to "premium quality" consisted of 34% flavor, while the change from "unsatisfactory" to "everyday quality" was 35%.

Results of the conditional logit model suggested that an increase in flavor created the largest increase in the probability of moving from "unsatisfactory" to "good everyday quality." However, for samples classified as "good everyday quality" or higher, flavor played a minimal role in increasing the probability of moving to the next level. This model also showed that tenderness had the largest impact on the probability of increasing to "premium quality" from "better than everyday quality."

The conditional logit model was also used to evaluate the impact of origin on consumer satisfaction (Table 15). Flavor was the most important factor for increasing satisfaction at all levels and origins, except for Australian grain-fed and USDA Select beef moving from "better than everyday quality" to "premium quality." As quality level increased more emphasis shifted to tenderness, but flavor was still the driving force behind satisfaction in most instances.

Table 15. Conditional logit model focusing on origin.

		Level 2 vs. Level 3 [a]			Level 3 vs. Level 4 [a]			Level 4 vs. Level 5 [a]		
Origin	Variables	Estimates	SE	Point Estimates	Estimates	SE	Point Estimates	Estimates	SE	Point Estimates
Australian Grass	Intercept	−4.85 *	0.336		−8.17 *	0.534		−13.29 *	1.326	
	Tenderness	0.05 *	0.005	1.047	0.05 *	0.005	1.048	0.07 *	0.014	1.039
	Juiciness	0.00	0.005	1.000	0.01 *	0.005	1.015	0.02 *	0.009	0.999
	Flavor	0.08 *	0.006	1.082	0.05 *	0.006	1.052	0.07 *	0.012	1.043
Australian Grain	Intercept	−4.50 *	0.233		−8.01 *	0.363		−10.82 *	0.797	
	Tenderness	0.03 *	0.003	1.035	0.04 *	0.003	1.041	0.06 *	0.009	1.061
	Juiciness	0.00	0.003	1.005	0.02 *	0.003	1.016	0.02 *	0.006	1.022
	Flavor	0.08 *	0.004	1.082	0.05 *	0.004	1.056	0.04 *	0.008	1.038
USDA Select	Intercept	−4.18 *	0.256		−8.51 *	0.432		−11.59 *	0.882	
	Tenderness	0.04 *	0.004	1.039	0.04 *	0.004	1.045	0.06 *	0.009	1.063
	Juiciness	0.00	0.004	1.000	0.01 *	0.004	1.010	0.01 *	0.006	1.015
	Flavor	0.07 *	0.005	1.076	0.06 *	0.005	1.064	0.05 *	0.009	1.056
USDA Choice	Intercept	−4.39 *	0.295		−7.85 *	0.390		−13.21 *	0.906	
	Tenderness	0.03 *	0.004	1.030	0.04 *	0.004	1.039	0.07 *	0.009	1.069
	Juiciness	0.01 *	0.004	1.009	0.01 *	0.004	1.011	0.01 *	0.006	1.010
	Flavor	0.08 *	0.005	1.080	0.06 *	0.005	1.062	0.07 *	0.009	1.072

* Denotes variables significant at $p < 0.05$. [a] Level 2 = Unsatisfactory; Level 3 = Good everyday quality; Level 4 = Better than everyday quality; Level 5 = Premium quality.

4. Conclusions

Perhaps the most important finding of this study was that flavor was the largest contributor to consumer satisfaction. This finding was consistent throughout all the models. In the base model, flavor represented 59% of the satisfaction rating. Additionally, results indicated domestic beef was preferred over Australian beef by US consumers. The use of our base model showed that US beef was preferred in all categories, but primarily due to flavor. The flavor generated by the US beef was at least 2% greater in the overall satisfaction model. Americans are accustomed to eating domestic grain-fed beef and may have acquired a preference for the flavor of US beef over beef from other countries. In addition, differences in feeding practices between Australia and the US may impact the flavor of beef.

Another important finding of the study was the impact of the demographic variables of age, income, and gender on satisfaction. The older group generally placed more emphasis on tenderness, while younger people preferred a juicier beef. Males were more responsive than females for all attributes, especially tenderness. Those with higher income were more responsive to tenderness for all quality levels, but the lower income group was more responsive to juiciness. In conclusion, this approach effectively integrated beef source (country and cattle finishing system) and sociodemographic factors of US consumers to generate insights into the drivers of beef satisfaction.

Supplementary Materials: The following are available online at http://www.mdpi.com/2304-8158/9/3/289/s1, Figure S1: Scatter plot illustrating the relationship between raw flavor data plotted against satisfaction. 0 = dislike flavor/overall extremely; 100 = like flavor/overall extremely, Figure S2: Scatter plot illustrating the relationship between flavor data cleaned with Cook's Distance method plotted against satisfaction. 0 = dislike flavor/overall extremely; 100 = like flavor/overall extremely, Figure S3: Scatter plot illustrating the relationship between raw tenderness data plotted against satisfaction. 0 = dislike flavor/overall extremely; 100 = like flavor/overall extremely, Figure S4: Scatter plot illustrating the relationship between tenderness data cleaned with Cook's Distance method plotted against satisfaction. 0 = dislike flavor/overall extremely; 100 = like flavor/overall extremely, Figure S5: Scatter plot illustrating the relationship between raw juiciness data plotted against satisfaction. 0 = dislike flavor/overall extremely; 100 = like flavor/overall extremely, Figure S6: Scatter plot illustrating the relationship between juiciness data cleaned with Cook's Distance method plotted against satisfaction. 0 = dislike flavor/overall extremely; 100 = like flavor/overall extremely.

Author Contributions: C.F., C.L., J.M., R.P., and M.M. contributed to the conceptualization and experimental design of the study; C.F., R.P., C.B., and M.M. undertook the study; C.F., C.L., and J.M. conducted the formal data analysis and presentation of the data; C.F. and A.G. wrote the original manuscript draft; reviews and editing of the completed manuscript were conducted by A.G., C.L., J.M., R.P., C.B., and M.M.; project administration was the responsibility of C.L., R.P., and M.M. All authors have read and agreed to the published version of the manuscript.

Funding: This research was funded by Meat and Livestock Australia P/L., North Sydney, NSW, Australia.

Conflicts of Interest: The authors declare no conflict of interest.

References

1. Font-i-Furnols, M.; Guerrero, L. Consumer preference, behavior and perception about meat and meat products: An overview. *Meat Sci.* **2014**, *98*, 361–371. [CrossRef] [PubMed]
2. Henchion, M.; McCarthy, M.; Resconi, V.C.; Troy, D. Meat consumption: Trends and quality matters. *Meat Sci.* **2014**, *98*, 561–568. [CrossRef] [PubMed]
3. Troy, D.J.; Kerry, J.P. Consumer perception and the role of science in the meat industry. *Meat Sci.* **2010**, *86*, 214–226. [CrossRef] [PubMed]
4. Chong, F.S.; Farmer, L.J.; Hagan, T.D.J.; Speers, J.S.; Sanderson, D.W.; Devlin, D.J.; Tollerton, I.J.; Gordon, A.W.; Methven, L.; Moloney, A.P.; et al. Regional, socioeconomic and behavioural-impacts on consumer acceptability of beef in Northern Ireland, Republic of Ireland and Great Britain. *Meat Sci.* **2019**, *154*, 86–95. [CrossRef] [PubMed]
5. Menkhaus, D.J.; Colin, D.M.; Whipple, G.D.; Field, R.A. The effects of perceived product attributes on the perception of beef. *Agribusiness* **1993**, *9*, 57–63. [CrossRef]
6. Lusk, L.L.; Fox, J.A. Consumer valuation of beef ribeye steak attributes. Presented at the American Agricultural Economics Association Annual Meeting, Tampa, FL, USA, 30 July–2 August 2000.

7.   Bentley, J.U.S. Per Capita Availability of Red Meat, Poultry, and Fish Lowest Since 1983. Available online: https://www.ers.usda.gov/amber-waves/2017/januaryfebruary/us-per-capita-availability-of-red-meat-poultry-and-fish-lowest-since-1983 (accessed on 12 February 2020).

8.   Purcell, W.D. The case of beef demand; A failure by the discipline. *Choices* **1989**, *2*, 16–19.

9.   Lusk, J.L.; Fox, J.A.; Schroeder, T.C.; Mintert, J.; Koohmaraie, M. In-store valuation of steak tenderness. *Am. J. Agric. Econ.* **2001**, *83*, 539–550. [CrossRef]

10. Lusk, J.L.; Schroeder, T.C. Are choice experiments incentive compatible? A test with quality differentiated beef steaks. *Am. J. Agric. Econ.* **2004**, *86*, 467–482. [CrossRef]

11. Miller, M.F.; Carr, M.A.; Ramsey, C.B.; Crockett, K.L.; Hoover, L.C. Consumer thresholds for establishing the value of beef tenderness. *J. Anim. Sci.* **2001**, *79*, 3062–3068. [CrossRef]

12. McIlveen, H.; Buchanan, J. The impact of sensory factors on beef purchase and consumption. *Nutr. Food Sci.* **2001**, *31*, 286–292. [CrossRef]

13. Boleman, S.J.; Boleman, S.L.; Miller, R.K.; Taylor, J.F.; Cross, H.R.; Wheeler, T.L.; Koohmaraie, M.; Shackelford, S.D.; Miller, M.F.; West, R.L.; et al. Consumer evaluation of beef of known categories of tenderness. *J. Anim. Sci.* **1997**, *75*, 1521–1524. [CrossRef] [PubMed]

14. Ron, O.S.; Garmyn, A.J.; O'Quinn, T.G.; Brooks, J.C.; Miller, M.F. Influence of production practice information on consumer eating quality ratings of beef top loin steaks. *Meat Muscle Biol.* **2019**, *3*, 90–104. [CrossRef]

15. Asem-Hiablie, S.; Rotz, C.A.; Stout, R.; Stackhouse-Lawson, K. Management characteristics of beef cattle production in the Northern Plains and Midwest regions of the United States. *Prof. Anim. Sci.* **2016**, *32*, 736–749. [CrossRef]

16. Gee, A. *Protocol Book 4: For the Thawing Preparation, Cooking and Serving of Beef for MSA (Meat Standards Australia) Pathway Trials*; Meat and Livestock Australia: North Sydney, Australia, 2006.

17. Watson, R.; Gee, A.; Polkinghorne, R.; Porter, M. Consumer assessment of eating quality—Development of protocols for MSA testing. *Aust. J. Exp. Agric.* **2008**, *48*, 1360–1367. [CrossRef]

18. Anderson, E.B. Diagnostics in categorical data analysis. *J. R. Stat. Soc.* **1992**, *54*, 781–791. [CrossRef]

19. Reicks, A.L.; Brooks, J.C.; Garmyn, A.J.; Thompson, L.D.; Lyford, C.L.; Miller, M.F. Demographics and beef preferences affect consumer motivation for purchasing fresh beef steaks and roasts. *Meat Sci.* **2011**, *87*, 403–411. [CrossRef]

20. Umberger, W.J.; Feuz, D.M.; Calkins, C.R.; Killinger-Mann, K.U.S. Consumer preference and willingness-to-pay for domestic corn-fed beef versus international grass-fed beef measured through an experimental auction. *Agribusiness* **2002**, *18*, 491–504. [CrossRef]

21. McFadden, D. The choice theory approach to market research. *Market. Sci.* **1986**, *5*, 275–297. [CrossRef]

22. O'Quinn, T.G.; Legako, J.F.; Brooks, J.C.; Miller, M.F. Evaluation of the contribution of tenderness, juiciness, and flavor to the overall consumer beef eating experience. *Trans. Anim. Sci.* **2018**, *2*, 26–36. [CrossRef]

23. Killinger, K.M.; Calkins, C.R.; Umberger, W.J.; Feuz, D.M.; Eskridge, K.M. Consumer sensory acceptance and value for beef steaks of similar tenderness, but differing in marbling level. *J. Anim. Sci.* **2004**, *82*, 3294–3301. [CrossRef]

24. Neely, T.R.; Lorenzen, C.L.; Miller, R.K.; Tatum, J.D.; Wise, J.W.; Taylor, J.F.; Buyck, M.J.; Reagan, J.O.; Savell, J.W. Beef customer satisfaction: Role of cut, USDA quality grade, and city on in-home consumer ratings. *J. Anim. Sci.* **1998**, *76*, 1027–1032. [CrossRef] [PubMed]

25. O'Quinn, T.G.; Brooks, J.C.; Polkinghorne, R.J.; Garmyn, A.J.; Johnson, B.J.; Starkey, J.D.; Rathmann, R.J.; Miller, M.F. Consumer assessment of beef strip loin steaks of varying fat levels. *J. Anim. Sci.* **2012**, *90*, 626–634. [CrossRef] [PubMed]

26. Corbin, C.H.; O'Quinn, T.G.; Garmyn, A.J.; Legako, J.F.; Hunt, M.R.; Dinh, T.T.N.; Rathmann, R.J.; Brooks, J.C.; Miller, M.F. Sensory evaluation of tender beef strip loin steaks of varying marbling levels and quality treatmets. *Meat Sci.* **2015**, *100*, 24–31. [CrossRef]

27. Hunt, M.R.; Garmyn, A.J.; O'Quinn, T.G.; Corbin, C.H.; Legako, J.F.; Rathmann, R.J.; Brooks, J.C.; Miller, M.F. Consumer assessment of beef palatability from four beef muscles from USDA Choice and Select graded carcasses. *Meat Sci.* **2014**, *98*, 1–8. [CrossRef]

28. Crownover, R.D.; Garmyn, A.J.; Polkinghorne, R.J.; Rathmann, R.J.; Bernhard, B.C.; Miller, M.F. The effects of hot vs. cold boning on eating quality of New Zealand grass fed beef. *Meat Muscle Biol.* **2017**, *1*, 207–217. [CrossRef]

29. Garmyn, A.J.; Polkinghorne, R.J.; Brooks, J.C.; Miller, M.F. Consumer assessment of New Zealand forage finished beef compared to US grain fed beef. *Meat Muscle Biol.* **2019**, *3*, 22–32. [CrossRef]

30. Garmyn, A.J.; Garcia, L.G.; Spivey, K.S.; Polkinghorne, R.J.; Miller, M.F. Consumer assessment of beef muscles from Australian and US production systems with or without enhancement. *Meat Muscle Biol.* **2020**. [CrossRef]

31. Sitz, B.M.; Calkins, C.R.; Feuz, D.M.; Umberger, W.J.; Eskridge, K.M. Consumer sensory acceptance and value of domestic, Canadian, and Australian grass-fed beef steaks. *J. Anim. Sci.* **2005**, *83*, 2863–2868. [CrossRef]

32. Bonny, S.P.F.; Gardner, G.E.; Pethick, D.W.; Allen, P.; Legrand, I.; Wierzbicki, J.; Farmer, L.J.; Polkinghorne, R.J.; Hocquette, J.F. Untrained consumer assessment of the eating quality of European beef: Demographic factors have only minor effects on consumer scores and willingness to pay. *Animal* **2017**, *11*, 1399–1411. [CrossRef]

33. Kubberød, E.; Ueland, Ø.; Rødbotten, M.; Westad, F.; Risvik, E. Gender specific preferences and attitudes towards meat. *Food Qual. Prefer.* **2002**, *13*, 285–294. [CrossRef]

34. Quagrainie, K.K.; Unterschultz, J.; Veeman, M. Effects of product origin and selected demographics on consumer choice of red meats. *Can. J. Agr. Econ.* **1998**, *46*, 201–219. [CrossRef]

35. Hwang, I.H.; Polkinghorne, R.; Lee, J.M.; Thompson, J.M. Demographic and design effects on beef sensory scores given by Korean and Australian consumers. *Aust. J. Exp. Agric.* **2008**, *48*, 1387–1395. [CrossRef]

Article

Exploring Meal and Snacking Behaviour of Older Adults in Australia and China

Behannis Mena *, Hollis Ashman, Frank R. Dunshea, Scott Hutchings, Minh Ha and Robyn D. Warner

School of Agriculture and Food, Faculty of Veterinary and Agricultural Sciences, University of Melbourne, Parkville, VIC 3010, Australia; hollis.ashman@unimelb.edu.au (H.A.); fdunshea@unimelb.edu.au (F.R.D.); scott.c.hutchings@outlook.com (S.H.); minh.ha@unimelb.edu.au (M.H.); robyn.warner@unimelb.edu.au (R.D.W.)
* Correspondence: bmena@student.unimelb.edu.au; Tel.: +61-424-741-029

Received: 12 February 2020; Accepted: 30 March 2020; Published: 3 April 2020

Abstract: Sensory perception and food preferences change as we age. This paper encompassed two studies with the aim being to investigate meal and snacking behaviour of older adults towards food, especially meat products, and understand the desirable characteristics of those products. A qualitative multivariate analysis (QMA) focus group with Australian and Chinese older (60–81 years old) adults was conducted. A conjoint concept database was used to determine older consumers' wants and needs for food in Australia and China. The QMA suggested that Australian consumers are not eating a proper breakfast or dinner but are 'snacking' throughout the day. In contrast, Chinese consumers are eating three regular meals through the day and occasionally snacks. For both groups, texture and flavour were key drivers for food choice. Difficulty in eating meat products was evident, e.g., beef jerky was found too dry and hard. Older consumers in China and Australia differed in responses to the four food categories investigated in terms of product traits and segmentation. Both the conjoint analysis and QMA showed that demographics have an impact on consumer preferences towards food. This research suggested that there is an opportunity to create ready-to-eat, nutrient dense products to enhance the wellness of older consumers.

Keywords: focus group; meat products; qualitative multivariate analysis; conjoint analysis; older adults

---

1. Introduction

In 2017, 8.7% of the population worldwide (654 million) were aged ≥65 [1]. This is expected to grow to nearly 17% (1.6 billion) of the worldwide population by 2050 [2]. With reduced fertility rates and people living longer, this is becoming a major concern for many European, Asian and North American countries [2]. For Australia, the group of people aged ≥65 was 15.7% in 2018 [3]. Life expectancy is increasing as the world population is getting older [4], thus lifestyle factors become increasingly relevant to improve the quality of the later years in life.

While Australia has a higher healthy life expectancy than China, it has been well established that general heath can decline as we age; and there are many elements that influence this fact. One of them is nutrition, as it is important to maintain healthy eating habits throughout life. Declining energy intake in older adults is an important phenomenon to address. Dietary patterns in older adults have received little attention [5]. Too often, older adults' nutrition practices are ignored, which impede health maximisation and quality of life improvement [5]. Good nutrition is essential to slow the mental and physical decline of ageing people. Sarcopenia, the degenerative loss of skeletal muscle mass and strength associated with ageing [6], is an important condition to consider. Research has demonstrated that increased protein in the diet of older people, coupled with exercise, can significantly delay sarcopenia [7]. One approach to increasing protein intake in older people is through the consumption of high protein meat-based snacks. To develop red meat-based snacks, the starting point is to understand 'snacking'. In particular, why people snack, the underlying motivations that lead them to snack and shape the choices they make and the most relevant channels through which a snack would be sourced to satisfy the need. Using these foundations will enable the design of superior red meat snack products that better meet these criteria [8].

Qualitative research is often used to develop and refine hypotheses in product development, which allows for quick, inexpensive probing of consumer demands in a natural and comfortable environment [9,10]. Qualitative multivariate analysis (QMA) and conjoint analysis are two powerful methods to investigate consumers´ choices. Studies applying those techniques to understand Australian and Chinese older people preferences towards meat products have not been conducted. QMA fosters a more 'bottom-up' approach to product development as new products can be more specifically tailored to the attributes seen as positive and negative by particular market segments. The advantage of conjoint analysis over other more direct approaches of analysis (i.e., asking consumers what they like and dislike) is that it provides an accurate analysis of the consumer mind rather than the factors that consumers 'think' affect their decisions.

The aim of this research was therefore to use QMA and conjoint analysis to investigate food preferences and behaviours of older Australian and Chinese adults and understand how they differ. Some of the questions we hoped to answer were; (i) how do snacks fit with elder consumers' preferences? (ii) is what we know regarding older adults eating behaviour correct? (iii) what else do we need to know? Another aim was to understand the meat market from older people's perspective, to understand what is similar/dissimilar among older adults and to find gaps where possible products could be developed that do not currently exist in the market.

2. Materials and Methods

2.1. QMA Study

QMA is a relatively new technique that has been used mainly in industry settings [11]. This is a reliable technique to perform research exploration. Drake et al. [11] found that the results obtained from quantitative mapping techniques (n = 110 consumers), i.e., landscape segmentation analysis with agglomerative hierarchical clustering, were similar to those obtained through QMA with a smaller sample size (n = 12 consumers). The QMA technique used by [11] (also known as perceptual mapping, landscape mapping, Napping®) uses a similar protocol to this study with group discussions and mapping exercise which brings advantages from integration of many techniques. Through QMA, researchers can discover insights about the products directly from consumers without any predetermined agenda, discussing only important matters identified in the discussion [11,12].

Exploratory studies were carried out in Australia at the Hastings Lifestyle Retirement Village with Australian participants in Hastings, Victoria and at the Chinese Senior Citizens Club with Chinese participants in Frankston, Victoria. QMA with focus group methodology was used, following a semi-structured protocol and was conducted under the University of Melbourne human ethics protocol number 1749295.

2.1.1. Participants

Sixteen (13 female, 3 male) Australian meat consumers, who were selected for having active lifestyles and being 65–79 years old, were recruited for two focus groups (n = 8 in each focus group). Twenty-one (17 female, 4 male) Chinese meat consumers with active lifestyles and being 60–81 years old, participated in two focus groups (n = 12 and 9). Due to limited number of participants, it was difficult to conduct this trial with balance for gender. The ratio of 4:1, female:male can be justified by the fact that women generally live longer than males [13] and they were more eager to participate in the focus groups. Participants declared they were not taking any strong medication and did not have any health conditions that would influence taste, smell or chewing of samples except for one participant that declared cream, onion and garlic allergies and abstained from tasting the bolognese. The sample size was determined according to saturation principles [14]. Participants were recruited by staff of the retirement village and senior club. All participants gave informed consent, signed allergen forms and received a small compensatory gift. A plain language statement was given with detailed information of the study.

2.1.2. Procedure

Data were collected in three stages following the sequence below and each focus group went through the three stages sequentially, over a period of 2 h. Both focus groups within a demographic, were conducted on the same day, one after the other. Sessions for each demographic group were conducted on separate dates. For the Chinese focus groups, a translator facilitated the activity and all paper materials were translated into Chinese. Sessions had video and voice recordings made. A discussion guide (Table S1) was followed to facilitate conversation and reach the learning objectives.

Stage 1: A 5Ws table (When, What, Why, Where, Who and What With) documenting the participant's meal/snacking behaviour was completed. Participants were asked to describe their eating habits throughout the day (early morning, breakfast, mid-morning, lunch, afternoon, dinner and after dinner) using a customised list of meat products sensory descriptors for flavour, appearance, texture and aroma (Table S2). These descriptors were given to enhance the flow of conversation while their responses were recorded on a flip chart.

Stage 2: Tasting of seven commercially available meat products (pictorial reference is shown in Figure 1) including meat bolognese (pasta and minced meat), cocktail sausage, prosciutto, meat sticks, meat floss (light and fluffy shredded dry pork common in China), liver pate and Chinese beef jerky (softer and more moist than traditional beef jerky produced and consumed in Australia). These products were chosen to provide stimuli across a range of textures (softest texture was liver pate and hardest was beef jerky) and familiarity (least familiar for Australians was meat floss, for Chinese was liver pate and most familiar for Australians was cocktail sausages, for Chinese was meat floss). Participants were presented with products, in a randomised sequence according to demographic group, everyone ate the same product at the same time. For each product, participants completed a paper-based questionnaire with open response questions regarding attributes of appearance, texture contrast, flavour/aroma. Products were purchased at local stores and kept under refrigeration (5 °C) where required. Cocktail sausages were cooked with boiling water and meat bolognese was heated in a microwave before serving. The remaining samples were ready-to-eat. Serving size was about 30 g per sample and they were served individually.

Stage 3: Perceptual mapping of commonly consumed snack foods was conducted, which consisted of providing pictures and placing them on a two-dimensional map with the x-axis being: 'Everyday' to 'Indulgent' and Y-axis being 'Likely to Eat' to 'Less Likely to Eat'. Products were selected from a wide range of texture, protein source and popularity (Figure 1). The map was drawn on a flipchart paper glued to a table with x- and y-axes taped on it using sticky notes. Participants sat around a large table in a private room, a moderator trained in QMA techniques stood in front of them to facilitate discussion and three assistants sat outside the group, taking notes. At the start of the session, a picture of cocktail sausages was shown and placed in the middle of the map as the reference point. This product was used as starting point for being generic (safe/popular), thus forcing participants to use the entire space and avoid clumping pictures in one quadrant and helping to separate out the differences between products more easily. The analysis involved foods being grouped by identifying common characteristics among them within a quadrant.

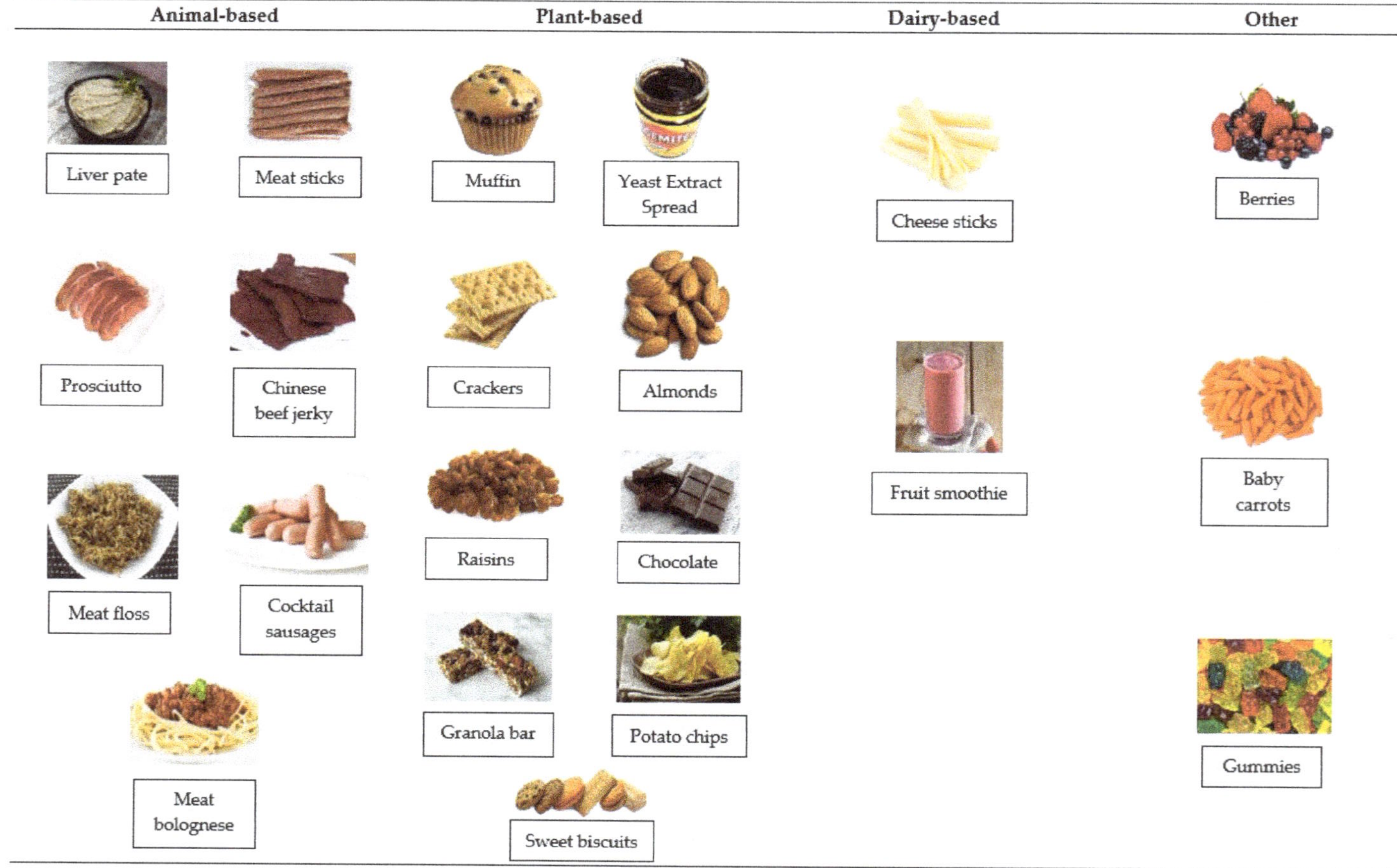

Figure 1. Images of the snacks presented in the perpetual mapping stage, and grouped into animal-, plant-, dairy-based and other.

2.2. Conjoint Study

Conjoint measurement is "a simple descriptive method, namely the use of experimental design to understand reactions to ideas by measuring reactions to mixtures of ideas" [15]. Conjoint measurement uses experimental design, mixing, together small components ('idealets'), generating combinations, acquiring subjective responses to those combinations and then deducing what components drive the reactions [16]. The point of view of conjoint measurement in particular, and statistical experimental design in general, is that the combination of independent variables allows each of them to affect the other in a way that could not be seen in the traditional one-at-a-time approach [17]. This technique allows us to better understand the mind of the consumer to generate insights and product ideas and, also provides various outputs for analysis, including part-worth utilities, counts, importance by ranking, shares of preference and purchase likelihood simulations [18]. Furthermore, conjoint analysis involves providing consumers with a set of product descriptions and requiring them to rate whether each description would or would not appeal to them [19]. These product descriptions comprise a core set of elements, which are systematically varied among the descriptions by a "mix and match" method. Once the behaviour of individuals is understood, consumers can be grouped according to their patterns of utility value also known as regression coefficients.

2.2.1. Selection of Foods and Respondents

The concept database comprised 18 food and beverage categories (wine, coffee, chocolate, yogurt, milk drinks, ice cream, cheese, baby formula, capsicum, pumpkin, tomato, banana, melon, tea, grains, beef, pork, olive oil) for consumers and was conducted in both Australia and China. The original objective of the database was to understand the key levers that drive 'premiumness' across those food groups as representations of the overall food industry. Both Australian and Chinese consumers were included in order to understand both the domestic market and potential export markets. In addition, an understanding of cultural bias is provided by comparing the two cohorts. For the current research, the focus was on beef, pork, chocolate and cheese categories, as according to the QMA study described above, these categories represent the key categories older consumers self-report eating. A total of 14,400 consumers (7200 in Australia; 7200 in China) were surveyed online and about 400 consumers per category were achieved. Recruitment was based on general population criteria, i.e., age, gender and geographical location that were representative of the overall demographic of the country. The survey had a male/female split of 43/57 for Australia and 51/49 for China to allow for gender segmentation of the data and was translated into Mandarin by an experienced translator for distribution in China and to ensure clear cross-country comparisons. Translation and data collection were conducted by Research Now Pty Ltd. (currently known as DynataTM). Respondents resided in both urban and rural areas and had familiarity with the food category evaluated. Each respondent only answered questions within one category.

2.2.2. Key Attributes

There were six key attributes evaluated for each category: product, packaging, ingredient uniqueness, provenance process, channel and occasion. The design of the experiment allowed for each of these attributes to have six levels ranging from a very basic everyday description to a very premium description (Table S3). Respondents were presented with different concepts, generated randomly from the possible combination of factors (6 attributes, 6 levels) and representing different statistically derived variations of the product concepts. They were required to choose which was considered most, or least, premium. The concepts were couched in language that goes beyond the simple one- or two-word statements into much more descriptive terms [20] and Table S3 shows a description of the terms used. To improve the consumers' response and achieve a better grasp of the consumers mind-set in response to the product, instead of using a simple one- or two-word statement, the concepts were more detailed to avoid confusion [20]. In addition, key questions about being the first to buy or adopt new products or ideas (Table S4), were asked for stratification of the consumers into segments based on adoption characteristics (see data analysis section for detail).

2.2.3. Conjoint Design

Conjoint analysis is a survey-based statistical technique used in market research that helps determine how people value different attributes that make up an individual product or service. The objective of conjoint analysis is to determine what combination of a limited number of attributes is most influential in respondent choice or decision making. In our study, six attributes along with pictures of packaging concepts for different products were shown to survey respondents (Figure S1). We then analysed how the respondents made preferences between these products, thus the implicit value (utility or part-worth) of the individual elements making up the product or service was determined. In this study, each of six attributes can be broken down into six levels. Both attributes and levels were defined based on previous marketing research and researcher's knowledge. For instance, the levels for ingredient uniqueness include basic, safe ingredients, high quality ingredients, chef inspired ingredients, bush ingredients, limited seasonal ingredients and healthy ingredients. Each of was specific to the food category under consideration and details are shown in Table S3.

2.2.4. Experimental Design

Similar to the method of Foley [20], we invited the respondents to participate, by sending an email. Those who agreed to participate simply clicked on a link embedded in the e-mail invitation. They were then taken to a survey 'wall', which listed the available food categories. Respondents ticked one box per food category, and the software kept track of how often different levels for each attribute were chosen. Then, with regression analysis, coefficients were obtained which is where the data presented in this paper came from.

Respondents were shown a set of 4 concepts to choose from (Figure S1) for each category. In total, 40 concepts per food category were used, created from a combination of levels from all of the constituent attributes. Respondents were asked to rate the products in terms of premiumness in a partial factorial design. Each concept was composed of a unique combination of levels. The data consisted of individual ratings among alternative combinations that were shuffled randomly for consumers. As the number of combinations of attributes and levels within food category increase, the number of potential concepts also increase exponentially. Consequently, fractional factorial from 6 × 6 design was used to reduce the number of concepts that had to be evaluated, while ensuring enough data are available for statistical analysis, resulting in a carefully controlled set of 'concepts' for the respondent to consider. Examples were similar enough that consumers see them as close substitutes, but dissimilar enough for them to clearly determine a preference. Each of these studies used the same overall design, so they could be compared to each other. This allowed for the research questions to be answered for an individual food category, a group of food categories, the entire set of food categories and cross-country comparisons.

2.3. Data Collection and Analysis

For the QMA study, no new topics emerged during the second focus group session, indicating that thematic saturation had been reached in the first focus group session and also indicating consistency between focus groups. After the QMA mapping, the participants generated groups of food categories that allowed for identification of linkages between different product and potential opportunity spaces to create new products. Data were transcribed from audio and video recordings by annotating important information given by participants and then used to identify recurring themes across sessions.

For the conjoint study, regression analysis of the 'Premiumness' classification was performed as described by Hughson [19]. The mathematical process illustrates how each element either adds to or detracts from the liking of a product as the effect of an element is based on its coefficient in the regression equation, with positive coefficients increasing acceptance and negative coefficients decreasing acceptance. The size of the coefficient reflects the strength of any effect. Those coefficients generated an equation similar to [18]: Premiumness classification = k0 + k1 (Element 1) + k2 (Element 2) + … k36 (Element 36). Where the constant, k0, is a theoretical value that functions as a correction factor and represents the likelihood that the respondent would like to consume a food concept that is not shown in the options. The element coefficients (k1–k36), represents the degree to which each level drives or reduces interest in a concept. Positive coefficients add to consumer interest while negative coefficients detract from interest. In total, there were 36 levels, with 6 within each of the 6 attributes. Thus, element also represents the frequency of selection for an individual level, within an attribute.

Methodology of [21] was applied to achieve segmentation of consumers into laggard (1–2), mainstream (3–4) and lead users (5–7). Segments were split by classifying the responses to the four questions shown in Table S4, into these three ranges and average of the regression coefficients across all levels for an attribute were calculated. This segmentation was determined through 'self-identification' and allowed an understanding of the predominant type of consumers in the given population, permitting targeting of products to a specify cohort. Percentage of attribute relative importance per segment was measured using the following equation which provided a way to represent how important an attribute was in a given segment relative to the other attributes.

$$\% = \frac{\sum \text{absolute average of regression coefficients per attribute}}{\text{Total } \sum \text{ of regression coefficients for all attributes}}$$

3. Results

3.1. QMA Study

3.1.1. Table of Meal/Snacking Behaviour

Participants provided detailed verbal descriptions of what they eat throughout the day. As shown in Table 1, Australians declared that they start eating around mid-morning (~11:00 a.m.) with a light snack, continue to snack across the day and consume the biggest portion of food in the afternoon (~4:00 p.m.). Often, they eat just for habit or a treat, not necessarily due to hunger. They rarely eat outside which is also true for Chinese. Most foods are ready-to-eat and consumed with their partner or accompanied with a beverage. Most foods consumed are shelf stable and the key protein source tends to be dairy oriented, e.g., cheese with biscuits.

Table 1. Key insights from the discussion during the construction of the meal/snacking behaviour table (Stage 1) of older Australian and Chinese adults.

Time of Consumption	Australian *	Chinese **
Early morning	No food consumed	Consumption of warm water first thing in the morning
Breakfast	Only tea or orange juice	Consumption of proper breakfast, lunch and dinner across the day for health, habit, hunger or share with family and friends
Mid-morning	A dry/crunchy food with coffee, consumed with partner	
Lunch	A dry/crunchy food, combination of sweet and salty, eaten at home, solo or with partner	
Afternoon	Most food consumption (cheese, biscuits, fruit/vegetables and sweets) due to hunger or habit with coffee or tea, eaten solo or with partner	Mid-morning and afternoon snacks only if hungry but unlikely
Dinner	No 'proper' dinner is consumed. Consumption of something sweet or crunchy for a treat	

Table 1. *Cont.*

Time of Consumption	Australian *	Chinese **
After dinner	The trend is to consume mostly sweet, crunchy foods for a treat, boredom or habit. Eaten at home with a hot beverage (coffee, tea, hot chocolate)	Snack for gut health or sleep better. Mainly dairy-based, i.e., plain yogurt, milk
Other comments	Do not cook very often, prefer ready-to-eat/convenient foods	No liquids are consumed during meals, just before or after, still cook by themselves, enjoy homemade and fresh food, warm water is consumed before bed

* $n = 16$; two sessions: $n = 8$ in each. ** $n = 21$; two sessions: $n = 12$ and 9 respectively.

The eating behaviour of the Chinese participants is totally different (Table 1) and almost the opposite to Australians. They have a healthier approach to food consumption as they are eating 3 regular meals during the day and snacking occasionally. This constitutes a habit along with warm water consumption as part of the traditional customs of their culture. For them, eating starts around early morning (~7:00 a.m.) and finishes by dinner (~6:00 p.m.) or after dinner (~8:00 p.m.).

3.1.2. Meat Products Tasting

The Australians and Chinese description of their eating experience of the meat products presented is shown in Table 2. Positive and negative attributes for each product were recorded by each participant in the questionnaire. After completion of questionnaires and tasting, the facilitator allowed open discussion. Beef jerky and meat sticks were recorded as being 'too hard' in texture. In contrast to initial assumptions, most products were described 'salty'. Surprisingly, even the skin of cocktail sausages and the fat layer of prosciutto were recorded as being difficult to eat. Meat floss was a very unfamiliar and undesirable product for the Australian participants.

Table 2. Results from the tasting of meat products (Stage 2) showing summary of responses from older Australian and Chinese adults, replicated across sessions in the qualitative multivariate analysis (QMA). The products with the softest texture are given at the top of the table and hardest at the bottom of the table.

Meat Product	Australian ($n = 16$)	Chinese ($n = 21$)
Liver pate	Spreadable, special occasions, too peppery, uninviting appearance (familiar)	Too soft, too peppery, uninviting appearance (not familiar)
Meat floss	Disgusting, hard to eat in terms of familiarity, 'fibrous' (not familiar)	Delicious, easy to eat, crispy and soft, umami (familiar)
Meat bolognese	Appetising, easy to eat, bland (familiar)	Meaty, moist, soft, easy to eat, umami (familiar)
Cocktail sausage	Party food, easy to eat filling, skin too hard (familiar)	Meaty, soft, easy to eat, too processed, umami (familiar)
Prosciutto	Appetising, snack food, fatty, chewy, salty (familiar)	Convenient, nutritious, too salty, not tasty, umami (not familiar)
Beef jerky	'Beefy' (not familiar)	'Beefy', meaty, hard to eat, chewy, umami (familiar)
Meat sticks	Party food, soft filling, hard skin, flavourless, difficult to bite (familiar)	Meaty, tasty, convenient, dry, tough skin, non-nutritious, umami (familiar)

Both Chinese and Australians described the beef jerky as the only 'Beefy' product, the liver pate had an unappealing appearance and too much pepper, the meat bolognese as easy to eat', prosciutto as salty and meat sticks had a hard skin to bite through. In spite of this, Chinese had a different response to the meat products compared to Australians. For instance, Chinese did not have difficulty eating the skin of the cocktail sausages and even enjoyed the chewiness of the beef jerky. They were happy to take the time to slowly chew on it, which the Australians described as frustrating. Apart from liver pate, all products were described as having umami flavour by the Chinese participants, which seemed to be important for these consumers. Additionally, relevant was how convenient and/or easy the product was to carry around in a handbag. Prosciutto, apart from being unfamiliar, was also rejected for being 'cured', which was considered unhealthy by the Chinese group.

3.1.3. QMA Perceptual Mapping

Insights regarding the snacks, for Australians are given in Figure 2. For instance, liver pate, prosciutto and potato chips were ranked as indulgent and easy to eat, except for the fat layers of the prosciutto which were removed before chewing. Dark chocolate was considered as 'likely to eat' and more indulgent for some than others, so it fell into two quadrants: Likely to Eat—Everyday and Likely to Eat—Indulgent. Cheese sticks were placed in the Everyday—Less Likely to Eat quadrant due to the negative perception the participants had of processed foods (even though consumers are eating cheese, they eat block cheese). Gummies and fruit smoothie were placed in the Less Likely to Eat—Indulgent as they were considered 'unhealthy' due to the perceived high amount of sugar.

At this stage, participants pointed out their negative attitude towards snacking, despite the fact that they snack throughout the day. This indicates that they are aware they are snacking and are also aware it is not a healthy daily eating habit. The ranking of snacks by the Chinese participants confirmed that they take a different approach to foods than Australians. Meat bolognese, berries, crackers, almonds, sultanas, potato chips, chocolate and gummies were the only foods that share the same location as for Australians, in the QMA map (Figure 3). It was interesting to find that dark chocolate was placed almost in the exact same spot by both groups. Beef jerky was placed in the Everyday—Less Likely to Eat quadrant due to the difficulty in eating because of its hard texture. Gummies and muesli bars were classified as 'unhealthy' mainly for the great amount of sugar some of those products have and placed in the Less Likely to Eat—Indulgent quadrant.

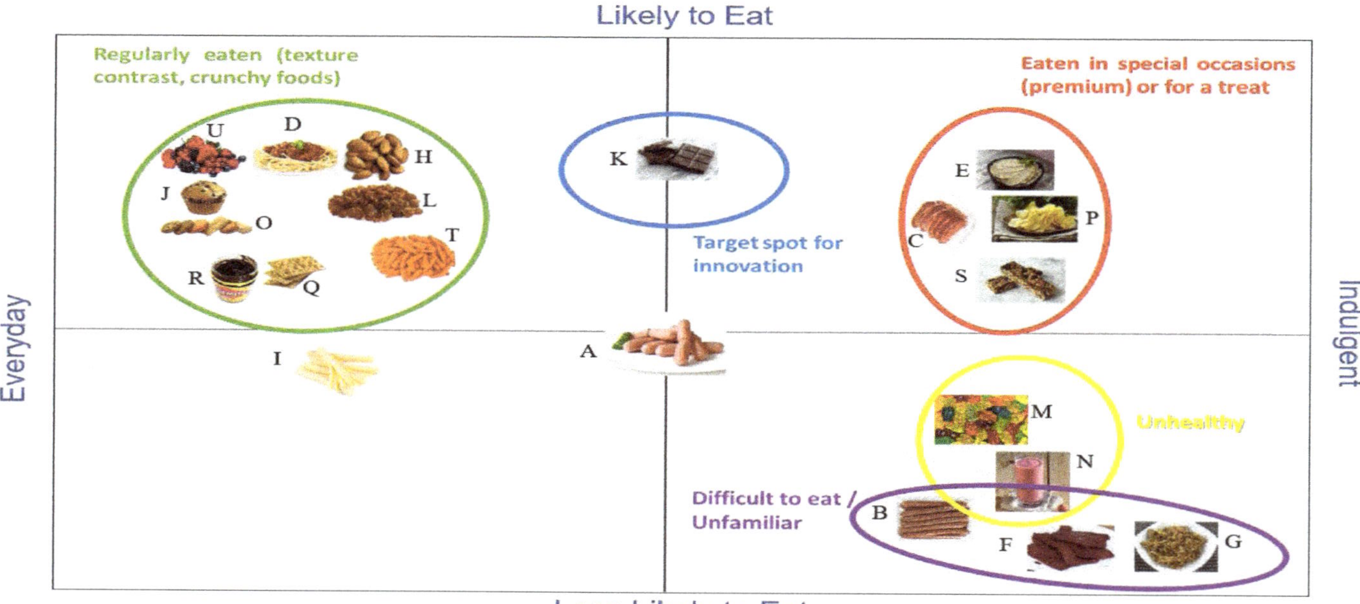

Figure 2. Perceptual map generated by Australian consumers in the qualitative multivariate analysis (QMA) discussions. The cocktail sausage image was the first shown, and everything else was mapped relative to it. (**A**) Cocktail sausages; (**B**) Meat sticks; (**C**) Prosciutto; (**D**) Meat bolognese; (**E**) Liver pate; (**F**) Chinese beef jerky; (**G**) Meat floss; (**H**) Almonds; (**I**) Cheese sticks; (**J**) Muffin; (**K**) Chocolate; (**L**) Raisins; (**M**) Gummies; (**N**) Smoothie; (**O**) Sweet biscuits; (**P**) Potato crisps; (**Q**) Crackers; (**R**) Vegemite; (**S**) Granola bar; (**T**) Baby carrots; (**U**) Mixed berries. The blue circle is drawn around the "target spot for innovation" and the foods in this target category would be frequently eaten with the desired texture contrast that can be consumed solo or shared with others. The green, red, yellow and purple shapes show groupings of products as indicated by the corresponding coloured text.

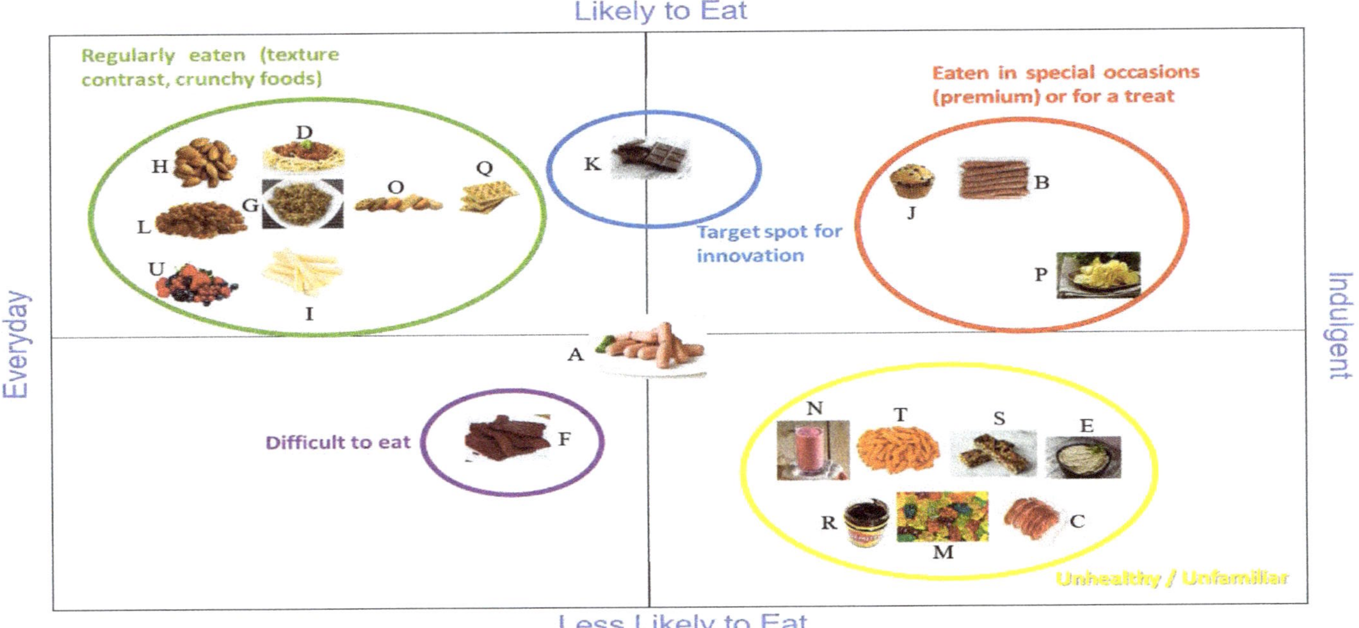

Figure 3. Perceptual map generated by Chinese consumers in the qualitative multivariate analysis (QMA) discussions. The cocktail sausage image was the first shown, and everything else was mapped relative to it. (**A**) Cocktail sausages; (**B**) Meat sticks; (**C**) Prosciutto; (**D**) Meat bolognese; (**E**) Liver pate; (**F**) Chinese beef jerky; (**G**) Meat floss; (**H**) Almonds; (**I**) Cheese sticks; (**J**) Muffin; (**K**) Chocolate; (**L**) Raisins; (**M**) Gummies; (**N**) Smoothie; (**O**) Sweet biscuits; (**P**) Potato crisps; (**Q**) Crackers; (**R**) Vegemite; (**S**) Granola bar; (**T**) Baby carrots; (**U**) Mixed berries. The blue circle is drawn around the "target spot for innovation" and the foods in this target category would be frequently eaten with the desired texture contrast that can be consumed solo or shared with others. The green, red, yellow and purple shapes show groupings of products as indicated by the corresponding coloured text.

3.2. Conjoint Study

3.2.1. Segments Classification into Laggard, Mainstream and Lead Users of Beef, Pork, Cheese and Chocolate Traits for Australians and Chinese Older Consumers

As shown in Figures 4 and 5, from the segment grouping, it was evident that the importance of each trait (product, package, ingredient, provenance, channel and occasion) varies depending on country of birth. For instance, for Australians, provenance and channel were the most important traits within beef as for Chinese, product, ingredients and channel seems to be more relevant. A similar trend was observed for pork (Figure 4). For cheese, Australians consumers care for ingredients and channel, whereas Chinese mainly for ingredients. For Australians, ingredients and channel were most important for chocolate and for Chinese, it was packaging (Figure 5). For all food categories, it was found that Chinese consumers tend to act more as mainstream and lead users in contrast to Australian consumers where there are also a great part of the population acting as laggards.

3.2.2. Utility Weight (Regression Coefficients) over Age and Gender for Most Important Beef, Pork, Cheese and Chocolate Traits for Australians and Chinese Consumers

It was interesting to look at how the importance of concepts for each food changes over age, gender and country. A subset table with the highest coefficients, i.e., >1, for each country is presented in this section (full-detailed tables for each food category can be found in Table S5). For beef, concepts for product and ingredient uniqueness, along with provenance and channel were the most relevant for both groups of consumers. As shown in Table 3, it is not surprising that in China, the concept of stir fry beef is more acceptable than in Australia for both male and female. However, a tender and juicy piece of beef sirloin was very appealing, especially for Australians. Preferences for beef ingredients differed between countries. Chinese preferred lean meat but flavoursome, and for Australians, a premium meat was better. For provenance, both groups liked an environmentally friendly meat production. Traceability was also important for Australians but not so much for Chinese. As for the place to buy beef, Australians picked "local store" and "supermarket" and Chinese mainly the later.

For pork, product, provenance and channel were important (Table 4). Interestingly, in product uniqueness, the concept of "authentic, tender and juicy Australian pork loin chop" was more appealing for both groups than just plain "pork loin chop". Important to note that the word 'authentic' made a massive difference for liking or not this food. Provenance concepts of pork mattered more for Australian consumers. Pork packaged and prepared in Australia and produced taking into consideration animal welfare and environmentally friendly practices, were most preferred. As for channel, Australians preferred to buy pork from a small market/specialty store, especially as people aged. In contrast, Chinese preferred to buy pork from a local seller.

In the cheese category, ingredient uniqueness and channel were important for both countries as shown in Table 5. It was interesting to see that as Australian women age, they become more concerned about animal wellbeing as "made with milk from a single cow free to roam on green pastures" was scored higher (Table 5). In contrast, the concept of "made with fresh Australian milk" had a negative score which increased with age. Importantly, Australians would buy cheese in any food store whilst Chinese are more likely to buy it from an Australian seller.

For chocolate, the most important attributes were package, ingredient and channel for both countries (Table 6). Individual wrap and gift pack were most preferred. The concept of "Australian cocoa beans blended with all Australian ingredients for a premium chocolate" was highly appealing for Australian women as they age (Table 6). The concept of healthiness was also important when buying chocolate. These findings can help to understand why chocolate was placed in the "likely to eat" area of the QMA maps (Figures 2 and 3) as it was thought to be a premium product and the majority of participants were women. As for channel, Australians would buy chocolate in any food store and Chinese mostly in the supermarket.

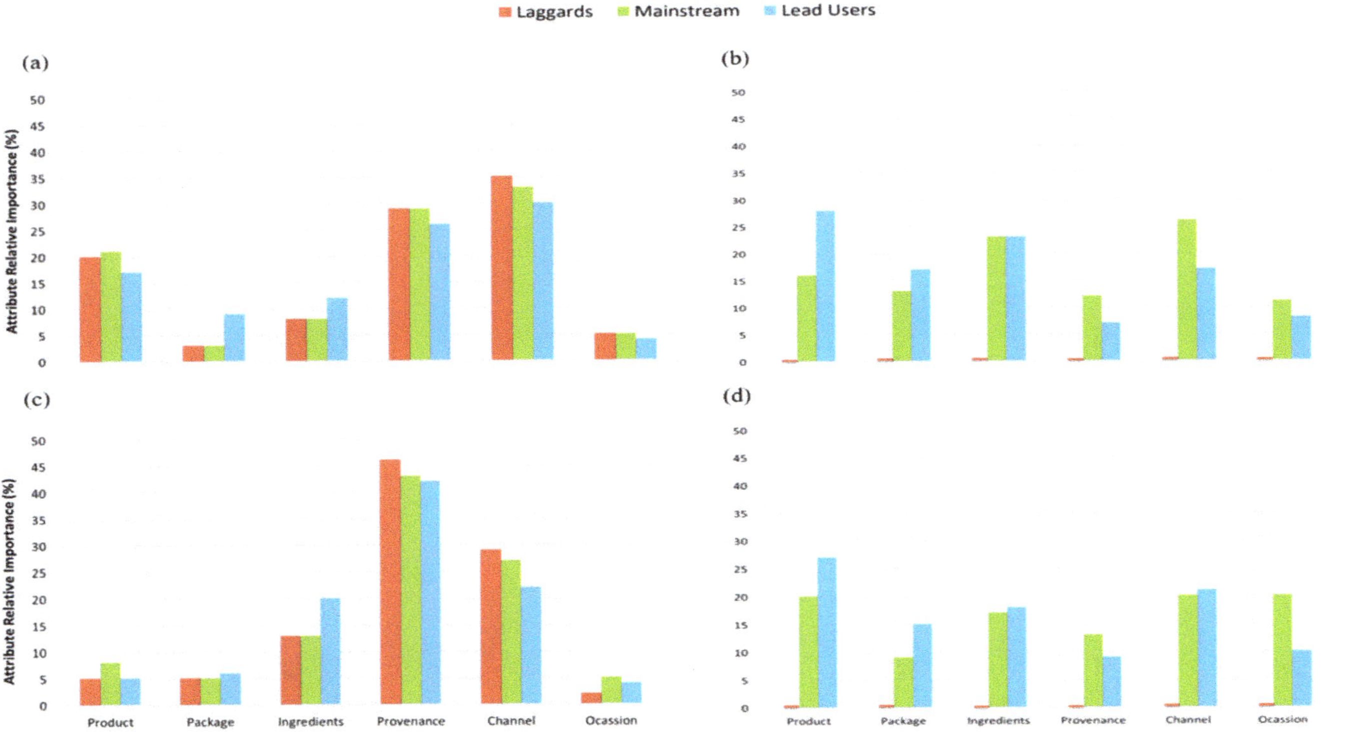

Figure 4. Average of the regression coefficients across all levels for an attribute for the three groupings of consumers being laggards (last to adopt a product), mainstream (general public) and lead users (fist to adopt a product) across the six attributes (product, package, ingredients, provenance, channel and occasion) for (**a**) beef Australia (*n* = 160, age = 55–84), (**b**) beef China (*n* = 32, age = 55–85), (**c**) pork Australia (*n* = 142 age = 55–84), (**d**) pork China (*n* = 36, age = 55–67). Regression coefficients of levels within attributes, within age groups and country, are given for selected food categories in Table 3 (for beef) and Table 4 (for pork).

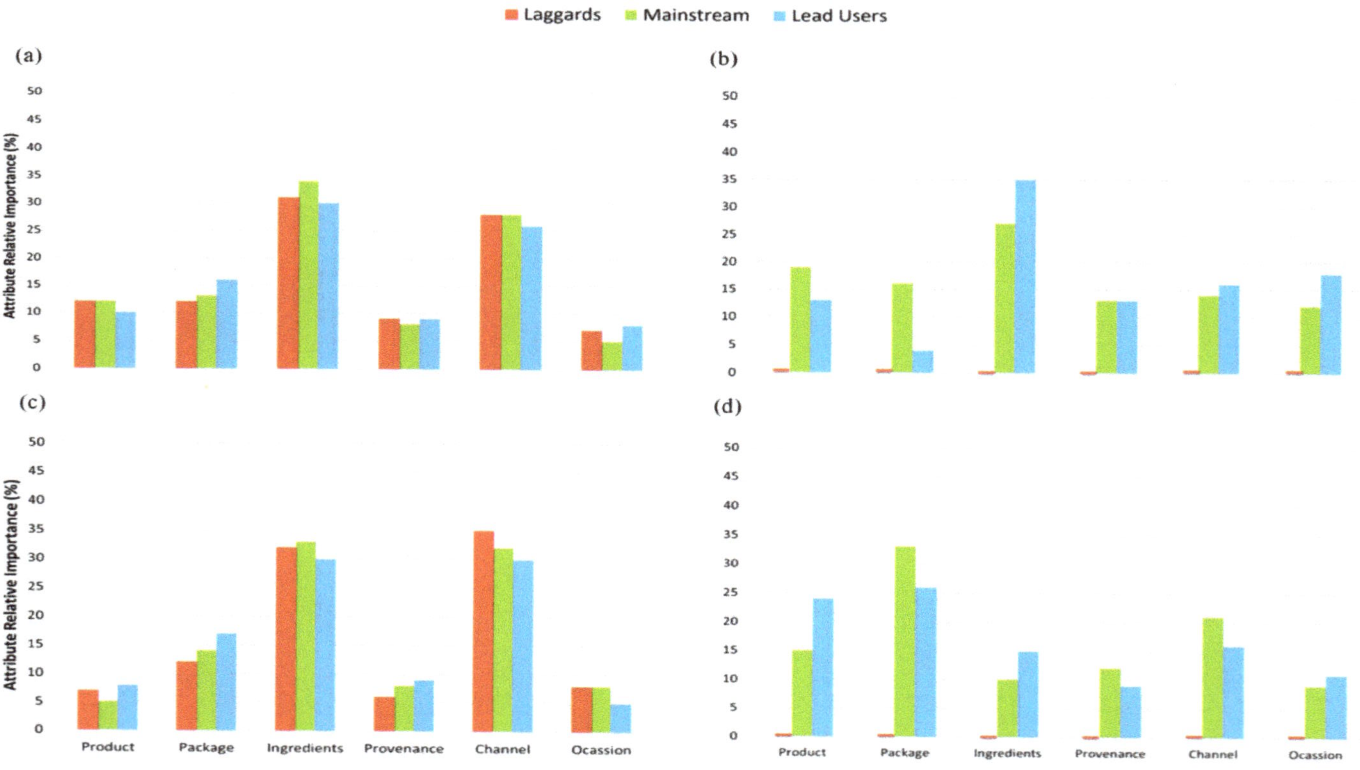

Figure 5. Average of the regression coefficients across all levels for an attribute for the three groupings of consumers being laggards (last to adopt a product), mainstream (general public) and lead users (fist to adopt a product) across the six attributes (product, package, ingredients, provenance, channel and occasion) for (**a**) cheese Australia ($n = 134$, age=55–80), (**b**) cheese China ($n = 31$, age = 55–80), (**c**) chocolate Australia ($n = 144$, age = 55–90) and (**d**) chocolate China ($n = 29$, age = 55–67). Regression coefficients of levels within attributes, within age groups and country, are given for selected food categories in Table 5 (for cheese) and Table 6 (for chocolate).

Table 3. Premiumness classification [1] (see footnote for equation) given to each level (1, 2), within each attribute (product, ingredient, provenance, channel) which makes an element for beef with the data divided for country (Australia, China), age group (25–44, 45–54, ≥55), sex (male, female). A high positive value indicates the consumer considered the element to be more premium within the attribute compared to other options and conversely negative values indicate less premiumness.

		Country:	AUSTRALIA						CHINA					
		Gender:	Male			Female			Male			Female		
		Age Group:	25–44	45–54	≥55	25–44	45–54	≥55	25–44	45–54	≥55	25–44	45–54	≥55
		Number of Respondents:	48	40	89	92	58	71	131	41	15	151	44	17
Attribute	**Element**													
PRODUCT	Beef stir fry, cut in just the right size		0.04	0.18	−0.20	−0.11	0.16	0.07	0.40	0.12	1.17	0.11	0.28	0.64
	Beef sirloin, tender and juicy every time		2.78	2.59	2.15	2.27	2.53	2.44	0.83	0.62	0.38	0.67	0.62	0.69
INGREDIENT	Lean heart healthy beef, raised to have monosaturated fats to lower your blood pressure and cholesterol, but still have lots of flavour		−0.59	−0.46	−0.50	−0.28	−0.66	−0.55	0.34	0.57	0.71	0.58	0.71	0.72
	Premium pasture fed beef from Blackmore's Wagyu, Cape Grim, or Minderoo		0.84	0.52	0.31	0.54	0.77	0.23	−0.23	−0.18	−0.26	−0.15	−0.20	−0.67
PROVENANCE	Authenticated, traceable back to the farm		0.74	0.96	1.10	0.64	1.00	0.98	−0.01	0.41	−0.03	0.15	0.17	0.14
	Raised on a small family farm, grass feed using biodiverse pastures, hormone free, using sustainable farming practices		1.50	1.36	1.58	1.61	1.74	1.99	0.33	0.20	0.11	0.39	0.44	0.10
CHANNEL	Available at my local store		1.45	1.43	1.75	1.69	1.50	1.86	0.10	0.03	0.96	−0.01	0.00	−0.14
	Available in the supermarket		1.34	1.33	1.57	1.37	1.62	1.83	0.59	0.40	1.01	0.67	0.71	0.34

[1] Premiumness classification = k0 + k1 (Element 1) + k2 (Element 2) + … k36 (Element 36). Where k0: constant and k1–k36 elements are concepts listed above within attributes. k0–k36 are the partial coefficients which are calculated in the equation and add up to zero.

Table 4. Premiumness classification [1] (see footnote for equation) given to each level (1, 2), within each attribute (product, provenance, channel) which makes an element for pork with the data divided for country (Australia, China), age group (25–44, 45–54, ≥55), sex (male, female). A high positive value indicates the consumer considered the element to be more premium within the attribute compared to other options and conversely negative values indicate less premiumness.

Country:		AUSTRALIA						CHINA					
Gender:		Male			Female			Male			Female		
Age Group:		25–44	45–54	≥55	25–44	45–54	≥55	25–44	45–54	≥55	25–44	45–54	≥55
Number of Respondents:		55	41	86	83	66	56	157	39	25	122	50	11
Attribute:	**Levels:**												
PRODUCT	Pork loin chop	−0.46	−0.64	−0.70	−0.58	−0.53	−0.47	−0.46	−0.18	−0.69	−0.38	−0.63	−0.19
	Authentic, tender and juicy Australian pork loin chop	0.47	0.10	0.01	0.13	0.17	0.07	1.25	0.79	1.10	1.16	1.50	1.10
PROVENANCE	Australian meat packaged and prepared in Australia	1.05	1.19	1.02	0.87	0.82	1.37	0.15	−0.01	0.39	0.08	−0.02	−0.18
	Raised on a small family farm, grass feed using biodiverse pastures, hormone free, using sustainable farming practices	2.05	2.01	2.20	2.39	2.07	2.24	0.12	0.15	0.32	0.15	−0.01	−0.11
CHANNEL	Available at a small market/specialty store	0.77	1.02	1.21	0.75	0.64	1.31	−0.16	−0.51	0.21	−0.51	−0.51	−0.83
	Available online from a Chinese seller	0.11	0.04	−0.12	0.15	0.22	−0.39	0.78	1.05	0.84	0.91	0.76	1.05

[1] Premiumness classification = k0 + k1 (Element 1) + k2 (Element 2) + ... k36 (Element 36). Where k0: constant and k1–k36 elements are concepts listed above within attributes. k0–k36 are the partial coefficients which are calculated in the equation and add up to zero.

Table 5. Premiumness classification [1] (see footnote for equation) given to each level (1, 2), within each attribute (ingredient, channel) which makes an element for cheese with the data divided for country (Australia, China), age group (25–44, 45–54, ≥55), sex (male, female). A high positive value indicates the consumer considered the element to be more premium within the attribute compared to other options and conversely negative values indicate less premiumness.

Country:		AUSTRALIA						CHINA					
Gender:		Male			Female			Male			Female		
Age group:		25–44	45–54	≥55	25–44	45–54	≥55	25–44	45–54	≥55	25–44	45–54	≥55
Number of respondents:		46	46	62	92	65	72	146	32	14	164	28	17
Attribute:	Levels:												
INGREDIENT	Made with fresh Australian milk	1.31	1.54	2.12	1.57	1.99	2.15	−0.11	−0.55	−0.61	−0.18	0.11	−0.51
	Made with milk from a single cow free to roam on green pastures	0.87	1.06	1.15	1.09	0.97	1.48	0.51	0.89	1.23	0.41	0.50	0.86
CHANNEL	Available at all stores where food and beverages are sold	1.05	1.32	1.39	1.22	1.57	1.36	0.11	−0.21	−0.46	0.15	0.32	0.30
	Available online from an Australian seller	0.02	−0.05	−0.26	0.00	0.04	0.17	0.42	0.18	0.31	0.65	0.07	0.47

[1] Premiumness classification = k0 + k1 (Element 1) + k2 (Element 2) + … k36 (Element 36). Where k0: constant and k1–k36 elements are concepts listed above within attributes. k0–k36 are the partial coefficients which are calculated in the equation and add up to zero.

Table 6. Premiumness classification [1] (see footnote for equation) given to each level (1, 2), within each attribute (package, ingredient, channel) which makes an element for chocolate with the data divided for country (Australia, China), age group (25–44, 45–54, ≥55), sex (male, female). A high positive value indicates the consumer considered the element to be more premium within the attribute compared to other options and conversely negative values indicate less premiumness.

	Country:	AUSTRALIA						CHINA					
	Gender:	Male			Female			Male			Female		
	Age Group:	25–44	45–54	≥55	25–44	45–54	≥55	25–44	45–54	≥55	25–44	45–54	≥55
	Number of Respondents:	136	44	66	155	49	78	144	40	15	143	40	14
Attribute:	**Levels:**												
PACKAGE	Individual wrap	0.42	0.80	0.56	0.64	0.68	0.68	0.52	0.68	0.42	0.64	0.59	0.80
	Gift pack	1.38	2.38	1.22	1.50	1.81	0.86	1.05	0.81	1.14	1.08	1.19	0.84
INGREDIENT	Made from wild bush grown cocoa beans that provide the maximum health benefits from antioxidants	0.52	0.20	0.56	0.72	0.68	0.73	0.31	0.45	0.49	0.35	0.27	0.42
	Australian cocoa beans blended with all Australian ingredients for a premium chocolate	2.01	2.03	2.35	2.14	2.22	2.74	0.14	0.20	0.13	0.33	−0.05	0.11
CHANNEL	Available at all stores where food and beverages are sold	1.07	1.06	1.30	1.04	1.18	1.26	0.05	0.21	0.03	0.16	0.00	0.36
	Available in the supermarket	0.92	1.34	1.34	1.02	1.48	1.50	0.57	0.33	0.69	0.57	0.61	0.35

[1] Premiumness classification = k0 + k1 (Element 1) + k2 (Element 2) + … k36 (Element 36). Where k0: constant and k1–k36 elements are concepts listed above within attributes. k0–k36 are the partial coefficients which are calculated in the equation and add up to zero.

4. Discussion

Opportunities for the food industry in general, include new products for older consumers. Hence, it is important to understand the thinking/mind of older consumers to be able to meet their product requirements.

Healthy life expectancy (HALE) considers the time spent living with disease and injury and is described by the World Health Organization (WHO) as "the average number of years that a person can expect to live in "full health" by taking into account "years lived in less than full health due to disease and/or injury" [22] and it is well known that a poor diet can cause disease. The HALE for Australia was 83.4 and for China 76.9 years for 2019 [23].

4.1. QMA Study

A snack is defined as foods that can be eaten in place of, or in between meals, that are convenient because they can be quick and easy to eat [24]. Snacking ('snackification') is one of the top five consumer trends in 2019 and is expected to gain further momentum in the future [25]. Snacking now makes up nearly half of all eating occasions and is one of the most profound changes in consumers' behaviour. Time-poor consumers, rising health consciousness, higher discretionary incomes and demand from grocery are drivers of snacking [26]. Snacking habits are no longer purely the domain of children and people under 30's, with 96% of Australians consuming some sort of snack on a regular basis [8]. Thus, snacking has now pervaded all segments of society, including 'seniors'. About 84% of non-institutionalised adults aged 65 years and older snack [27]. Snacking on foods and beverages between meals can be an effective way to increase daily calories in older adults [27]. Older adult non-snackers consumed 134 fewer kcal than the 1600 kcal recommended by the US Department of Agriculture [28] for sedentary older women, while snackers consumed 1718 kcal daily [5]. Food intake in general is affected as people get older since texture and flavour of foods, particularly meat, is perceived differently. Products such as beef jerky, that once were easy to eat can become difficult, as people age. It is also believed that taste bud sensitivity might decrease with age, causing reduced perception of certain flavours, such as saltiness.

The qualitative data collected on snacking behaviour shows that older Australian consumers are eating small portions of food throughout the day in place of meals. Mostly, they are consuming ready-to-eat products. Baby boomers (people born between 1946 and 1965) snack to avoid consuming larger meals, often alone or just in a group of two [26]. This was not true for older Chinese who eat regular meals throughout the day.

Although the recommended daily intake for adults is 0.8 g protein/kg body weight, some aged care nutritionists have recommended that older people should increase their daily intake of protein to 1.2 g per body weight [29]. For example, a person weighing 70 kg will need to increase their intake of protein from 56 to 84 g per day. An Australian Health Survey in 2011–2012 showed that older males are only getting 1.6 servings of lean meat and poultry, fish, eggs, tofu, nuts and seeds, and legumes/beans vs. the recommended 2.5 servings while older women are getting only 1 serving vs. the recommended 2 servings [30]. Protein intake should be increased in older adults, given its importance in delaying sarcopenia. Snacking or eating protein dense foods, such as meat products, can create a positive impact in increasing protein intake.

When considering the importance of texture vs. flavour, the assumption that older adults have difficulty eating beef jerky and meat sticks due to their dry and hard nature [31,32] turned out to be confirmed by the Australian participants. Many of them had consumed these products at a younger age and were now missing the textures and their flavours. Interestingly, eating the skin of cocktail sausages was also a difficult task for most of the Australian participants. In contrast, Chinese participants enjoyed the chewiness of beef jerky and ate the cocktail sausages without peeling off the skin. It is clear that while age can have an effect on eating quality of meat, there is also a cultural or geographical overlay.

Age-related changes in taste and chemosensory acuity may be the result of impaired swallowing and difficulty chewing, because of associated problems with teeth and gums and potential issues arising from the need to take medications for various health conditions [31]. This needs to be considered when developing food products for older adults. In our study, no participants were taking strong medication, hence this was not relevant. In a relevant study, authors fed subjects beef with contrasting textures (either tough and dry or tender and juicy, obtained by varying post-mortem aging and cooking temperature) [32]. Regardless of the beef texture, chewing duration before swallowing was longer, and bolus residual strength was greater for older compared to younger adults. Although there have been several studies developing meat products for the elderly [33–36], there is still insufficient information on a wide range of meat products that are easy to eat and consequently increase protein consumption in older people. These overall difficulties with texture, drives older consumers to consider more softer textures in their food choices and avoid eating meat. If consumption of meat and meat products is reduced, significant nutritional advantages are lost, and total intake of energy and high-quality protein may fall below dietary requirements [33].

Understanding what the appropriate texture range is for a product becomes critical to creating a novel red meat snack that will be easily consumed, while also providing the satisfaction of meat consumption. When eating meat products, particle size was found to be important for ease of break down in the mouth. For instance, in the present study, fine particles in liver pate were preferred by older Australians over 'fibrous' pieces of meat bolognese as the term 'fibrous' can include 'stringiness', which can make the product difficult to eat. A soft texture with a 'little bit of bite', 'melty' but 'not fibrous mouthfeel' and 'moisture', during consumption for Australians and in the product itself for Chinese, were also important traits for older consumers.

As we age, the way our senses (hearing, vision, taste, smell, touch) function changes and gustatory dysfunction may indeed be related to the normal ageing process [37]. Participants, especially older Australians, found cocktail sausages too salty. Again, this may present a gap in the marketplace for older consumers, as they are perceiving a salty flavour, which goes against their desire for low-salt products for health reasons.

In the QMA mapping and group discussion, the moderator explored how participants felt towards processed foods and what they considered as 'unhealthy'. Interestingly, older Australians defined 'unhealthy' as a high content of fat, salt and sugar, which agrees with the global definition [24]. However, for older Chinese, a high amount of sugar seemed to be the biggest reported trigger for unhealthiness, with less focus on fat and salt. Both groups had a negative perception of processed foods, which they classified as too artificial, e.g., cheese sticks vs. cheese block. The latter product is preferred as it is considered closer to a whole food, and less processed.

4.2. Conjoint Study

Country of birth was a defining factor for classifying importance of each product trait as shown in the segmentation results (Figures 4 and 5). Ethnicity is often found to be important for food perception and consumption, possibly due to differences in oral physiology and anatomy between consumers belonging to different ethnicities, which might cause differences in food oral processing behaviour [38,39]. Others have also found ethnicity to be important when investigating the consumer perception of meat alternatives in USA, India and China [40].

Laggards are the last users to adopt a product and generally they prefer simple products [41]. They are the last to pick up on innovations and only buy an innovative product when it works completely flawlessly, and all traditional alternatives are no longer available [42]. Mainstream users, as the name implies, refers to the general public or majority of consumers [43]. They rely on security, trying the products after they have been proven and the system "forced them to" [42]. Lead users are users whose present strong needs will become general in a marketplace months or years in the future [21]. Lead users are defined as open to innovation and pioneers among all users in a population. They actively participate in product development, often creating their own prototypes based on their need [42].

Older Chinese consumers seemed to be more 'adventurous' when choosing food products compared to older Australian consumers who behave more as mainstream and laggards. In contrast, the older Australian QMA participants showed less interest in trying new products unless they were highly recommended by trusted people. Chinese culture differs significantly from Western and even other Asian cultures, so consumers have different values and a different perception of product attributes [44–46].

Product and ingredient uniqueness, provenance and channel concepts of beef were important in terms of premiumness for both older Chinese and Australians consumers. However, the importance of each level within attributes differed between Australia and China. According to [47], Chinese consumers may desire similar product features (e.g., brand name, quality and flavour) to Western consumers, but the value that consumers attach to the same product may differ cross-nationally. For instance, whole meat cuts and premium meat are highly preferable for Australians in comparison with Chinese.

Product uniqueness, provenance and channel concepts were the most relevant attributes for premiumness for both older Australians and Chinese when buying pork. Chinese like to buy their pork form a Chinese seller, but if the meat is from Australia, their interest increased. The same cannot be said for Australians who said they preferred to by pork from a specialty store (e.g., local butcher). In a study investigating consumer preference for pork in different Asian countries, including China, specific consumer preferences differed for meat cut choices, provenance, price and sensory characteristics, and the authors suggested this should be considered carefully to increase product consumption [48] by offering the specific products each cohort is demanding.

For cheese, ingredient and channel were important for both countries. Using Australian milk for cheese manufacture was polarising for Australians and Chinese, as were other elements. This might be due to the low cheese consumption by older Chinese consumers. Furthermore, older Chinese preferred dry dairy products, including those from Australian origin [49], which is reflected in our findings.

For chocolate, channel was important for both countries. "Bush ingredients", which implies native Australian or natural food, was defined as important for respondents in the conjoint study and agrees with the finding from the QMA of participants wanting to consume less processed cheese, as ingredients play a key role for a product to be considered as more or less processed. In a study on consumer views on "healthier" processed meat, authors reported findings from seven focus groups where participants considered that, in order to improve the (perceived) healthiness of those products, the focus should be on the use of better-quality ingredients, e.g., natural ingredients, and less salt, fat, preservatives and other additives [50].

5. Conclusions

This research revealed that older Chinese are eating in a healthier way than older Australians. It was shown that older Chinese are eating three regular meals during the day and not snacking much, whilst Australians are doing the opposite. In the older population, snacking appeared to be more important for Australians than Chinese. Both Australians and Chinese reported that they avoid overly processed and sugary foods. These findings demonstrated that older adults, especially Australians, need more healthy choices for snacks, that fit their needs for the proper texture and familiar flavour. A marked difference was observed among participants depending on their country of origin. These results were reinforced with the findings from the conjoint analysis which supported the importance of demographics for perception and preference of, in this case, premium foods. For older Chinese, one possibility might be to create a product, with similar characteristics to meat floss, that could be eaten as a stand-alone product, or used as filling/ingredient in other dishes. Clearly, there is opportunity to create a nutrient dense product to better fit their need to a larger range of healthy snack foods. Our research suggest that the novel product should be 'ready to eat' and bite-sized, to align with the snacking behaviour of older people. The ability to describe the product space characteristics and how it fits with the behaviours, health rules and emotions of older consumers provides a starting point for the meat industry to create a differentiated snack that will fit their lifestyles. For the product to be successful, it is important to consider demographics as a factor that is important for food choices. However, older adults in general are not a homogeneous group. Depending on the age range and other factors, they have different nutritional needs. Further research with specific older age subgroups is needed to determine the most effective approaches for promoting health and wellbeing.

Supplementary Materials: The following are available online at http://www.mdpi.com/2304-8158/9/4/426/s1, Table S1: Moderator guide used for group discussion in QMA. Table S2: Lexicon for sensory attributes of meat products [51]. Table S3: Conjoint design for product concepts per food category (36 levels). Table S4: Questions included for segments stratification, Table S5: Utility weight (regression coefficient) for beef, pork, cheese and chocolate concepts per attributes according to age, gender and country of origin (Australia or China), Figure S1: Example per food category of how the respondent saw the concepts in the survey, Table S5: Utility weight (regression coefficient) for concepts per food category and attributes according to age, gender and country of origin (Australia or China).

Author Contributions: Conceptualization, B.M., H.A. and F.R.D.; methodology, B.M. and H.A.; formal analysis, B.M., H.A., S.H., M.H. and R.D.W.; investigation, B.M.; resources, B.M., H.A., F.R.D. and R.D.W.; data curation, B.M. and H.A.; writing—original draft preparation, B.M.; writing—review and editing, B.M, H.A, S.H, F.R.D., M.H. and R.D.W.; visualization, B.M.; supervision, H.A., and R.D.W.; project administration, R.D.W.; funding acquisition, R.D.W. and F.R.D. All authors have read and agreed to the published version of the manuscript.

Funding: This research was funded by The University of Melbourne through granting the Melbourne Research Scholarship to undertake the PhD degree under this study was conducted. The conjoint study was supported by the Australian Government through the AUSTRALIAN RESEARCH COUNCIL (Grant number IH120100053) "Unlocking the Food Value Chain: Australian industry transformation for ASEAN markets".

Acknowledgments: The authors wish to acknowledge Veronica Welburn-Brown from Hastings Retirement Village and Sarah Yujie from the Chinese Senior Citizens Club for the recruitment support to carry out this research project. Special thanks to Xinyu Miao for translation services. Thanks to Evan Bittner, Allison McNamara, Melindee Hastie, Paula González and Colette Day for helping with study design/data collection, and all the participants who joined the studies.

Conflicts of Interest: The authors declare no conflict of interest. The funders had no role in the design of the study; in the collection, analyses, or interpretation of data; in the writing of the manuscript, or in the decision to publish the results.

References

1. World Bank. Population Ages 65 and Above, Total. Available online: https://data.worldbank.org/indicator/SP.POP.65UP.TO (accessed on 18 January 2019).

2. He, W.; Goodkind, D.; Kowal, P. *An Aging World: 2015*; US Census Bureau; U.S. Government Publishing Office: Washington, DC, USA, 2016. [CrossRef]

3. ABS (Australian Bureau of Statistics). Population by Age and Sex, Australia: States and Territories. Available online: http://www.abs.gov.au/ausstats/abs@.nsf/0/1cd2b1952afc5e7aca257298000f2e76?opendocument (accessed on 20 January 2019).

4. Van der Zanden, L.D.; Van Kleef, E.; de Wijk, R.A.; Van Trijp, H.C. Knowledge, perceptions and preferences of elderly regarding protein-enriched functional food. *Appetite* **2014**, *80* (Suppl. C), 16–22. [CrossRef]

5. Zizza, C.A.; Tayie, F.A.; Lino, M. Benefits of snacking in older Americans. *J. Am. Diet. Assoc.* **2007**, *107*, 800–806. [CrossRef] [PubMed]

6. McNeill, S.; Van Elswyk, M.E. Red meat in global nutrition. *Meat Sci.* **2012**, *92*, 166–173. [CrossRef] [PubMed]

7. Evans, W.J. Protein nutrition, exercise and aging. *J. Am. Coll. Nutr.* **2004**, *23*, 601S–609S. [CrossRef]

8. Jenkinson, D. *Red Meat Innovation Insights Report: The Snacking Opportunity*; Meat and Livestock Australia Limited: North Sydney, NSW, Australia, 2015.

9. Jervis, M.; Drake, M. The use of qualitative research methods in quantitative science: A review. *J. Sens. Stud.* **2014**, *29*, 234–247. [CrossRef]

10. Hastie, M.; Ashman, H.; Torrico, D.; Ha, M.; Warner, R. A Mixed Method Approach for the Investigation of Consumer Responses to Sheepmeat and Beef. *Foods* **2020**, *9*, 126. [CrossRef]

11. Drake, S.; Lopetcharat, K.; Drake, M. Comparison of two methods to explore consumer preferences for cottage cheese. *J. Dairy Sci.* **2009**, *92*, 5883–5897. [CrossRef]

12. Beckley, J.; Paredes, D.; Lopetcharat, K. Future Trends and Directions. *Prod. Innov. Toolbox* **2012**, 374. [CrossRef]

13. WHO (World Health Organization). Female Life Expectancy. Available online: https://www.who.int/gho/women_and_health/mortality/situation_trends_life_expectancy/en/ (accessed on 16 March 2020).

14. Glaser, B.; Strauss, A.; Strauss, F. *The Discovery of Grounded Theory: Strategies for Qualitative Research*; Routledge: New York, NY, USA, 2017.

15. Moskowitz, H.R.; Gofman, A.; Beckley, J.; Ashman, H. Founding a new science: Mind genomics. *J. Sens. Stud.* **2006**, *21*, 266–307. [CrossRef]

16. Box, G.E.; Hunter, W.G.; Hunter, J.S. *Statistics for Experimenters*; John Wiley and Sons: New York, NY, USA, 1978; Volume 664.

17. Anderson, N.H. Functional measurement and psychophysical judgment. *Psychol. Rev.* **1970**, *77*, 153. [CrossRef]

18. Orme, B.K. *Getting Started with Conjoint Analysis: Strategies for Product Design and Pricing Research*; Research Publishers LLC: Madison, WI, USA, 2010.

19. Hughson, A.; Ashman, H.; De la Huerga, V.; Moskowitz, H. Mind-sets of the wine consumer. *J. Sens. Stud.* **2004**, *19*, 85–105. [CrossRef]

20. Foley, M.; Beckley, J.; Ashman, H.; Moskowitz, H.R. The mind-set of teens towards food communications revealed by conjoint measurement and multi-food databases. *Appetite* **2009**, *52*, 554–560. [CrossRef] [PubMed]

21. Von Hippel, E. Lead users: A source of novel product concepts. *Manag. Sci.* **1986**, *32*, 791–805. [CrossRef]

22. WHO (World Health Organization). Healthy Life Expectancy (HALE): Data by Country. Available online: https://apps.who.int/gho/data/node.main.HALE?lang=en (accessed on 28 January 2019).

23. Roser, M.; Ortiz-Ospina, E.; Ritchie, H. Life Expectancy. Available online: https://ourworldindata.org/life-expectancy (accessed on 16 March 2020).

24. Fellows, P.; Hampton, A. *Small-Scale Food Processing: A Guide to Appropriate Equipment*; Intermediate Technology Publications: London, UK, 1992.

25. NZMP (New Zealand Milk Products). Top 5 Consumer Trends 2019. Available online: https://www.nzmp.com/global/en/news/top-5-consumer-trends-2019.html (accessed on 17 March 2020).

26. AIFST (Australian Institute of Food Science and Technology). Top five trends in 2017 by Sarah Hyland. *Digital Magazine*, 31 January 2017; 28.

27. Higgins, M.M. Food and Nutrition Professionals Can Help Older Adults Improve Dietary Practices. *J. Am. Diet. Assoc.* **2007**, *107*, 806–807. [CrossRef]

28. USDA (US Department of Agriculture). MyPyramid. Available online: https://www.fns.usda.gov/mypyramid (accessed on 5 May 2017).

29. Castellanos, V.H.; Litchford, M.D.; Campbell, W.W. Modular protein supplements and their application to long-term care. *Nutr. Clin. Pract.* **2006**, *21*, 485–504. [CrossRef] [PubMed]

30. ABS (Australian Bureau of Statistics). Australian Health Survey: Consumption of food groups from the Australian Dietary Guidelines, 2011–2012. Available online: https://www.abs.gov.au/ausstats/abs@.nsf/Lookup/by%20Subject/4364.0.55.012~{}201112~{}Main%20Features~{}Measuring%20the%20consumption%20of%20food%20groups%20from%20the%20Australian%20Dietary%20Guidelines~{}10000 (accessed on 17 January 2019).

31. Baugreet, S.; Hamill, R.M.; Kerry, J.P.; McCarthy, S.N. Mitigating nutrition and health deficiencies in older adults: A role for food innovation? *J. Food Sci.* **2017**, *82*, 848–855. [CrossRef]

32. Mioche, L.; Bourdiol, P.; Monier, S.; Martin, J.; Cormier, D. Changes in jaw muscles activity with age: Effects on food bolus properties. *Physiol. Behav.* **2004**, *82*, 621–627. [CrossRef]

33. Farouk, M.M.; Yoo, M.J.Y.; Hamid, N.S.A.; Staincliffe, M.; Davies, B.; Knowles, S.O. Novel meat-enriched foods for older consumers. *Food Res. Int.* **2018**, *104*, 134–142. [CrossRef]

34. Reyes-Padilla, E.; Valenzuela-Melendres, M.; Camou, J.P.; Sebranek, J.G.; Alemán-Mateo, H.; Dávila-Ramírez, J.L.; Cumplido-Barbeitia, G.; González-Ríos, H. Quality evaluation of low fat bologna-type meat product with a nutritional profile designed for the elderly. *Meat Sci.* **2018**, *135*, 115–122. [CrossRef]

35. Baugreet, S.; Kerry, J.P.; Botineştean, C.; Allen, P.; Hamill, R.M. Development of novel fortified beef patties with added functional protein ingredients for the elderly. *Meat Sci.* **2016**, *122*, 40–47. [CrossRef]

36. Rothenberg, E.; Ekman, S.; Bülow, M.; Möller, K.; Svantesson, J.; Wendin, K. Texture-modified meat and carrot products for elderly people with dysphagia: Preference in relation to health and oral status. *Scand. J. Food Nutr.* **2007**, *51*, 141–147. [CrossRef]

37. Boyce, J.M.; Shone, G. Effects of ageing on smell and taste. *Postgrad. Med J.* **2006**, *82*, 239–241. [CrossRef] [PubMed]

38. Xue, S.A.; Hao, J.G. Normative standards for vocal tract dimensions by race as measured by acoustic pharyngometry. *J. Voice* **2006**, *20*, 391–400. [CrossRef] [PubMed]

39. Ketel, E.C.; Aguayo-Mendoza, M.G.; de Wijk, R.A.; de Graaf, C.; Piqueras-Fiszman, B.; Stieger, M. Age, gender, ethnicity and eating capability influence oral processing behaviour of liquid, semi-solid and solid foods differently. *Food Res. Int.* **2019**, *119*, 143–151. [CrossRef] [PubMed]

40. Bryant, C.J.; Szejda, K.; Deshpande, V.; Parekh, N.; Tse, B. A survey of consumer perceptions of plant-based and clean meat in the USA, India, and China. *Front. Sustain. Food Syst.* **2019**, *3*, 11. [CrossRef]

41. Jahanmir, S.F.; Lages, L.F. The lag-user method: Using laggards as a source of innovative ideas. *J. Eng. Technol. Manag.* **2015**, *37*, 65–77. [CrossRef]

42. Hengsberger, A. What Distinguishes Lead User from Customers? Available online: https://www.lead-innovation.com/english-blog/what-distinguishes-lead-user-from-customers (accessed on 26 January 2020).

43. Gauthier, V. Three Reasons Why Extreme Users Boost Your Innovation. Available online: https://www.linkedin.com/pulse/3-reasons-why-extreme-users-boost-your-innovation-vivien-gauthier (accessed on 25 January 2019).

44. Hasimu, H.; Marchesini, S.; Canavari, M. A concept mapping study on organic food consumers in Shanghai, China. *Appetite* **2017**, *108* (Suppl. C), 191–202. [CrossRef]

45. Del Giudice, T.; Caracciolo, F.; Cicia, G.; Grunert, K.G.; Krystallis, A. Consumatori cinesi e cibo: Tra tradizione millenaria e influenze culturali occidentali. *Econ. Agro-Aliment.* **2012**, *14*, 85–89.

46. Hasimu, H.; Marchesini, S.; Canavari, M. Chinese distribution practitioners' attitudes towards Italian quality foods. *J. Chin. Econ. Foreign Trade Stud.* **2008**, *1*, 214–231. [CrossRef]

47. Lee, P.Y.; Lusk, K.; Mirosa, M.; Oey, I. The role of personal values in Chinese consumers' food consumption decisions. A case study of healthy drinks. *Appetite* **2014**, *73*, 95–104. [CrossRef]

48. Oh, S.H.; See, M.T. Pork preference for consumers in China, Japan and South Korea. *Asian-Australas. J. Anim. Sci.* **2012**, *25*, 143. [CrossRef] [PubMed]

49. Xu, J.; Wu, Y. A Comparative Study of the Role of Australia and New Zealand in Sustainable Dairy Competition in the Chinese Market after the Dairy Safety Scandals. *Int. J. Environ. Res. Public Health* **2018**, *15*, 2880. [CrossRef] [PubMed]

50. Shan, L.C.; Regan, A.; Monahan, F.J.; Li, C.; Murrin, C.; Lalor, F.; Wall, P.G.; McConnon, A. Consumer views on "healthier" processed meat. *Br. Food J.* **2016**, *118*, 1712–1730. [CrossRef]

51. Hodder Education. Tasting Word Bank—Sensory Descriptors. OCR Design & Technology For GCSE: Food Technology. Available online: https://www.john-spendluffe.lincs.sch.uk/documents/food/word%20bank%20-%20Examples%20of%20words%20to%20use%20when%20filling%20in%20sensory%20sheets.pdf (accessed on 15 May 2017).

Article

Influence of Demographic Factors on Sheepmeat Sensory Scores of American, Australian and Chinese Consumers

Rachel A. O'Reilly [1,2,*], **Liselotte Pannier** [1,2], **Graham E. Gardner** [1,2], **Andrea J. Garmyn** [3], **Hailing Luo** [4], **Qingxiang Meng** [4], **Markus F. Miller** [3] and **David W. Pethick** [1,2]

[1] Australian Cooperative Centre for Sheep Industry Innovation, Armidale, New South Wales 2351, Australia; L.Pannier@murdoch.edu.au (L.P.); G.Gardner@murdoch.edu.au (G.E.G.); D.Pethick@murdoch.edu.au (D.W.P.)

[2] College of Science, Health, Engineering and Education, Murdoch University, Perth, Western Australia 6150, Australia

[3] Department of Animal and Food Sciences, Texas Tech University, Lubbock, TX 79409, USA; andrea.garmyn@ttu.edu (A.J.G.); markus.miller@ttu.edu (M.F.M.)

[4] State Key Laboratory of Animal Nutrition, College of Animal Science and Technology, China Agricultural University, Beijing 100083, China; hailingluo_cau@126.com (H.L.); qxmeng@cau.edu.cn (Q.M.)

* Correspondence: r.oreilly@murdoch.edu.au

Received: 14 February 2020; Accepted: 13 April 2020; Published: 22 April 2020

Abstract: Along with animal production factors, it is important to understand whether demographic factors influence untrained consumer perceptions of eating quality. This study examined the impact of demographic factors and sheepmeat consumption preferences on eating quality scores of American, Australian and Chinese untrained consumers. M. *longissimus lumborum* (LL) and m. *semimembranosus* (SM) were grilled according to sheep Meat Standards Australia protocols and evaluated by 2160 consumers for tenderness, juiciness, flavour and overall liking. Linear mixed effects models were used to analyse the impact of demographic factors and sheepmeat consumption habits on eating quality scores. Consumer age, gender, number of adults in a household and income had the strongest effect on sensory scores ($P \leq 0.05$), although, the impact was often different across countries. Frequency of lamb consumption had an impact on sensory scores of American, Australian and Chinese consumers but larger sample sizes in some underrepresented subclasses for Australian and Chinese consumers are needed. Results suggest it is important to balance sensory panels for demographic factors of age, gender, number of adults and income to ensure sensory preferences are accurately represented for these particular populations.

Keywords: consumer; demographic; sensory; lamb; yearling; longissimus; semimembranosus; cross-cultural

1. Introduction

As in beef, the development of a sheep Meat Standards Australia (MSA) grading system is underpinned by sensory evaluations of vast numbers of untrained consumers [1–3]. The MSA prediction model forecasts consumer eating quality of a final cooked product, and is based on consumer scoring of a variety of muscles using descriptors of tenderness, juiciness, liking of flavour and overall liking [1,4]. This prediction model is successfully implemented for beef and is being developed for sheep [3]. While the merit of using untrained consumers is that they are by definition unbiased, some potential drawbacks can include high variance of consumer scores, and the risk of sampling from discreet sectors of the community not representative of an entire population. Previously, O'Reilly et al. [5] demonstrated minor differences in sensory scores of grilled Australian sheepmeat between

American, Australian and Chinese consumers, with a consistent response across consumer groups to production factors of muscle type, animal age and sire type. However, the influence of demographic factors and meat consumption habits on sensory scores particular to these culturally unique groups remained unexplored. Lamb consumption rates in the USA are markedly lower than for Australian consumers, 0.6 kg compared to 8.6 kg per person annually [6], therefore many American consumers are unfamiliar with this protein. Similarly, Chinese consumers would largely be unaccustomed to Western style grilled lamb with the top three sheepmeat cooking styles in China being stewing, roasting or hotpot [7,8].

Previous research has shown that consumer demographic factors and meat consumption habits have some impact on eating quality scores in sheep [9] and beef [10], however the magnitude of these effects are often low, inconsistent across eating quality traits, and differ across countries. Australian, French, Northern Irish and Polish consumers with a higher appreciation of red meat in their diet tend to score lamb and beef more favourably than those who classify themselves as indifferent to red meat [9,10]. This was not observed for Irish or Korean consumers rating grilled beef samples [10,11]. A preferred higher degree of cooking doneness has also been shown to positively influence sensory scores compared to those who prefer medium doneness for Australian, Northern Irish and Irish consumers, however no effect was observed in French or Korean consumers, with Polish consumers demonstrating the opposite effect [9–11].

Similar to meat consumption habits, demographic factors of gender, consumer age, occupation, income level and household size have an inconsistent impact on eating quality scores across the literature [9–12]. For example, in European comparisons, Northern Irish, Irish and Polish males scored grilled beef up to two units higher than females [10]. Similarly, Kubberød et al. [13] demonstrated higher scores for Norwegian males than females when assessing lamb and beef. Both studies attributed this effect to males, placing a higher degree of importance of meat in their diet compared to females [10,13]. In contrast, Thompson et al. [9] found Australian females to score lamb up to two units higher than males. While Huffman et al. [12] and Hwang et al. [11] found no effect of gender on sensory scores for American, Korean and Australian consumers scoring beef grilled and barbequed Korean style. Older consumers within Australia have demonstrated higher sensory scores than younger consumers [9,14], while an inconsistent response to age was observed in the USA [12]. On the other hand, a small negative relationship in eating quality scores was observed with increasing consumer age for tenderness in France and Poland, and for juiciness in Ireland, Northern Ireland and Poland [10]. Huffman et al. [12] found as income increased, sensory scores decreased, with the assumption that participants on higher incomes have greater expectations of the products they consume. In addition, a greater number of adults in the household has also yielded slightly higher eating quality scores in some studies [9,10].

Largely, these studies imply that for sensory evaluations of beef and sheepmeat using untrained consumers, demographic factors pose a relatively unimportant source of bias on sensory scores [9–13]. However, to accurately assess American and Chinese consumer perceptions of sheepmeat, it is important to assess whether their demographic factors play a role in sensory scoring as this has not been reported before. In addition, lambs have demonstrated greater palatability than yearlings and mutton for Australian, American and Chinese consumers [5,15,16]. Hence, it is important to investigate whether demographic factors have an effect on animal age preferences. Therefore, this study examined the effect of consumer demographic factors on eating quality scores of American, Australian and Chinese consumers testing grilled Australian lamb and yearling m. *longissimus lumborum* (LL) and m. *semimembranosus* (SM) muscles. We hypothesised that demographic factors would have a small and inconsistent effect on eating quality scores across countries, and that higher appreciators of meat and those that prefer their meat medium-well done will score more favourably. In addition, given similarities between Korean and Chinese culinary habits, we hypothesised that Chinese eating quality scores would be less responsive to changes in meat consumption preferences compared to American and Australian counterparts.

2. Materials and Methods

2.1. Animal and Muscle Collection

Carcasses used in this study were described in detail by O'Reilly et al. [5]. In brief, a total of 164 lambs (no erupted permanent incisors; average age 368 days) and 168 yearlings (2–4 erupted permanent incisors; average age 726 days) were included in this experiment. Lambs (females and wethers) were the progeny of Maternal (Border Leicester, and Dohne Merino), Merino (Merino, and Poll Merino), and Terminal (Poll Dorset, Suffolk, Texel and White Suffolk) sires, whereas yearlings (wethers) were the progeny of Merino (Merino, and Poll Merino) sires only. Animals were sourced from the Meat and Livestock Australia genetic resource flock at Kirby (NSW, Australia) [17]. Animals were commercially slaughtered at an abattoir licensed for international export to China and the USA. Medium voltage electrical stimulation [18] was applied to all carcasses, before being trimmed according to AUS-MEAT specifications [19], and chilled for 24 h at 3–4 °C prior to sampling. Left and right LL (AUS-MEAT 5150) and SM (AUS-MEAT 5077) muscles were collected from each carcass at 24 h postmortem, vacuum packed and aged for ten days at 2–3 °C prior to frozen transport to China and the USA or retained within Australia. All samples were transported according to commercial processing plant specifications with temperatures maintained below minus 10 °C. A total of 648 LL, 648 SM muscles were collected with every carcass represented in two of three countries at any one time through allocation of muscles collected.

2.2. Sample Preparation and Sensory Testing

Consumer assessment of the sheepmeat was performed according to MSA testing protocols detailed by Thompson et al. [2] and Watson et al. [20], with minor updates described by O'Reilly et al. [5]. In brief, 24 h prior to each session, muscle cuts were thawed at 2–5 °C, each muscle was then sliced into five 15 mm thick steaks (approximately 5 × 5 × 1.5 cm), and allocated to consumers using a 6 × 6 Latin square design. Steaks were grilled on a Silex griller (electric S-Tronic Single Grill S161GR, Hamburg, Germany) with top and bottom grill plates set to 180 °C and 195 °C to obtain a medium doneness (2.25 min; 65 °C internal temperature). Steaks were rested (1.5 min) before being halved and served to consumers seated in individual booths, resulting in 10 samples per muscle being served and tasted. The same model of grill was used in Australia, China and the USA.

A total of 2160 untrained consumers were recruited across Australia, China and the USA to participate in one of the 12 tasting sessions conducted per country (n = 720 consumers per country, 60 per session). Each consumer scored six samples (3 LL and 3 SM) for tenderness, juiciness, liking of flavour, and overall liking on a hedonic scale line of 0–100. Scale lines were anchored with 'not' and 'very' preceding the eating quality trait for tenderness and juiciness, and 'dislike extremely' and 'like extremely' for liking of flavour and overall liking. In Australia and the USA, participants largely consisted of community organisations and clubs recruited through a market research company and Texas Tech University. While in China, participants were recruited though China Agricultural University and localised to the community surrounding the university. Tasting sessions in Australia were conducted in the outer Melbourne suburbs of Victoria, at the China Agricultural University in Beijing, China and across ten American states: California, Colorado, Florida, Kansas, Ohio, Oklahoma, Pennsylvania, South Carolina, Texas and Utah.

2.3. Consumer Demographics

In addition to scoring sheepmeat samples, consumers were asked to fill in a questionnaire on their demographic details and meat consumption preferences. Recruitment requested participants aged between 18 and 70 years old, preferred to be consumers of sheepmeat at least once per fortnight and have a preference for their meat to be cooked to a medium level of doneness. The questions asked were:

1. Consumer Age group, based on 6 categories: *(a)* <20, *(b)* 20–25, *(c)* 26–30, *(d)* 31–39, *(e)* 40–60, *(f)* 61–70;
2. Gender, based on two categories: *(a)* male, *(b)* female;
3. Number of adults in the household, based on 5 categories: 1, 2, 3, 4, ≥5;
4. Number of children in the household, based on 6 categories: 0, 1, 2, 3, 4, ≥5;
5. Income per annum, based on 5–8 categories depending on country: Australia in AUD: *(a)* <\$25,000, *(b)* \$25,000–\$50,000, *(c)* \$50,001–\$75,000, *(d)* \$75,001–\$100,000, *(e)* \$100,001–\$125,000, *(f)* \$125,001–\$150,000, *(g)* >\$150,000 *(h)* Prefer not to say; China in yuan: *(a)* ≤¥24,000, *(b)* ¥24,000–¥36,000, *(c)* ¥36,001–¥60,000, *(d)* ¥60,001–¥96,000, *(e)* ¥96,001–¥120,000, *(f)* >¥120,000, *(g)* Prefer not to say; and the USA in USD: *(a)* <\$20,000, *(b)* \$20,000–\$50,000, *(c)* \$50,001–\$75,000, *(d)* \$75,001–\$100,000, *(e)* >\$100,000.
6. Occupation, based on 8–11 categories depending on country: for Australia and China *(a)* manager, *(b)* professionals, *(c)* technician and trade workers, *(d)* community and personal service, *(e)* administration, *(f)* sales and service, *(g)* machinery operators and drivers, *(h)* labourers, *(i)* home duties, *(j)* student, *(k)* other; and for the USA *(a)* professionals, *(b)* tradesperson, *(c)* administration, *(d)* sales and service, *(e)* labourers, *(f)* home duties, *(g)* student, *(h)* other;
7. Consumption frequency, based on seven categories: for Australia and China *(a)* daily, *(b)* 4–5 times per week, *(c)* 2–3 times per week, *(d)* weekly, *(e)* fortnightly, *(f)* monthly, *(g)* never eat lamb; and for the USA *(a)* daily, *(b)* weekly, *(c)* fortnightly, *(d)* monthly, *(e)* every other month, *(f)* 2–3 times per year, *(g)* never eat lamb;
8. Preference for lamb, based on four categories: *(a)* Appreciator: 'I enjoy lamb. It's an important part of my diet', *(b)* Lamb is important: 'I like lamb well enough. It's a regular part of my diet', *(c)* Indifferent: 'I do eat some lamb although, truthfully it wouldn't worry me if I didn't', *(d)* Rarely/never eat: 'I rarely/never eat lamb';
9. Preferred degree of cooking doneness, based on 6 categories: *(a)* blue, *(b)* rare, *(c)* medium/rare, *(d)* medium, *(e)* medium/well done, *(f)* well done.

2.4. Statistical Analysis

The effect of demographic factors on eating quality scores (tenderness, juiciness, liking of flavour and overall liking) were examined using linear mixed effects models in SAS (SAS Version 9.1, SAS Institute, Cary, NC, USA). All ten consumer responses for each muscle were included in the analysis to account for the consumer variation. Initially, base models were established for tenderness, juiciness, liking of flavour and overall liking. These included fixed effects for country (Australia, China, USA), muscle (LL, SM), sex within animal age class (female lamb, wether lamb and wether yearling), sire type within animal age class (Maternal lamb, Merino lamb, Terminal lamb and Merino yearling), and kill group (animals were slaughtered in two separate groups), which were only retained if significant ($P < 0.05$). These base models have previously been reported in O'Reilly et al. [5] and enabled the models to account for any imbalance in animal type inadvertently occurring across demographic factor categories. Demographic factors were then included together in each of the base models as fixed effects. These included consumer age group, gender, income, occupation, adults in household, children in household, frequency of lamb consumption, preferred cooking doneness and importance of lamb in diet. Income level, occupation and frequency of lamb consumption were fitted within-country because their category options were defined differently between countries. In addition, preferred degree of doneness was also fitted within country, in this case because there were missing subclasses of 'blue' for the USA and 'blue and rare' for China. All demographic factors were interacted with muscle and with country, except those nested within country (income level, occupation, consumption frequency and preferred doneness). Non-significant terms ($P > 0.05$) were removed in a stepwise manner. Random terms included consumer identification within tasting session by country, and animal identification within sire identification. The Satterthwaite function was used in all models to approximate the degrees of freedom.

3. Results

3.1. Demographic Factor Distribution

Table 1 shows the demographic distribution of consumers recruited within Australia, China and the USA. Of the 2160 participants that attended tasting sessions, 2117 were included in the analysis as they had data for all demographic categories. Consumer age group varied by country with Australian consumers generally being older with more than half over 40 years of age, while American and Chinese consumers were largely aged between 20 and 25 years old. Gender was consistent for Australia and the USA with 10% more males than females, while China had 20% more females than males.

Table 1. Percentage distribution of consumers who scored sheepmeat samples (and number of consumers) within each demographic and meat consumption category for each country.

Consumer Age	<20	20–25	26–30	31–39	40–60	>60		
Australia	7.64 (55)	8.33 (60)	6.94 (50)	19.86 (143)	49.72 (358)	7.50 (54)		
China	6.39 (46)	48.75 (351)	19.31 (139)	4.31 (31)	12.50 (90)	8.75 (63)		
USA	10.39 (70)	41.25 (278)	9.20 (62)	11.57 (78)	23.00 (155)	4.60 (31)		
Gender	**Male**	**Female**						
Australia	54.17 (390)	45.83 (330)						
China	37.22 (268)	62.78 (452)						
USA	55.04 (371)	44.96 (303)						
Adults	**1**	**2**	**3**	**4**	**≥5**			
Australia	7.36 (53)	51.11 (368)	20.14 (145)	14.72 (106)	6.67 (48)			
China	2.50 (18)	15.00 (108)	42.22 (304)	25.14 (181)	15.14 (109)			
USA	16.34 (110)	45.77 (308)	21.4 (144)	10.40 (70)	6.09 (41)			
Children	**0**	**1**	**2**	**3**	**4**	**≥ 5**		
Australia	34.03 (245)	20.28 (146)	26.67 (192)	12.64 (91)	3.75 (27)	2.64 (19)		
China	69.31 (499)	23.89 (172)	5.00 (36)	1.39 (10)	0.28 (2)	0.14 (1)		
USA	62.17 (419)	15.43 (104)	11.42 (77)	7.27 (49)	2.37 (16)	1.34 (9)		
Income level [1]	*a*	*b*	*c*	*d*	*e*	*f*	*g*	*h*
Australia	6.39 (46)	10.42 (75)	14.58 (105)	17.36 (125)	13.61 (98)	10.97 (79)	15.28 (110)	11.39 (82)
China	48.33 (348)	15.97 (115)	12.08 (87)	8.19 (59)	2.64 (19)	4.44 (32)	8.33 (60)	
USA	19.88 (134)	19.88 (134)	19.14 (129)	17.95 (121)	23.15 (156)			
Occupation	**Manager**	**Professional**	**Trade Worker**	**Community Service**	**Administration**	**Sales and Service**		
Australia	17.36 (125)	22.22 (160)	13.06 (94)	4.58 (33)	8.75 (63)	3.06 (22)		
China	3.75 (27)	20.97 (151)	9.86 (71)	1.81 (13)	5.42 (39)	3.06 (22)		
USA		6.23 (42)	6.82 (46)		28.64 (193)	0.74 (5)		
Occupation	**Machinery Operators**	**Labourer**	**Home Duties**	**Student**	**Other**			
Australia	5.56 (40)	3.47 (25)	5.00 (36)	5.28 (38)	11.67 (84)			
China	1.81 (13)	10.69 (77)	0.83 (6)	28.75 (207)	13.06 (94)			
USA		10.83 (73)	32.94 (222)	8.61 (58)	5.19 (35)			
Frequency of Lamb Consumption [2]	*a*	*b*	*c*	*d*	*e*	*f*	*g*	
Australia	0.69 (5)	1.94 (14)	14.31 (103)	36.11 (260)	23.61 (170)	22.64 (163)	0.69 (5)	
China	0.83 (6)	1.39 (10)	12.36 (89)	21.81 (157)	24.44 (176)	37.08 (267)	2.08 (15)	
USA	0 (0)	1.48 (10)	4.01 (27)	9.50 (64)	8.61 (58)	42.88 (289)	33.53 (226)	
Importance of Lamb in the Diet	**Appreciator of lamb**	**Lamb is important**	**Indifferent to lamb**	**Rarely/never eat lamb**				
Australia	29.44 (212)	44.17 (318)	24.31 (175)	2.08 (15)				
China	5.69 (41)	29.03 (209)	62.08 (447)	3.19 (23)				
USA	7.27 (49)	14.54 (98)	35.16 (237)	43.03 (290)				
Preferred Degree of Doneness	**Blue**	**Rare**	**Medium/Rare**	**Medium**	**Medium/Well Done**	**Well Done**		
Australia	0.28 (2)	3.33 (24)	32.78 (236)	28.19 (203)	25.28 (182)	10.14 (73)		
China	0 (0)	0 (0)	0.28 (2)	1.94 (14)	23.75 (171)	74.03 (533)		
USA	0 (0)	4.01 (27)	30.12 (203)	27.00 (182)	29.23 (197)	9.64 (65)		

[1] Income categories different for each country. In all countries, income level per annum. Australia (AUD) *(a)* <$25,000, *(b)* $25,000–$50,000, *(c)* $50,001–$75,000, *(d)* $75,001–$100,000, *(e)* $100,001–$125,000, *(f)* $125,001–$150,000, *(g)* > $150,000 *(h)* Prefer not to say; China (yuan): *(a)* ≤¥24,000, *(b)* ¥24,000–¥36,000, *(c)* ¥36,001–¥60,000, *(d)* ¥60,001–¥96,000, *(e)* ¥96,001–¥120,000, *(f)* > ¥120,000, *(g)* Prefer not to say; and the USA (USD): *(a)* < $20,000, *(b)* $20,000–$50,000, *(c)* $50,001–$75,000, *(d)* $75,001–$100,000, *(e)* >$100,000'; [2] Consumption frequency categories different for the USA. For Australia and China (a) daily, (b) 4–5 times per week, (c) 2–3 times per week, (d) weekly, (e) fortnightly, (f) monthly, (g) never eat lamb; for the USA (a) daily, (b) weekly, (c) fortnightly, (d) monthly, (e) every other month, (f) 2–3 times per year, (g) never eat lamb.

The majority of American and Australian consumers came from households of one to three adults, while Chinese participants mainly came from households of three or above. All three countries leant towards less children (≤2), but this was more prominent for Chinese and American consumers. Australian and American consumers were evenly distributed across income brackets while Chinese consumers were largely within the lowest income category. Australians classified themselves primarily as managers, professionals and trade workers, Americans as administrators and homemakers, and Chinese as professionals and students. Lamb consumption habits were varied across the three countries. Most Australian and Chinese consumers eat lamb weekly to monthly, with a tendency towards higher and lower consumptions rates respectively. American consumers largely ate little to no lamb with over 70% of participants in the '2–3 times per year' to 'never eat' categories. Australians were the highest appreciators of lamb, with the majority indicating lamb is important in their diet. Chinese consumers were largely indifferent to inclusion of lamb in their diet, and the majority of American consumers were either indifferent to lamb or never eat it. Preferred degree of cooking doneness was similar for American and Australian consumers with close to 90% of participants evenly distributed across medium/rare to medium/well done categories. In contrast, almost all Chinese consumers preferred their meat cooked from medium to well done.

3.2. The Impact of Demographic Factors on Sensory Scores

Table 2 presents significance levels of demographic factors included in base models for tenderness, juiciness, liking of flavour and overall liking. Consumer age group, number of adults in the household, income level and frequency of lamb consumption reached significance most often across the four sensory traits while occupation, consumer appreciation of lamb and preferred degree of doneness were not significant for any sensory trait. Consumer gender and number of children in a household had a limited effect on sensory traits (Table 2).

3.2.1. Consumer Age Group

On average, across the three countries and both muscles, scores for tenderness, flavour and overall liking scores varied between consumer age group ($P < 0.05$; Table 2). Consumers ≥40 years old scored tenderness 3.0 to 6.3 units higher than younger counterparts, while 26 to 30 year olds scored flavour and overall liking 3.0 to 5.7 units lower than both younger and older age groups.

The age group effect varied between countries and muscle types for tenderness, juiciness, flavour and overall liking ($P < 0.05$; Table 3). As a general trend, younger Australian consumers tended to score eating quality highest, while the reverse was observed for Chinese and American consumers with older participants scoring more favourably. In Australia, younger consumers tended to score tenderness and juiciness higher. Consumers ≤25 years old scored 5.0 to 9.1 tenderness units higher than 26 to 60 year olds. For juiciness, this trend remained for the SM muscle, with consumers ≤25 years old scoring samples 5.9 to 10.6 units higher than those aged 26 to 60 years old. However, for the LL muscle, juiciness scores differed across the older consumer groups ($P < 0.05$; Table 3).

Table 2. *F*-values, numerator and denominator degrees of freedom (NDF and DDF) for base linear mixed effects models for predicted eating quality scores of tenderness, and juiciness, liking of flavour and overall liking.

Effects	Tenderness		Juiciness		Flavour		Overall Liking	
	NDF, DDF	*F*-Value	NDF, DDF	*F*-Value	NDF, DDF	*F*-Value	NDF, DDF	*F*-Value
Country	2, 2107	2.50	2, 2225	2.48	2, 2083	3.42 *	2, 2235	1.41
Muscle	1, 9944	69.12 **	1, 10,000	105.97 **	1, 10,000	104.85 **	1, 9978	146.96 **
Country*muscle	2, 9963	1.29	2, 9978	3.39 *	2, 9997	0.84	2, 9984	0.43
Age	5, 2029	5.45 **	5, 2060	2.01	5, 2055	2.80 *	5, 2054	3.49 *
Gender	–	–	1, 2063	0.14	1, 2061	2.49	1, 2058	2.45
Adults	4, 2021	3.25 *	4, 2050	3.90 *	4, 2052	3.52 *	4, 2048	3.46 *
Children	5, 2001	0.38	–	–	–	–	–	–
Income level (country)	17, 2021	2.68 **	–	-	17, 2052	1.88 *	17, 2047	2.13 *
Consumption frequency (country)	17, 2024	0.96	17, 2053	1.21	17, 2054	1.06	17, 2053	1.25
Country*age	10, 2032	2.58 *	10, 2063	3.01 **	10, 2058	2.55 *	10, 2058	2.36 *
Country*gender	–	–	–	–	2, 2062	4.71 *	2, 2058	3.88 *
Country*adults	8, 2023	1.11	8, 2051	2.76 *	–	–	–	–
Country*children	10, 2024	0.90	–	–	–	–	–	-
Muscle*age	5, 9961	2.67 *	5, 9990	3.36 *	5, 10,000	1.89 *	5, 9938	1.70
Muscle*gender	–	–	1, 10,000	6.77 *	–	–	–	–
Muscle*adults	4, 9957	0.88	–	–	–	–	–	–
Muscle*children	5, 9942	1.17	–	–	–	–	–	–
Muscle*income level (country)	–	–	–	–	17, 10,000	1.69 *	17, 9929	1.94 *
Muscle*consumption frequency (country)	17, 9952	2.71 **	17, 9989	1.71 *	17, 10,000	2.11 *	17, 9935	2.37 *
Country*muscle*adults	8, 9961	4.15 **	–	–	–	–	–	–
Country*muscle*children	10, 9957	2.96 *	–	–	–	–	–	–
Country*muscle*age	–	–	10, 9987	2.02 *	10, 10,000	2.67 *	10, 9934	3.30 **

*: *P* < 0.05; **: *P* < 0.001; –: effect not in final base model after stepwise regression.

Table 3. Least square means ± standard error (on a scale of 0–100) of tenderness, juiciness, flavour and overall liking scores of m. *longissimus lumborum* (LL) and m. *semimembranosus* (SM) samples by a consumer's country and age group. Values within brackets signify difference from mean of country and muscle type (*italics = greater than 3 units from the mean*).

Sensory Trait	Tenderness		Juiciness		Flavour		Overall Liking	
Muscle Type	LL	SM	LL	SM	LL	SM	LL	SM
Age (years)				*Australian consumers*				
<20	69.0 ± 2.8 [a] (1.1)	64.4 ± 2.8 [a] *(7.8)*	70.0 ± 2.6 [abc] (0.5)	65.7 ± 2.6 [b] *(5.6)*	71.4 ± 2.6 [ab] (2.3)	66.5 ± 2.6 [a] *(5.7)*	72.4 ± 2.6 [b] (2.3)	68.2 ± 2.6 [d] *(7.0)*
20–25	66.9 ± 2.9 [ac] (−1.0)	59.2 ± 2.9 [ac] (2.6)	69.6 ± 2.6 [abc] (0)	64.3 ± 2.6 [b] *(4.2)*	69.3 ± 2.5 [ab] (0.2)	65.9 ± 2.5 [ab] *(5.1)*	69.9 ± 2.6 [ab] (−0.2)	65.5 ± 2.6 [cd] *(4.4)*
26–30	65.4 ± 3.0 [b] (−2.5)	49.9 ± 3.0 [b] *(−6.7)*	67.9 ± 2.7 [ab] (−1.6)	55.1 ± 2.7 [a] *(−5.0)*	65.6 ± 2.6 [a] *(−3.6)*	54.0 ± 2.7 [d] *(−6.8)*	66.2 ± 2.7 [a] *(−3.9)*	53.6 ± 2.7 [a] *(−7.5)*
31–39	69.3 ± 2.3 [bc] (1.4)	53.8 ± 2.3 [bc] *(−2.9)*	70.4 ± 2.1 [bc] (0.9)	58.5 ± 2.1 [a] (−1.7)	69.9 ± 2.0 [ab] (0.7)	59.2 ± 2.0 [c] (−1.6)	70.9 ± 2.0 [ab] (0.8)	59.5 ± 2.1 [b] (−1.7)
40–60	68.0 ± 2.1 [bc] (0.2)	55.5 ± 2.1 [bc] (−1.2)	66.3 ± 1.8 [a] *(−3.2)*	57.3 ± 1.8 [a] (−2.8)	66.9 ± 1.7 [a] (−2.2)	58.9 ± 1.8 [c] (−2.0)	68.2 ± 1.8 [ab] (−1.9)	59.7 ± 1.8 [b] (−1.4)
>60	68.6 ± 3.1 [ac] (0.7)	57.0 ± 3.2 [ac] (0.4)	73.0 ± 2.7 [c] (3.4)	59.9 ± 2.7 [ab] (−0.2)	71.7 ± 2.7 [b] (2.6)	60.4 ± 2.7 [bc] (−0.5)	72.7 ± 2.7 [b] (2.7)	60.4 ± 2.7 [bc] (−0.7)
				Chinese consumers				
<20	60.5 ± 4.6 [a] *(−3.5)*	47.6 ± 4.6 [a] *(−4.2)*	61.0 ± 2.7 [ab] (−1.8)	54.7 ± 2.7 [b] (−2.5)	63.4 ± 2.9 (−1.6)	49.7 ± 2.9 [a] *(−6.3)*	66.4 ± 2.9 (−0.9)	51.2 ± 2.9 [a] *(−5.5)*
20–25	63.2 ± 4.1 [a] (−0.8)	51.0 ± 4.1 [a] (−0.7)	61.8 ± 1.7 [a] (−1.0)	56.6 ± 1.7 [b] (−0.6)	64.2 ± 1.9 (−0.8)	53.5 ± 1.9 [ac] (−2.6)	67.2 ± 2.0 (−0.1)	54.4 ± 2.0 [a] (−2.2)
26–30	63.7 ± 4.0 [a] (−0.2)	51.0 ± 4.0 [a] (−0.8)	64.2 ± 2.0 [ab] (1.4)	57.9 ± 2.0 [ab] (0.7)	65.4 ± 1.9 (0.4)	55.5 ± 1.9 [bc] (−0.5)	67.1 ± 2.0 (−0.2)	56.6 ± 2.0 [ab] (0)
31–39	61.9 ± 4.7 [a] (−2.1)	47.4 ± 4.7 [a] *(−4.4)*	59.9 ± 3.2 [ab] (−2.9)	53.2 ± 3.2 [b] (−4)	65.0 ± 3.1 (−0.1)	55.9 ± 3.1 [acd] (−0.1)	66.7 ± 3.1 (−0.6)	56.1 ± 3.1 [ab] (−0.6)
40–60	67.1 ± 4.0 [b] (3.2)	55.6 ± 4.0 [b] (3.9)	66.0 ± 2.1 [b] (3.2)	59.6 ± 2.1 [ab] (2.4)	65.5 ± 2.1 (0.4)	60.3 ± 2.1 [d] (4.2)	67.9 ± 2.2 (0.6)	60.4 ± 2.2 [b] (3.7)
>60	67.4 ± 4.3 [b] (3.4)	58.0 ± 4.3 [b] (6.2)	63.8 ± 2.4 [ab] (1)	61.0 ± 2.5 [ac] (3.9)	66.7 ± 2.5 (1.6)	61.4 ± 2.5 [d] (5.4)	68.6 ± 2.5 (1.3)	61.4 ± 2.5 [b] (4.7)
				American consumers				
<20	72.0 ± 2.5 [a] (−1.2)	56.4 ± 2.6 [a] (−1.1)	67.9 ± 2.2 [abc] (−0.4)	57.8 ± 2.2 [ab] (−0.3)	69.5 ± 2.1 [ab] (1.5)	64.4 ± 2.1 [ac] (3.2)	70.0 ± 2.2 [ab] (1.4)	62.0 ± 2.2n[b] (2.9)
20–25	69.6 ± 2.0 [a] (−3.7)	55.1 ± 2.0 [a] (−2.4)	65.3 ± 1.4 [a] (−2.9)	57.6 ± 1.4 [a] (−0.4)	65.6 ± 1.4 [a] (−2.3)	59.7 ± 1.4 [bd] (−1.4)	65.8 ± 1.4 [a] (−2.7)	57.4 ± 1.4 [ac] (−1.7)
26–30	72.2 ± 2.8 [a] (−1.1)	52.2 ± 2.8 [a] *(−5.4)*	65.6 ± 2.3 [ab] (−2.6)	52.8 ± 2.3 [b] *(−5.2)*	65.5 ± 2.2 [ab] (−2.5)	56.4 ± 2.2 [b] *(−4.8)*	65.9 ± 2.3 [a] (−2.7)	53.1 ± 2.3 [a] *(−6.0)*
31–39	72.5 ± 2.5 [a] (−0.7)	56.8 ± 2.5 [a] (−0.7)	68.4 ± 2.2 [abc] (0.2)	59.1 ± 2.2 [a] (1.1)	67.2 ± 2 [ab] (−0.7)	60.5 ± 2.0 [bcd] (−0.6)	68.1 ± 2.1 [ab] (−0.5)	59.3 ± 2.1 [bc] (0.2)
40–60	76.4 ± 2.3 [b] (3.2)	60.5 ± 2.3 [b] (3.0)	70.9 ± 1.8 [bc] (2.6)	59.3 ± 1.8 [a] (1.2)	70.0 ± 1.7 [b] (2.0)	61.8 ± 1.7 [acd] (0.7)	70.9 ± 1.7 [b] (2.4)	59.7 ± 1.7 [bc] (0.6)
>60	76.7 ± 3.6[b] (3.5)	64.3 ± 3.7 [b] (6.7)	71.3 ± 3.1 [c] (3.1)	61.5 ± 3.2 [a] (3.5)	70.0 ± 3.0 [ab] (2.0)	64.1 ± 3.0 [acd] (3.0)	70.6 ± 3.1 [ab] (2.0)	63.1 ± 3.1 [bc] (4.0)

[a, b, c, d] Values within a column, country and muscle type with different superscript letters, differ significantly at $P < 0.05$.

In contrast, Chinese consumers ≥40 years old and American consumers ≥31 years old scored tenderness and juiciness 3.6 to 8.6 units higher than most younger age groups. Though, for Chinese consumers, improved scoring of juiciness (4.2 units) was only observed in the LL for 40 to 60 year olds compared to 20 to 25 years, whereas in the SM, the effect extended to those above 60 years compared to most age group categories under 40 years old differing by about 4.4–7.8 units. American consumers ≥40 years old scored LL samples 5.2 to 6.0 juiciness units greater than 20 to 30 year olds, while for the SM, 26 to 30 year olds scored juiciness 4.8 to 8.7 units lower than almost every other age group ($P < 0.05$; Table 3).

Aligning with tenderness and juiciness, when averaged across both muscles, younger Australian consumers scored flavour (4.7–6.1 units) and overall liking (5.1–6.3 units) more favourably than older consumers. Notably, those aged 26 to 30 years old scored flavour and overall liking on average 4.7 to 10.2 units lower than all other age groups. This trend was evident in the SM muscle for which flavour and overall liking scores were highest amongst those ≤ 25 years old, with increases of 6.2 to 14.5 units compared to categories upwards of 26 years old. However, this trend differed within the LL, as Australian consumers over 60 scored flavour higher than those aged 26 to 30 and 40 to 60 years old (6.2 and 4.8 units). Similarly, overall liking scores in the LL were significantly greater for those over 60 and under 20 compared to the 26 to 30 years-old category (6.6 and 6.2 units).

Older Chinese consumers scored flavour (≥40 years) and overall liking (≥60 years) 5.2 to 7.5 units higher than younger consumers (≤25 years old). There was no impact of age group seen for LL samples, however, in the SM muscle, older Chinese consumers (generally ≥40 years old) scored flavour and overall liking more favourably than those in categories ≤25 years old (4.7–11.7 units).

Similar to some Australian participants, American consumers aged 20 to 30 years old scored flavour and overall liking lower than younger and most older age groups by 3.2 to 7.3 units ($P < 0.05$; Table 3). This effect was seen in the LL samples with consumers aged 20 to 25 years old and 20 to 30 years old scoring flavour (4.4 units) and overall liking (~5 units) lower than consumers aged 40 to 60 years old. Similarly, for the SM muscle, 20 to 30 year olds scored significantly lower for flavour and overall liking than all other age groups (4.6–9.0 units) ($P < 0.05$; Table 3).

Country differences were also observed within consumer age groups ($P < 0.05$; Table 3). Across muscle types, American tenderness scores were greater than Chinese consumers for the age groups under 20, and 31–39 years old (~10.1 units), and Australian consumers 40–60 years old (6.8 units). In addition, Australian tenderness scores were higher than Chinese in the under 20-year-old category (12.7 units). Australian consumers scored juiciness higher than Chinese consumers ≤25 years old (10 units), and 31 to 39 years old (7.7 units) and American consumers 20 to 25 years old (5.4 units). American consumers within the 31 to 39 years-old category also scored juiciness 7.2 units higher than Chinese consumers. No differences between countries was observed for tenderness in age categories 20 to 30 years old and above 60 years old, and for juiciness in groups 26 to 30 and above 40 years old. Country differences across age groups were generally consistent for flavour and overall liking with Australian and American consumer scores greater than Chinese scores for those under 20 years old (7.2 to 12.4 units). Within consumer group 20 to 25 years old, Australian flavour and overall liking scores were greater than American and Chinese consumers by 6.1–8.8 units. There were no country differences for flavour and overall liking within age groups above 26 years old ($P < 0.05$; Table 2).

3.2.2. Number of Adults and Children in a Household

Number of adults in a household when averaged across all countries and muscle types had an impact on tenderness, juiciness, flavour and overall liking scores ($P < 0.05$; Table 2; individual data not shown). In general, households with more adults present scored higher eating quality scores than those containing less adults. Households with 3 adults scored tenderness, and juiciness higher than households with 1 and 2 adults (3.0 units; $P < 0.05$), and 1, 2 and 5 adults (3.0 to 3.3 units; $P < 0.05$) respectively. Similarly, flavour and overall liking was also scored highest by households of 3 adults compared to 2 and 4 adult households (1.8 to 3.2 units; $P < 0.05$). These trends differed across country and muscle for tenderness and juiciness ($P < 0.05$; Table 2).

Within Australia, LL samples scored 6.6 tenderness units higher in households of 4 adults compared to 5, while households of 3 and 5 adults scored SM samples 6.2 to 7.2 units higher than 1 and 2 adults. Within China, households of 5 adults scored LL tenderness about 4.9 units higher than 1 and 4 adults, and within the USA households with 1 and 3 adults scored LL tenderness 4.7 and 5 units higher than 2 adults. Number of adults within a household did not have an impact on tenderness scoring of the SM for American and Chinese consumers. For juiciness, on average across both muscles, Australian consumers with households of 3 and 4 adults scored 3.8 to 5.8 units higher than all other categories. Within China juiciness scores did not differ based on number of adults in a household. American consumers of households with 1 and 3 adults scored juiciness 4.6 to 6.5 units higher than households of 2 and 5 adults.

Comparing within adult categories (Table 2; individual data not shown), American tenderness LL scores were greater than Australian for 1 and 5 adults (7.4 and 10.4 units; $P < 0.05$) and Chinese in 1, 3 and 4 adults in a household (9.3 to 12.4 units; $P < 0.05$). SM scores largely did not differ by country. Within adult categories, Australian consumer juiciness scores were greater than Chinese for 2, 3 and 4 adults (4.7 to 5.8; $P < 0.05$) and American for 4 adults in a household. American juiciness scores were also higher than Chinese for 1 and 3 adults in a household (9.6 and 4.7 units). There was no difference in countries for 5 adults in a household.

Number of children within a household only had an impact on tenderness scores ($P < 0.05$; Table 2; individual data not shown). American consumers reported no tenderness differences in LL samples across the number of children in a household, whereas for the SM households with 5 children scored tenderness 12.9 to 16.1 units higher than those with 0 to 3 children ($P < 0.05$). Within Australia and China, the number of children within a household had no detectable impact on tenderness scores. Differences across the three countries were similar to those found for other demographic traits. For households with 0 and 1 children, Australian and American consumers scored tenderness 7.0 to 10.3 units higher than China for the LL and SM. For households of 2 and 3 children, Americans scored the LL higher than Chinese consumers (9.5 and 12.3; $P < 0.05$), and for households of 5 children, Americans score the SM 13.9 units greater than Australians ($P < 0.05$). No country difference was reported for SM samples in 3 children, LL and SM samples in 4 children and LL samples in households of 5 children.

3.2.3. Consumer Gender

Consumer gender had a significant impact on flavour and overall liking, but only for American consumers ($P < 0.05$; Table 2). American males scored flavour and overall liking 3.6 and 3.5 units higher than females ($P < 0.05$; individual data not shown) However, country differences were detected with American and Australian males scoring flavour and overall liking higher than Chinese males (4.0–6.4 units; $P < 0.05$; individual data not shown), while female scores across the three countries were not significantly different.

3.2.4. Consumer Income

Within each country, consumer income bracket had a significant impact on sensory scores of tenderness, flavour and overall liking ($P < 0.05$; Tables 2 and 4). This effect varied within each country, though a cross-country comparison could not be made as the local currency was used in questionnaires. Overall, income had the greatest influence on sensory scores for Australian consumers of low income, and middle-income Chinese consumers.

Table 4. Least square means ± standard error (on a scale of 0–100) of tenderness, juiciness, flavour and overall liking scores of sheepmeat samples by a consumer's income bracket. Values within brackets signify difference from mean of country (*italics = greater than 3 units from the mean*).

Sensory Trait	Tenderness	Juiciness	Flavour	Overall Liking
Income				
AUD	*Australian consumers*			
<$25,000	63.9 ± 2.6 [abc] (1.6)	67.3 ± 2.5 (2.8)	66.0 ± 2.4 [ab] (1.0)	67.4 ± 2.4 [bc] (1.8)
$25,000–$50,000	67.9 ± 2.3 [a] *(5.7)*	66.6 ± 2.2 (2.1)	68.4 ± 2.1 [b] *(3.4)*	70.3 ± 2.1 [c] *(4.7)*
$50,001–$75,000	64.1 ± 2.1 [ab] (1.8)	65.8 ± 2.0 (1.3)	65.8 ± 1.9 [ab] (0.8)	66.6 ± 2.0 [bc] (1.0)
$75,001–$100,000	63.1 ± 2.1 [bc] (0.9)	63.5 ± 2.0 (−1.0)	63.8 ± 1.9 [a] (−1.2)	64.4 ± 1.9 [ab] (−1.2)
$100,001–$125,000	58.6 ± 2.3 [d] *(−3.6)*	61.9 ± 2.1 (−2.7)	62.1 ± 2.0 [a] (−2.9)	62.2 ± 2.1 [a] *(−3.4)*
$125,001–$150,000	60.1 ± 2.4 [bcd] (−2.1)	63.1 ± 2.3 (−1.5)	64.4 ± 2.2 [ab] (−0.6)	64.1 ± 2.2 [ab] (−1.5)
>$150,000	62.4 ± 2.2 [bc] (0.1)	64.4 ± 2.1 (−0.1)	65.4 ± 2.0 [ab] (0.4)	65.7 ± 2.0 [ab] (0.1)
Prefer not to say	57.9 ± 2.3 [d] *(−4.4)*	63.6 ± 2.1 (−0.9)	64.0 ± 2.1 [a] (−1.0)	64.1 ± 2.1 [ab] (−1.5)
Yuan	*Chinese consumers*			
≤¥24,000	57.3 ± 3.3 [ab] (−0.6)	59.5 ± 1.8 (−0.2)	60.4 ± 1.7 [ab] (−0.2)	61.3 ± 1.7 [a] (−0.7)
¥24,000–¥36,000	59.3 ± 3.5 [ab] (1.4)	62.3 ± 1.9 (2.6)	62.8 ± 1.8 [a] (2.3)	64.7 ± 1.8 [b] (2.7)
¥36,001–¥60,000	57.5 ± 3.6 [ab] (−0.4)	59.4 ± 2 (−0.3)	58.0 ± 1.9 [b] (−2.5)	60.0 ± 1.9 [a] (−2.0)
¥60,001–¥96,000	61.4 ± 3.7 [a] (3.5)	63.5 ± 2.3 *(3.8)*	63.4 ± 2.2 [a] (2.8)	64.1 ± 2.2 [ab] (2.1)
¥96,001–¥120,000	55.5 ± 4.6 [ab] (−2.4)	57.7 ± 3.5 (−2.0)	57.8 ± 3.4 [ab] (−2.7)	61.0 ± 3.5 [ab] (−1.0)
>¥120,000	58.7 ± 4.1 [ab] (0.8)	56.5 ± 2.9 *(−3.2)*	61.5 ± 2.8 [ab] (0.9)	62.8 ± 2.8 [ab] (0.8)
Prefer not to say	55.4 ± 3.6 [b] (−2.5)	59.0 ± 2.3 (−0.7)	59.9 ± 2.2 [ab] (−0.6)	60.2 ± 2.3 [ab] (−1.8)
USD	*American consumers*			
<$20,000	67.3 ± 2.2 [ab] (1.9)	65.2 ± 1.8 (2.0)	66.4 ± 1.7 [a] (1.9)	66.3 ± 1.7 [a] (2.5)
$20,000–$50,000	64.6 ± 2.0 [ab] (−0.8)	63.2 ± 1.7 (0.1)	66.0 ± 1.5 [a] (1.5)	65.2 ± 1.6 [a] (1.4)
$50,001–$75,000	64.0 ± 2.0 [a] (−1.5)	62.0 ± 1.7 (−1.2)	62.0 ± 1.6 [b] (−2.5)	61.6 ± 1.6 [bc] (−2.2)
$75,001–$100,000	67.4 ± 2.0 [b] (2.0)	64.0 ± 1.7 (0.8)	66.1 ± 1.6 [a] (1.5)	64.6 ± 1.6 [ac] (0.7)
>$100,000	63.8 ± 1.9 [a] (−1.6)	61.4 ± 1.6 (−1.8)	62.2 ± 1.6 [b] (−2.3)	61.4 ± 1.6 [bc] (−2.4)

[a, b, c, d] Values within a column and country with different superscript letters, differ significantly at $P < 0.05$.

Australian consumers within lower income brackets tended to score tenderness, flavour and overall liking higher than those within higher income brackets with the exception of those within the >\$150,000 category (Table 4). For tenderness, on average, consumers with income <\$100,000 scored 3.9 to 9.3 units higher than those in higher income brackets. For flavour, the two income brackets ranging from \$25,000 to \$75,000 demonstrated scores 3.7 to 6.3 units higher compared to categories \$75,000 to \$120,000. Similarly, average overall liking scores were greatest for lower income earners (<\$75,000) with increases ranging from 4.3 to 8.1 units compared to higher income categories. This income difference also varied slightly between muscle types for flavour and overall liking ($P < 0.05$; Table 2; individual data not shown) but typically lower income earners scored flavour and overall liking more favourably than higher income earners both in LL and SM muscle ($P < 0.05$; Tables 2 and 4; individual data not shown).

Within China, there was no consistent trend observed for income bracket. Tenderness only differed amongst those who preferred not to divulge their income and earners within the ¥60,001–¥96,000 category. For flavour and overall liking, on average, consumers within the income bracket ¥36,001 to ¥60,000 scored 4.7 to 5.3 units lower than the income brackets immediately above and below. This varied between the muscles, flavour LL scores increased with higher income up to ¥96,000, while improvements in scoring were limited to bracket ¥24,000 to ¥36,000 for the SM. Similarly, greater overall liking scores of the LL were restricted to ¥24,000 to ¥36,000 ($P < 0.05$; Tables 2 and 4; individual data not shown) with no detectable difference for the SM.

Similar to Chinese consumers, there was no consistent impact of income on sensory scores for American consumers. Participants within income bracket US\$75,001 to \$100,000 scored tenderness on average 3.5 units higher than income brackets immediately above and below. Flavour scores were on average 3.8 to 4.4 units higher amongst the income brackets <US\$50,000 and US\$75,001 to \$100,000, following the same trend for LL and SM muscles (3.8 to 4.8 units higher). Overall liking scores were on average, 3.5 to 4.9 units higher in earners <US\$50,000 per annum, again with LL and SM following the same trend ($P < 0.05$; Tables 2 and 4; individual data not shown).

3.2.5. Meat Consumption Habits

Of all meat consumption habits analysed, only frequency of consumption had a significant impact on eating quality scores. This was observed for tenderness, juiciness, flavour and overall liking and varied across muscles ($P < 0.05$; Tables 2 and 5). The general trend was similar across countries with higher consumption frequency having a positive influence on sensory scores, however statistical comparisons between countries was not possible given different scales were used. Typically, more frequent lamb consumption habits increased eating quality scores, however for LL samples, Australian and Chinese scoring was largely unaffected by frequency of consumption, with significant differences only observed for juiciness. Daily consumers of lamb within Australia scored LL juiciness 16.8 to 22.5 units higher than those in lower frequency categories, similarly Chinese consumers who more frequently eat lamb scored juiciness 11.4 to 11.7 units higher than those that never eat lamb ($P < 0.05$; Tables 2 and 5). As for the SM, the more frequent consumption habits (daily consumers) of Australian consumers resulted in higher scoring for tenderness (15.1–17.7 units), flavour (13.9–14.3 units) and overall liking (20.4–26.5 units) than those who had lower consumption habits. Similarly, Chinese consumers with greater consumption frequency habits scored tenderness (9.6–19.7 units), juiciness (3.1–17.2 units), flavour (11.7–20.4 units) and overall liking (4.3–17.2 units) higher than those in lower consumption categories. Within the USA, consumers that eat lamb monthly consistently scored LL tenderness, juiciness, flavour and overall liking (4.9–8.4 units) higher than those in categories of more and less frequent consumption. In contrast to Australian and Chinese consumers, the highest American consumption group (once per week) scored LL flavour and overall liking (10.1–18 units) lower than those that eat lamb less frequently. SM scoring was unaffected by frequency of consumption across all eating quality traits.

Table 5. Least square means ± standard error (on a scale of 0–100) of tenderness, juiciness, flavour and overall liking scores of LL and SM samples by a consumer's consumption frequency of sheepmeat. Values within brackets signify difference from mean of country and muscle type (*italics = greater than 3 units from the mean*).

| Sensory Trait | Tenderness | | Juiciness | | Flavour | | Overall Liking | |
Muscle Type	LL	SM	LL	SM	LL	SM	LL	SM
Consumption Frequency				*Australian consumers*				
Daily	72.5 ± 7.4 (4.5)	69.6 ± 7.3 [b] (13.1)	85.1 ± 7.2 [b] (15.5)	70.0 ± 7.1 (9.9)	77.6 ± 7.1 (8.4)	72.8 ± 7.0 [b] (12.0)	79.7 ± 7.2 (9.6)	81.0 ± 7.1 [b] (19.9)
4–5/week	63.7 ± 4.5 (−4.3)	53.4 ± 4.5 [a] (−3.1)	62.6 ± 4.3 [a] (−7.0)	55.2 ± 4.3 (−5.0)	63.3 ± 4.2 (−5.9)	60.3 ± 4.2 [ab] (−0.5)	65.7 ± 4.3 (−4.4)	60.6 ± 4.3 [a] (−0.5)
2–3/week	68.7 ± 1.9 (0.8)	55.4 ± 1.9 [a] (−1.1)	68.3 ± 1.7 [a] (−1.3)	59.6 ± 1.7 (−0.5)	69.7 ± 1.7 (0.6)	60.4 ± 1.7 [ab] (−0.4)	70.8 ± 1.7 (0.7)	59.4 ± 1.7 [a] (−1.7)
1/week	67.7 ± 1.6 (−0.2)	51.9 ± 1.6 [a] (−4.6)	67.3 ± 1.2 [a] (−2.3)	57.1 ± 1.2 (−3)	69.6 ± 1.2 (0.5)	58.5 ± 1.1 [a] (−2.3)	70.3 ± 1.2 (0.2)	56.8 ± 1.2 [a] (−4.4)
Fortnightly	68.6 ± 1.7 (0.6)	51.9 ± 1.7 [a] (−4.6)	67.6 ± 1.4 [a] (−1.9)	57 ± 1.4 (−3.1)	68.4 ± 1.3 (−0.8)	58.9 ± 1.3 [a] (−2.0)	70 ± 1.4 (0)	56.9 ± 1.4 [a] (−4.2)
Monthly	66.9 ± 1.7 (−1.1)	54.5 ± 1.7 [a] (−2.0)	67.1 ± 1.4 [a] (−2.4)	59.6 ± 1.4 (−0.5)	69.2 ± 1.3 (0.1)	59.5 ± 1.3 [ab] (−1.3)	69.6 ± 1.4 (−0.4)	58.7 ± 1.4 [a] (−2.4)
Never eat	67.8 ± 7.3 (9.5)	58.7 ± 7.6 [ab] (2.2)	68.9 ± 7.1 [ab] (7.6)	62.4 ± 7.3 (2.2)	66.2 ± 6.9 (3.3)	55.4 ± 7.2 [ab] (−5.4)	64.3 ± 7.0 (1.1)	54.6 ± 7.2 [a] (−6.6)
				Chinese consumers				
Daily	58.2 ± 7.5 (−5.5)	54.2 ± 7.6 [abc] (2.2)	55.1 ± 6.5 [ab] (−7.7)	56.4 ± 6.6 [ab] (−0.7)	65.1 ± 6.4 (0.1)	60.0 ± 6.5 [a] (4.0)	69.1 ± 6.4 [ab] (1.8)	62.3 ± 6.5 [ac] (5.6)
4–5/week	59.4 ± 6.3 (−4.3)	62.2 ± 6.3 [c] (10.2)	64.1 ± 5.1 [ab] (1.3)	67.0 ± 5.1 [a] (9.8)	62.6 ± 5.0 (−2.4)	63.6 ± 5.0 [a] (7.6)	67.2 ± 5.1 [ab] (−0.1)	64.0 ± 5.1 [a] (7.3)
2–3/week	67.6 ± 4.0 (3.9)	52.9 ± 3.9 [bc] (0.9)	66.5 ± 1.9 [a] (3.7)	57.6 ± 1.9 [ab] (0.4)	66.1 ± 1.8 (1.0)	57.9 ± 1.8 [a] (1.8)	68.0 ± 1.9 [ab] (0.6)	58.1 ± 1.9 [a] (1.4)
1/week	68.0 ± 3.7 (4.3)	49.2 ± 3.7 [ab] (−2.8)	66.6 ± 1.5 [a] (3.8)	55.5 ± 1.5 [bc] (−1.7)	67.2 ± 1.5 (2.1)	54.9 ± 1.5 [a] (−1.2)	68.7 ± 1.5 [ab] (1.3)	53.8 ± 1.5 [bc] (−2.9)
Fortnightly	67.0 ± 3.7 (3.3)	52.1 ± 3.7 [bc] (0.1)	66.3 ± 1.5 [a] (3.4)	58.5 ± 1.5 [ac] (1.4)	66.7 ± 1.4 (1.6)	57.6 ± 1.4 [a] (1.6)	69.1 ± 1.5 [b] (1.8)	56.6 ± 1.5 [ac] (−0.1)
Monthly	66.5 ± 3.7 (2.8)	51.0 ± 3.7 [b] (−1.0)	66.3 ± 1.3 [a] (3.4)	55.5 ± 1.3 [b] (−1.7)	66.6 ± 1.2 (1.6)	55.2 ± 1.2 [a] (−0.9)	68.4 ± 1.3 [ab] (1.1)	55.3 ± 1.3 [ac] (−1.4)
Never eat	59.3 ± 5.5 (−4.4)	42.5 ± 5.5 [a] (−9.6)	54.9 ± 4.3 [b] (−7.9)	49.7 ± 4.3 [b] (−7.4)	61 ± 4.2 (−4.1)	43.2 ± 4.2 [b] (−12.8)	60.8 ± 4.2 [a] (−6.5)	46.8 ± 4.2 [b] (−9.9)
				American consumers				
Daily	-	-	-	-	-	-	-	-
1/week	70.8 ± 5.3 [ac] (−2.4)	59.0 ± 5.3 (1.3)	62.8 ± 5.0 [ab] (−5.5)	57.7 ± 5.0 (−0.3)	59.4 ± 4.9 [ac] (−8.6)	60.8 ± 4.9 (−0.4)	56.3 ± 5.1 [a] (−12.2)	58.5 ± 5.0 (−0.6)
Fortnightly	72.9 ± 3.5 [ac] (−0.3)	57.5 ± 3.4 (−0.2)	69.1 ± 3.2 [ab] (0.9)	57.5 ± 3.2 (−0.5)	70.5 ± 3.1 [bd] (2.5)	60.3 ± 3.1 (−0.8)	72.3 ± 3.1 [bc] (3.7)	58.1 ± 3.1 (−1.0)
Monthly	78.1 ± 2.6 [a] (4.9)	58.8 ± 2.6 (1.1)	72.6 ± 2.1 [a] (4.4)	58.4 ± 2.2 (0.4)	72.8 ± 2.1 [b] (4.9)	61.4 ± 2.1 (0.3)	74.3 ± 2.1 [b] (5.8)	59.8 ± 2.1 (0.7)
Every other month	69.6 ± 2.7 [bc] (−3.5)	57.4 ± 2.7 (−0.3)	66.6 ± 2.3 [b] (−1.6)	57.1 ± 2.3 (−0.9)	68.7± 2.1 [abc] (0.8)	61.1 ± 2.2 (−0.1)	69.6 ± 2.2 [bc] (1.1)	59.8 ± 2.2 (0.7)
2–3/year	74.5 ± 1.8 [a] (1.4)	55.4 ± 1.8 (−2.2)	69.4 ± 1.3 [ab] (1.1)	58.2 ± 1.3 (0.2)	69.5 ± 1.1 [b] (1.6)	62.5 ± 1.1 (1.4)	70.3 ± 1.2 [bc] (1.8)	59.4 ± 1.2 (0.3)
Never eat	73.2 ± 1.8 [bc] (0)	57.9 ± 1.8 (0.3)	69.0 ± 1.3 [ab] (0.8)	59.1 ± 1.3 (1.1)	66.8± 1.2 [acd] (−1.2)	60.8 ± 1.2 (−0.4)	68.5 ± 1.2 [c] (−0.1)	59.0 ± 1.2 (−0.1)

[a, b, c, d] Values within a column and country with different superscript letters, differ significantly at $P < 0.05$.

4. Discussion

4.1. The Effect of Consumer Age Group

In agreement with our hypothesis, consumer age had an impact on all eating quality traits, with older consumers generally scoring samples more favourably than younger consumers, particularly American and Chinese. For Chinese consumers, the tenderness scores of those that were 40 and above were higher across both cuts compared to younger age groups. For juiciness, flavour and overall liking scores a similar pattern was observed, although in this case, it did show some variation across cuts. American consumer tenderness scoring was similar to the Chinese, with those aged above 30 scoring tenderness higher than younger American consumers across both cuts. This trend was also evident for juiciness, flavour and overall liking, yet similar to the Chinese, this also varied across both cuts. In contrast, Australians scored tenderness lower amongst older consumers in both cuts, with the same trend present for juiciness, flavour and overall liking scores in the SM. The reverse was observed for the LL, whereby older consumers tended to score higher than some younger age groups (Table 3).

The results for American and Chinese consumers generally align with previous research in sheepmeat [14], where Australian consumers ≥31 years old scored tenderness 11.5 units, and juiciness 9.5 units higher than younger counterparts. In addition, Thompson et al. [9] also found Australians 40 to 50 years old scored juiciness 3.5 units higher than 20 to 25 year olds. The magnitude of difference between age categories within this study was comparable to the findings of Hastie et al. [14] and Thompson et al. [9], however, in this study, the impact of age also extended to flavour and overall liking. The consistency of the age group effect across tenderness, juiciness, flavour and overall liking was expected as these traits are all highly correlated, an association shown in this dataset [5] and numerous previous studies [21–23]. Nonetheless, demographic factors do not routinely impact on all eating quality traits [9,11,14]. Interestingly, Australian consumer scores did not align neatly with previous Australian findings of Hastie et al. [14] and Thompson et al. [9]. Discrepancies with previous results could be due to the larger sample size used in the current study compared to Hastie et al. [14] (n = 720 versus n = 75), or due to generational changes in consumer attitudes over time, given our testing was conducted at least ten years after Thompson et al. [9]. Alternatively, in contrast to the current findings, there are also studies where there was negligible or no effect of consumer age on eating quality scores in beef for Australian, French, Irish, Korean, Northern Irish and Polish consumers [10,11]. The differences found between age groups in the current study suggests tasting sessions should be balanced across age groups to ensure the eating quality preferences represented for the population of interest are not biased. Notably, differences found between Australian consumers in the current study and those of previous studies highlight the importance of testing large numbers of consumers, as well as continued testing over time to capture current population preferences.

4.2. The Effect of Number of Adults and Children in a Household

The number of adults in a household had a significant impact on average scores of all sensory traits, while the number of children in a household specifically affected tenderness ($P < 0.05$; Table 2). Participants from American and Australian households largely comprised of one to three adults, while Chinese consumers mainly came from households of three or above, which is consistent with traditional Chinese living arrangements of housing shared with extended family. In each country, a large proportion of consumers had 2 or less children, particularly in America and China, which is perhaps unsurprising given a large proportion of participants were aged 20–25 years old in these countries. When averaged across countries and muscle types, households with three adults consistently scored more favourably than households of more and less adults (up to 3.3 units; $P < 0.05$). This did however differ by country and muscle for tenderness and juiciness, for example higher tenderness scores corresponded with more adults in a household for Australian and Chinese consumers but not American (Table 2). Results partially align with previous research, whereby more adults in a household yielded higher sensory scores, however in the current study the effect on sensory traits varied somewhat and the magnitude of difference was much higher compared to Bonny et al. [10] (tenderness and overall liking, range 0.5–1.0 units) and Thompson et al. [9] (juiciness, range 1.7 to 3.7 units). The effect of the number of children in a household, was restricted to tenderness scores of the SM samples for American consumers, where consumers of households of 5 children scored higher than those of 0 to 3 children (up to 16.1 units; $P < 0.05$). This particular sub-class consisted of only 9 consumers, and as such should be interpreted with caution. Country differences were also observed in scoring for tenderness and juiciness within the adult and children sub-classes. Overall, American and Australian scores were higher than Chinese consumers ($P < 0.05$), these particular subgroup differences may help explain some drivers of the country effect observed for tenderness in the production factor analysis of the same dataset [5].

4.3. The Effect of Consumer Gender

In support of our hypothesis, gender had an impact on eating quality scores but was only evident in American consumers for flavour and overall liking traits. American males scored flavour and overall liking around 3.5 units higher than females. This aligns with previous research of Bonny et al. [10] and Kubberød et al. [13] with higher scores observed for males within some countries tested, however the magnitude of effect was almost double in this study compared to Bonny et al. [10]. Some speculation surrounding gender scoring differences, has attributed higher scores to a greater appreciation of meat in the diet, for males in particular [10,13]. Our findings may support this, as there were ten percent more American males than females (Table 1), and a higher proportion of males were more frequent lamb consumers and higher appreciators of lamb compared to females (data not shown). Notably, previous research by Huffman et al. [12] found no impact of gender on sensory scoring for American consumers. Similarly, within our study, there was no difference in scoring between the genders for Australian and Chinese consumers, which aligns with Hwang et al. [11] testing Australian and Korean sensory responses to grilled and Korean barbequed beef. Chinese females tended towards lower consumption and appreciation of lamb in the diet compared to males, and with around twenty percent more females recruited within the China cohort, a possible negative gender effect was expected but was not apparent. These results are in contrast to findings of Thompson et al. [9] and Hastie et al. [14] demonstrating an improvement in scoring for Australian females compared to males. Lack of a consistent gender effect across the countries suggests the differences found may be a reflection of other societal or inherent red meat consumption preferences, rather than gender alone. However, regardless of the driver, eating quality preferences particular to gender in America were demonstrated and as such require consideration when evaluating palatability of sheepmeat.

4.4. The Effect of Consumer Income

Validating our hypothesis, income bracket had an impact on average sensory scores for tenderness, flavour and overall liking, and differed by country and muscle for flavour and overall liking ($P < 0.05$; Table 2). The most prominent trend was within Australian consumers, with lower income earners scoring more favourably than higher income earners except for those earning above $150,000 per annum (up to 9.3 units higher; $P \leq 0.05$; Table 4). This result is consistent with Huffman et al. [12] who demonstrated higher income earners to score sensory traits lower when conducting in home evaluations of beef, with the suggestion that product expectations may be greater in higher income earners. It stands to reason that higher income earners may have easier access to protein sources and premium products, therefore be more frequent consumers and thus more discerning. However, for Australian consumers, distribution of participants across consumption frequency and lamb appreciation categories was very consistent across all income brackets (data not shown), demonstrating Australian income groups were homogenous in regards to consumption preferences.

In contrast to Australian results, there was no consistent impact of income on sensory scoring for Chinese consumers. A tenderness difference was detected for the group who did not want to divulge their income compared to one income bracket, as thus no conclusion can be drawn. Flavour scores increased with rising income up to ¥96,000 for the LL and were higher only for income bracket ¥24,000 to ¥36,000 in the SM, similarly, overall liking increases were restricted to the ¥24,000 to ¥36,000 bracket for the LL and no difference in the SM. Mao et al. [8] reported that in urban areas, income is a strong driver of consumption frequency of sheepmeat with positive attitude changes to sheepmeat accompanying increased income categories. However, within this study, there were no marked differences in frequency of consumption distribution across the income brackets measured (data not shown). Following the same trend as Chinese scoring, there was no consistent pattern of scoring for American consumers across income brackets. Tenderness was scored most favourably by those earning US$75,000 to $100,000 per annum, higher flavour scores were reported for those earning under US$50,000 and US$75,000 to $100,000, while overall liking increases were restricted to those earning under US$50,000 (up to 4.9 units higher; $P \leq 0.05$; Table 4). This is in partial agreement with Huffman et al. [12] whereby lower income earners scored tenderness, juiciness and flavour more favourably than higher income earners. Income had a significant influence on eating quality scores for all three countries. As such, sensory panels should strive for a balanced representation of different incomes, achieved through sampling of large numbers of consumers and inclusion of different geographic locations.

4.5. The Effect of Meat Consumption Habits

Contrary to our hypothesis, appreciation of lamb in the diet and preferred degree of cooking doneness did not have an impact on eating quality scores within this study. In contrast to previous studies [9–11,21], the only meat consumption habit to influence sensory scores was frequency of consumption, which was significant for all sensory traits and varied across muscles (Tables 1 and 2). Examination of the number of lamb appreciators spread across consumption frequency categories, showed those that selected highest consumption rates were largely the highest appreciators and vice versa (data not shown). As such, it appears consumption habits and attitudes are largely intertwined. Higher frequency of consumption had positive effect on juiciness scores in the LL for Australian and Chinese consumers ($P \leq 0.05$; Table 5), and tenderness, juiciness, flavour and overall liking in the SM ($P \leq 0.05$; Table 5). These results should be interpreted with caution as significant differences were detected mainly between the highest and lowest consumption categories within Australian and Chinese consumer groups (these being, daily consumers within Australia ($n = 5$) and those that never eat lamb in China ($n = 15$); Table 1). Therefore, these results are unlikely to be representative of the wider population.

For American consumers, frequency of consumption had no impact on scores of the SM, however monthly consumers of lamb ($n = 64$) scored LL tenderness, juiciness, flavour and overall liking up to 8.4 units greater than more and less frequent consumers ($P \leq 0.05$; Table 5). Given previous studies have demonstrated a greater appreciation of red meat to impact on sensory scores [9,10,21], a suggestion for more favourable scores in the American once a month category is that they have a greater proportion of higher appreciators than every other category. Monthly consumption of lamb would likely be considered a higher frequency category in the USA, given annual per capita consumption rates are quite low [6]. Therefore, while close to 70% of monthly consumers identified lamb as important in their diet (data not shown), the weekly consumption category who scored LL flavour and overall liking lower, actually had a higher proportion of appreciators, negating this theory. This negative response observed in the most frequent consumers of lamb in the American cohort (weekly; Table 1) could be attributed to a preference for locally sourced lamb, as more favourable eating quality scores have been observed for domestic products compared to imported Australian lamb [24]. However, similar to Australia and China, consumer numbers were very low ($n = 10$) for this group and as such should be interpreted with a degree of caution.

5. Conclusions

Confirming our hypothesis, demographic factors had a variable impact on eating quality scores of Australian, Chinese and American consumers testing Australian lamb and yearling samples grilled according to MSA protocols. Demographic attributes of consumer age, gender and number of adults in a household, and income bracket had a significant but different effect within the three countries, occasionally varying by muscle and sensory trait. There was minimal effect of the number of children in a household and no effect of consumer occupation on sensory scores. Contrary to our hypothesis, the importance of sheepmeat and preferred degree of cooking doneness did not impact on eating quality scores for these consumer groups. Frequency of consumption had a significant effect in all three countries, contrasting with previous studies examining the effects of meat consumption habits on sensory scores. However, under-represented consumption categories at either end of the spectrum were largely driving results in this study. Overall, the magnitude of effect for significant attributes in this study were generally greater than those previously reported. This suggests that sensory panels within Australia, China and the USA should be balanced where possible, for demographic factors of age, gender, number of adults in a household and income, with the aim of reducing any bias on sheepmeat sensory scores. Challenges of bias can be overcome through recruitment of sufficiently large populations, and inclusion of geographically diverse locations to help ensure the most accurate representation of sensory preferences is captured for the population of interest.

Author Contributions: L.P., G.E.G., and D.W.P. contributed to the conceptualisation and methodology of the study; R.A.O., and L.P., conducted the investigation; resources were provided by A.J.G., H.L., Q.M., M.F.M.; R.A.O., L.P. and G.E.G. conducted the statistical analysis of the data and data visualisation; R.A.O. prepared the original manuscript draft; review and editing provided by R.A.O., L.P., G.E.G., A.J.G., H.L., Q.M., M.F.M., and D.W.P.; project administration was steered by R.A.O., and L.P.; funding acquisition was obtained through D.W.P., L.P., and G.E.G. All authors have read and agreed to the published version of the manuscript.

Funding: This research was funded by the Cooperative Research Centre for Sheep Industry Innovation, supported by the Australian Government's Cooperative Research Centre Program, and Meat and Livestock Australia. Grant number: R.2.2.3.2 'Eating quality prediction of yearling sheep meat'.

Acknowledgments: The authors wish to gratefully acknowledge both staff and resources provided by Murdoch University, NSW Department of Primary Industries, University of New England, WA Department of Primary Industries and Regional Development, China Agricultural University, and Texas Tech University. The authors would also like to thank Rod Polkinghorne for his consumer sensory expertise.

Conflicts of Interest: The authors declare no conflict of interest.

References

1. Pleasants, A.; Thompson, J.; Pethick, D. A model relating a function of tenderness, juiciness, flavour and overall liking to the eating quality of sheep meat. *Aust. J. Exp. Agric.* **2005**, *45*, 483–489. [CrossRef]

2. Thompson, J.; Gee, A.; Hopkins, D.; Pethick, D.; Baud, S.; O'Halloran, W. Development of a sensory protocol for testing palatability of sheep meats. *Anim. Prod. Sci.* **2005**, *45*, 469–476. [CrossRef]

3. Pannier, L.; Gardner, G.; O'Reilly, R.; Pethick, D. Factors affecting lamb eating quality and the potential for their integration into an MSA sheepmeat grading model. *Meat Sci.* **2018**, *144*, 43–52. [CrossRef] [PubMed]

4. Polkinghorne, R.; Thompson, J.; Watson, R.; Gee, A.; Porter, M. Evolution of the Meat Standards Australia (MSA) beef grading system. *Anim. Prod. Sci.* **2008**, *48*, 1351–1359. [CrossRef]

5. O'Reilly, R.; Pannier, L.; Gardner, G.; Garmyn, A.; Luo, H.; Meng, Q.; Miller, M.; Pethick, D. Minor differences in perceived sheepmeat eating quality scores of Australian, Chinese and American consumers. *Meat Sci.* **2020**. [CrossRef] [PubMed]

6. Meat and Livestock Australia. MLA Market Snapshot: Beef and Sheepmeat, North America. 2019; pp 1–8.

7. Meat and Livestock Australia. MLA Market Snapshot: Beef and Sheepmeat, Greater China. 2019; pp 1–10.

8. Mao, Y.; Hopkins, D.L.; Zhang, Y.; Luo, X. Consumption Patterns and Consumer Attitudes to Beef and Sheep Meat in China. *Amer. J. Food Nutrit.* **2016**, *4*, 30–39.

9. Thompson, J.; Pleasants, A.; Pethick, D. The effect of design and demographic factors on consumer sensory scores. *Anim. Prod. Sci.* **2005**, *45*, 477–482. [CrossRef]

10. Bonny, S.P.F.; Gardner, G.E.; Pethick, D.W.; Allen, P.; Legrand, I.; Wierzbicki, J.; Farmer, L.J.; Polkinghorne, R.J.; Hocquette, J.F. Untrained consumer assessment of the eating quality of European beef: 2. Demographic factors have only minor effects on consumer scores and willingness to pay. *Animal* **2017**, *11*, 1399–1411. [CrossRef] [PubMed]

11. Hwang, I.; Polkinghorne, R.; Lee, J.; Thompson, J. Demographic and design effects on beef sensory scores given by Korean and Australian consumers. *Anim. Prod. Sci.* **2008**, *48*, 1387–1395. [CrossRef]
12. Huffman, K.; Miller, M.; Hoover, L.; Wu, C.; Brittin, H.; Ramsey, C. Effect of beef tenderness on consumer satisfaction with steaks consumed in the home and restaurant. *J. Anim. Sci.* **1996**, *74*, 91–97. [CrossRef] [PubMed]
13. Kubberød, E.; Ueland, Ø.; Rødbotten, M.; Westad, F.; Risvik, E. Gender specific preferences and attitudes towards meat. *Food Qual. Pref.* **2002**, *13*, 285–294. [CrossRef]
14. Hastie, M.; Ashman, H.; Torrico, D.; Ha, M.; Warner, R. A Mixed Method Approach for the Investigation of Consumer Responses to Sheepmeat and Beef. *Foods* **2020**, *9*, 126. [CrossRef] [PubMed]
15. Pannier, L.; Gardner, G.; Pethick, D. Effect of Merino sheep age on consumer sensory scores, carcass and instrumental meat quality measurements. *Anim. Prod. Sci.* **2018**, *59*, 1349–1359. [CrossRef]
16. Pethick, D.; Hopkins, D.; D'Souza, D.; Thompson, J.; Walker, P. Effects of animal age on the eating quality of sheep meat. *Aust. J. Exp. Agric.* **2005**, *45*, 491–498. [CrossRef]
17. Fogarty, N.; Banks, R.; Van der Werf, J.; Ball, A.; Gibson, J. The information nucleus–a new concept to enhance sheep industry genetic improvement. In Proceedings of the Association for the Advancement of Animal Breeding and Genetics, Armidale, Australia, 23–26 September 2007; pp. 29–32.
18. Pearce, K.; Van De Ven, R.; Mudford, C.; Warner, R.; Hocking-Edwards, J.; Jacob, R.; Pethick, D.; Hopkins, D. Case studies demonstrating the benefits on pH and temperature decline of optimising medium-voltage electrical stimulation of lamb carcasses. *Anim. Prod. Sci.* **2010**, *50*, 1107–1114. [CrossRef]
19. Anonymous. *Handbook of Australian Meat—International Red Meat Manual*, 7th ed.; AUS-MEAT: Sydney, Australia, 2005.
20. Watson, R.; Polkinghorne, R.; Thompson, J. Development of the Meat Standards Australia (MSA) prediction model for beef palatability. *Aust. J. Exp. Agric.* **2008**, *48*, 1368–1379. [CrossRef]
21. Bonny, S.P.F.; Hocquette, J.F.; Pethick, D.W.; Legrand, I.; Wierzbicki, J.; Allen, P.; Farmer, L.J.; Polkinghorne, R.J.; Gardner, G.E. Untrained consumer assessment of the eating quality of beef: 1. A single composite score can predict beef quality grades. *Animal* **2016**, *11*, 1389–1398. [CrossRef] [PubMed]
22. Polkinghorne, R.; Nishimura, T.; Neath, K.; Watson, R. Japanese consumer categorisation of beef into quality grades, based on Meat Standards Australia methodology. *Anim. Sci. J.* **2011**, *82*, 325–333. [CrossRef] [PubMed]
23. Pannier, L.; Gardner, G.; Pearce, K.; McDonagh, M.; Ball, A.; Jacob, R.; Pethick, D. Associations of sire estimated breeding values and objective meat quality measurements with sensory scores in Australian lamb. *Meat Sci.* **2014**, *96*, 1076–1087. [CrossRef] [PubMed]
24. Phelps, M.; Garmyn, A.; Brooks, J.; Martin, J.; Carr, C.; Campbell, J.; McKeith, A.; Miller, M. Consumer assessment of lamb loin and leg from Australia, New Zealand, and United States. *Meat Muscle Biol.* **2018**, *2*, 64–74. [CrossRef]

MDPI

St. Alban-Anlage 66

4052 Basel

Switzerland

Tel. +41 61 683 77 34

Fax +41 61 302 89 18

www.mdpi.com

Foods Editorial Office

E-mail: foods@mdpi.com

www.mdpi.com/journal/foods